T0201738

PHYSICS AND DYNAMICS OF CLOUDS AND PRECIPITATION

What does a cloud contain and how does it form? How and why does lightning occur? How might clouds change the Earth's climate?

This key textbook provides a state-of-the-art view of the physics of cloud and precipitation formation, covering the most important topics in the field: the microphysics, thermodynamics, and cloud-scale dynamics. Highlights include: the condensation process explained with new insights from chemical physics studies; the impact of particle curvature (the Kelvin equation) and solute effect (the Köhler equation); homogeneous and heterogeneous nucleation from recent molecular dynamics simulations; and the hydrodynamics of falling hydrometeors and their impact on collision growth. 3D cloud-model simulations demonstrate the dynamics and microphysics of the formation of deep convective clouds and cirrus, and each chapter contains problems that enable students to review and implement their new learning.

Packed with detailed mathematical derivations and cutting-edge stereographic illustrations, this is an ideal text for graduate and advanced undergraduate courses, and also serves as a reference for academic researchers and professionals working in atmospheric science, meteorology, climatology, remote sensing, and environmental science.

PAO K. WANG is a Professor in the Department of Atmospheric and Oceanic Sciences, University of Wisconsin-Madison, where he has been teaching and conducting research for more than 30 years. He has won much recognition for his contributions to atmospheric science, including the Alexander von Humboldt Award, an S. C. Johnson Distinguished Fellowship, and election as a Fellow of the American Meteorological Society and the Meteorological Society of Taiwan. Professor Wang has been the principal investigator of numerous research grants sponsored by NSF, EPA, NASA, and DOE, covering topics in cloud physics, cloud dynamics, aerosol physics, air pollution, and historical climatology. He is an associate editor of *Atmospheric Research* and *European Journal of Physics-Plus*, a member of the international advisory board of *Terrestrial, Atmospheric and Oceanic Sciences*, and is also currently an advisory committee member of the Research Center for Environmental Change (RCEC), Academia Sinica-Taiwan.

Praise for this book:

"Finally a comprehensive textbook, filling an empty slot between mainly descriptive and encyclopaedic cloud physics books. It is carefully written, covering all relevant aspects, and starts from first principles in a pedagogic way: invaluable for cloud physics teachers and graduate students."

– **Professor Dr. Andrea Flossmann**, *Université Blaise Pascal de Clermont Ferrand*

"Without hesitation I am endorsing this book. It will be a great addition to atmospheric science."

– **Professor Dr. Hans R. Pruppacher**, *author of Microphysics of Clouds and Precipitation*

PHYSICS AND DYNAMICS
OF CLOUDS AND PRECIPITATION

PAO K. WANG

University of Wisconsin-Madison

CAMBRIDGE UNIVERSITY PRESS
Cambridge, New York, Melbourne, Madrid, Cape Town,
Singapore, São Paulo, Delhi, Mexico City

Cambridge University Press
The Edinburgh Building, Cambridge CB2 8RU, UK

Published in the United States of America by Cambridge University Press, New York

www.cambridge.org
Information on this title: www.cambridge.org/9781107005563

First published 2013

Printed and bound in the United Kingdom by the MPG Books Group.

A catalogue record for this publication is available from the British Library

Library of Congress Cataloguing in Publication data
Wang, Pao K.
Physics and dynamics of clouds and precipitation / Pao K. Wang, University of Wisconsin, Madison.
pages   cm
Includes bibliographical references and index.
ISBN 978-1-107-00556-3
1. Cloud physics.   2. Precipitation (Meteorology)   I. Title.
QC921.5.W363   2013
551.57′6–dc23
2012027751

ISBN 978-1-107-00556-3 Hardback

To my parents
and
to Libby, Lawrence and Victor

Contents

Color plates can be found between pages 304 and 305

Preface

Clouds are magic in the sky. On a perfectly clear day, clouds may suddenly appear from nowhere, literally out of the blue, and soon cover the whole sky with inconceivable rapidity. And they can be gone just as quickly, as you will know if you have ever encountered a sudden rainstorm, sometimes even with lightning and thunder, while hiking in the mountains. When you are scurrying to find a hiding place, the rain suddenly stops, clouds disappear, and the sun shines brightly over the freshly washed cypress trees. Such was my experience when I was a juvenile in Taiwan, a subtropical volcanic island in the Western Pacific where mountains, sunshine, and water vapor are all abundant. The stir-fry of these ingredients is surely an excellent recipe for cloud making.

Ever since I was a child, I have been intrigued by clouds and had wondered where clouds lived and how they can appear mysteriously in the sky and climb up mountain slopes, and how such seemingly solid blocks can stay afloat in air. Nobody seemed to know the answers. Some adults cited an old saying from ancient Chinese literature: "clouds come out from mountains," but I had never been able to find any "central storage house" of clouds in any mountain in Taiwan.

So one day I was hiking on a mountain and saw a patch of cloud resting on the slope not too far ahead, up the winding path, and I decided to find out what there was in the cloud. I rushed to it as I was afraid it would disappear. But the closer I was, the more it didn't look like a cloud, but rather like fog. And finally I was very sure I was in it, but I couldn't see anything tangible. It was even more dilute than many fogs I had experienced before. Later, I read a poem *Reply to the Imperial Inquiry* by the noted Taoist hermit scholar and alchemist Tao Hongjing (AD 456–536) who wrote it to decline the emperor's invitation to serve in the court:

> What is there in the mountain?
> Over the ranges there are copious white clouds
> But I can only enjoy them by myself
> And am unable to offer them as a gift to you.

Truly, you cannot even hold a piece of cloud in your hand.

But what intrigued me most are those tall cumulus clouds (towering cumulus). In the summer time of Taiwan, the tropopause is very high and these clouds can easily go above 10 or 12 kilometers. Viewed from a distance, they look like silvery mountain peaks topped with intricate palaces or castles. Old fables maintain that there are immortals living in these palaces. The fable notion fades quickly with age, of course, but one question remained in my mind: How would these tall clouds look if viewed from the top?

In college I majored in meteorology but there was no cloud physics course in the first three years, so I had to satisfy my curiosity by reading Horace Byers' (1965) *Elements of Cloud Physics*, the only textbook dedicated to cloud physics in the library there. Only in my senior year was the first-ever cloud physics course offered. It was a one-semester course but was taught at a rather elementary level. Meanwhile, another cloud physics book, *The Physics of Rainclouds* by N. H. Fletcher (1966), became available in the library.

In 1973 I went to UCLA to pursue my graduate study and was fortunate to be taken into Hans Pruppacher's group. Ken Beard, who later became a Professor at the University of Illinois, was a research associate with Hans at the time. It was Ken who got me involved in some experimental work with aerosol particles that marked the beginning of my career in cloud physics. Hans taught his cloud physics class using his hand-written notes and I was looking for a textbook for additional information. I found a used copy of B. J. Mason's (1971) *The Physics of Clouds* for sale in the UCLA student book store and purchased it. Both this and Fletcher's books were my major references for cloud physics in addition to the lecture notes.

There is an interesting anecdote about this copy of Mason's book. In the summer of 2010 I went to NCAR in Boulder, Colorado to collaborate with Bob Sharman, also a UCLA graduate, on some research on turbulence generated by deep convection. One day Bob lamented that he shouldn't have sold his copy of Mason's book after he finished the cloud physics course. He now felt that he needed to consult the book from time to time. It turned out that the copy he sold is precisely the copy I bought. Of course, I didn't really know Bob at the time. So, some 37 years later I finally discovered the origin of that book.

In 1978, Hans and Jim Klett together published their classic *Microphysics of Clouds and Precipitation* (Pruppacher and Klett, 1978), which was the most comprehensive cloud physics textbook available at the time. Their second revised and enlarged edition came out in 1996 and has remained the standard reference in this field. When I became a faculty member in the University of Wisconsin-Madison in 1980, I began teaching cloud physics using Pruppacher and Klett's book as the textbook.

But for beginners, Pruppacher and Klett's book can be overwhelming, as it covers such a vast amount of information in great detail. Thus I used only parts of the book that I deemed necessary for beginners. I re-derived many equations in somewhat different ways. I also added some materials not covered in their book, such as the summary of cloud observational methods. Over time, I developed a set of lecture notes that eventually evolved to become this book.

Coming to Madison has added new perspectives to my vision of cloud physics. First of all, it is here that my childhood dream of "seeing a towering cumulus from above" has been realized. UW-Madison is a hub of satellite meteorology, as it is where Vern Suomi and his team pioneered early satellite meteorology. One cannot but be exposed to chats about satellite observations all the time in such an environment. And what a satellite sees (via the visible channel) in the atmosphere is the top of clouds which, of course, include large towering cumulus and thunderclouds. Viewing clouds from above over the globe certainly gives a different perspective and stimulates new threads of thoughts about them, especially their role in global atmospheric processes.

Wisconsin is in the upper Midwest of the US where "real" thunderstorms occur. When I was in Los Angeles, watching the passage of a frontal system usually amounted to seeing a band of not-particularly-thick clouds moving over the otherwise blue sky. Severe storms usually don't occur in southern California. It is here in the US Midwest that I developed a true appreciation of deep extra-tropical cyclonic systems and the severe weather they spawn that are the textbook cases of the Bergen school of meteorology. Though the present book is not about satellite observations or thunderstorms specifically, I kept these two subjects in mind when writing.

Cloud physics as a branch of atmospheric science started relatively late compared with other branches such as synoptic or dynamic meteorology. Initially, the term "cloud physics" usually meant "cloud microphysics," i.e. the study of the initiation, growth, and dissipation of individual cloud and precipitation particles, and topics such as cloud dynamics and large-scale cloud processes (such as cloud impact on the radiative budget of the atmosphere) were not included. However, the boundaries between these disciplines have become blurred as the resolutions of observations and theoretical models improved, and I now feel a necessity to include at least some brief discussions of such non-traditional topics.

We are just beginning to understand the important impact of clouds on our atmosphere. The strange ability of water substance to switch quickly from invisible vapor to visible clouds, and vice versa, in our atmosphere makes the accurate prediction of atmospheric behavior extremely difficult. Such an ability has a strong impact on the atmospheric optical (hence radiative) properties and atmospheric chemistry. It is now recognized that the cloud factor (especially that related to aerosol) is the single largest source of uncertainties in climate model predictions because we just don't have adequate knowledge of how, when, where, how much, and what kind of clouds will form under certain environmental conditions. It is precisely the purpose of cloud physics studies that will address these issues, and I hope this book helps to clarify some of these issues at least to a certain extent.

This book is divided into 15 chapters. Chapter 1 deals with the observation and classification of clouds based largely on the conventional methods, but remote sensing techniques are also discussed. This chapter serves as the introduction. Beginning in Chapter 2 we start to enter the proper realm of cloud microphysics and examine the internal structure of clouds and precipitation – the size, shape, and concentration of cloud and precipitation particles; all are important properties of the cloud as a colloid system. In Chapter 3 we look into the molecular structure of water substance – individual water molecules and their aggregation states of vapor, liquid, and solid. There are some recent discoveries in this area, such as the presence of quasi-liquid layer on an ice crystal surface and the status of cubic ice in tropical tropopause that may be important when assessing the impact of clouds on climate, and they are briefly summarized here.

Chapter 4 reviews the classical bulk thermodynamics of water substance that eventually leads to the well-known Clausius–Clapeyron equation, and the associated phase change phenomena are discussed from the classical point of view. At the end, we point out the inadequacy of such classical bulk systems in explaining cloud particle thermodynamic behavior and this leads to Chapter 5, which expounds on the necessity of including the surface properties of the condensed phase and the effect of solutes on cloud particles. Two major products of these discussions are the Kelvin and Köhler equations.

Up to this point, all discussions are about the thermodynamic equilibrium among different components of cloud particles. In preparation of what will come next, we need to discuss the aerosol particles in the atmosphere, as they play such an important role in cloud formation and they themselves have a great impact on the global climate. This is presented in Chapter 6.

Chapter 7 is about nucleation – the initiation process of condensed phases. Unlike previous chapters, we discuss the kinetics of molecules instead of their equilibrium behavior. Both homogeneous and heterogeneous nucleation processes are discussed in some detail. In addition, recent molecular level imaging techniques has made it possible to visualize the nucleation on the crystal surface and some results are summarized here.

Up to Chapter 7, discussions are about the physics of cloud particles in the static state. But cloud particles are moving relative to air and such motions generate complicated flow fields around the particles and impact the growth of particles and the overall development of clouds. Hence in Chapter 8, we examine the flow fields around various hydrometeors falling in air from both the theoretical and experimental aspects. We need the knowledge gained in this chapter to understand the discussions in the next few chapters.

Chapter 9 discusses the diffusion growth, namely the growth of cloud and precipitation particles by the diffusion of water vapor towards the particle surface. This process dominates the early stage of cloud growth and we discuss the cases of water drops and ice crystals separately. But to develop large precipitation particles, diffusion growth is too slow and the collision and coalescence of hydrometeors take over as the dominant mechanism. This is discussed in Chapter 10 where the collision and coalescence between two individual particles are considered. In this chapter, we also examine the related process of break-up of particles, especially the drops, and the melting of ice particles. Chapter 11 examines the population dynamics of cloud drops to illustrate the impact of the stochastic nature of the collision growth process in clouds.

All the above microphysical processes occur in the cloud whose behavior is determined by the joint play of these processes and the environmental dynamic condition. To put the microphysics in a proper perspective, we need to understand some fundamental dynamics of the cloud formation, which is the main subject of Chapter 12. Here we review some basic cloud thermodynamics and convection physics. I include some observations by Lidar on the boundary layer processes prior to the formation of clouds. The mathematical expressions used for illustrating these basic processes are analytical which, while useful, cannot handle the complicated non-linear interactions between microphysics and dynamics. Thus in Chapter 13, I introduce numerical cloud models, which can simulate more realistically the time evolution of clouds if the microphysical processes are represented by suitable parameterizations. Simulation results of a deep convective storm and some thin cirrus clouds are shown to illustrate applications of such cloud models.

Chapter 14 is about the cloud electricity, which announces its existence visibly and audibly during a thunderstorm. This is a field that still needs more research to understand better both qualitatively and quantitatively. Earlier works in this field were rather speculative, but recently substantial progress has been made in both observational and modeling aspects, and I have summarized some of these studies.

The last chapter, Chapter 15, serves to illustrate some aspects of the impact of clouds on the global atmospheric processes. I selected the cloud scavenging of aerosol particles and sulfur dioxide, the radiative impact of cirrus due to cloud microphysics, and the cross-tropopause transport of water substance by deep convective storms as examples. All these phenomena strongly impact the global atmospheric process. Naturally, the discussions presented here are confined to the cloud scale processes.

This book is not meant to be comprehensive. The topics I selected for inclusion here are what I thought essentials for a coherent but uncomplicated understanding of cloud and precipitation processes in the atmosphere. There are so many beautiful pearls of research in the vast sea of cloud physics literature that I couldn't include here owing to space limitation. Undoubtedly the term "essentials" contains personal bias as others may have different opinions. Similarly, some topics get more detailed treatments and others much less for the same reason.

The main targeted readers of this book are advanced undergraduate and graduate students of atmospheric sciences. The prerequisite is quite general for college students in physical sciences, namely, a solid background in calculus-based general physics with some exposure to partial differential equations. Readers familiar with physical chemistry will find earlier chapters easy to read as they are essentially the chemical thermodynamics of clouds. But the book is sufficiently self-contained and can also serve as a self-reading reference for research scientists who are not specialized in cloud physics but need to gain some familiarity in this area.

The book contains more materials than can be covered in a one-semester cloud physics course. For a one-semester course, the instructor can use Chapters 1–10 at normal pace. By adding Chapters 11–15, and going through derivations step by step, the book can be used for a two-semester or tri-quarter cloud physics course. I have used mostly the same mathematical symbols as in Pruppacher and Klett's book, so readers can easily cross-check the two books for similar subjects. A few mostly simple problems are given at the end of each chapter, which are designed to get readers familiarized with the subjects discussed.

Book writing is also a self-education process. During the writing, I often found something I thought I knew well a long time ago, but in the process of trying to explain it clearly to readers I discovered that my previous understanding of it was rather superficial. The rewriting, re-derivation, and searching for original papers to clarify these subjects have afforded me to understand them more deeply and I have tried to convey these subtleties in my writing. Of course, the readers are the ultimate judge of how successful I am on this point.

Writing this book also caused me to recall the fond memory of my journey into the field of cloud physics and the people who have helped me along the way, and it is a pleasure to acknowledge them here. Foremost, I would like to thank Prof. Dr. Hans R. Pruppacher who took me into his group and got me started in this field. It was a real pleasure and indeed a privilege for me to participate in the Pruppacher Symposium in Mainz, Germany to celebrate Hans's 80th birthday. The hospitality Hans and his wife Monica have shown to me and my wife Libby every time we visit their house is simply unforgettable. Ken Beard introduced me to the experimental aspects of cloud and aerosol physics. Fellow graduate students and colleagues of Hans at the time, among them Bill Hall, John Pflaum, Sol Grover, Subir Mitra, and John Topalian, have enriched my study experience of cloud physics as well as my social life in a new country. Sol was also a very patient teacher who tried to improve my English.

My former students in the University of Wisconsin-Madison had worked out some excellent results of fundamental importance to cloud physics that have contributed significantly to the contents of this book. Among them, Jerry Straka single-handedly worked out the 3-D prognostic cloud model that has been used profitably by others to produce many model-derived studies by Dan Johnson and Hsin-Mu Lin. Emily Hui-Chun Liu developed the cirrus model that we used to study the radiative impact of ice microphysics. The radiation code for this cirrus model was graciously provided by Dr. Dave Mitchell of the Desert Research Institute, at Reno, Nevada. Norman Miller, Wusheng Ji, and Mihai Chiruta contributed to the ice scavenging, ice flow fields, and ice capacitance, respectively. My colleague Bob Schlesinger, one of the pioneers in 3-D cloud modeling, has provided much valuable advice to our cloud model studies.

In 1993, I spent half a year in Mainz, Germany under the support of a Senior Research Award bestowed on me by the Alexander von Humboldt Foundation of Germany. During this time and later when I was invited back, I had fruitful exchanges with many scientists in the Department of Atmospheric Physics, University of Mainz, and the Max-Planck Institute for Chemistry, especially Professors Stephan Borrmann, Andrea Flossmann, Paul Crutzen, Ruprecht Jaenicke, Peter Warneck, and many other younger scientists.

I have been also benefited from extensive exchanges with Prof. Franco Prodi and Dr. Vincenzo Levizzani of the Institute of Atmospheric Sciences and Climate, National Research Council, Italy (ISAC-CNR), during my several visits to Bologna. Franco arranged support for my visits to Bologna, Ferrara and Venice. Franco's comment that "a book should deliver a vision" alerted me to re-examine the contents of this book more carefully in that respect. I also met Dr. Martin Setvak of the Czech Institute of Hydrometeorology (CHMI), Prague, Czech Republic, via Vincenzo and we have since had very active collaborations in thunderstorm research.

UW-Madison supported my sabbatical leaves to the Massachusetts Institute of Technology and the National Taiwan University. I was also supported by the National Science Council of Taiwan for several summer visits in Taiwan. This support contributed directly or indirectly to the contents of this book.

My research in cloud physics has been supported for a long time by grants sponsored by US National Science Foundation (NSF) to which I am very grateful. Other federal agencies including EPA, NASA, NOAA and DOE have also contributed substantially. In 1992, the Samuel C. Johnson Foundation of Racine, Wisconsin, conferred upon me a S. C. Johnson Distinguished Fellowship that afforded me additional support for my research.

I am grateful to the American Meteorological Society and the American Geophysical Union for allowing researchers to reproduce figures and tables from their publications for use with academic books. Many fellow researchers have also generously granted permission for me to reproduce figures from their research and I have acknowledged their courtesy individually in the figure captions.

The publisher Cambridge University Press has played a big role in the production of this book. Although writing a cloud physics textbook had been in my mind for quite a while, it was Dr. Susan Francis of CUP who urged me several times to get it really started. Ms. Laura Clark, Abigail Jones, Kirsten Bot and Mr. Geoff Amor went through various aspects of the book production to make its completion so professionally done. Kai-Yuan Cheng and Dierk Polzin contributed by drawing or redrawing many figures.

Finally, I want to thank my wife Libby for her faithful support, without which the completion of this book could not have been achieved so smoothly.

Pao K. Wang
Fall, 2012

1

Observation of clouds

1.1 Water vapor in the atmosphere

The clouds in our atmosphere are a condensed form of water (water droplets and ice particles) suspended in air. Such a system is called an aerosol. Naturally, the necessary constituent in air for forming clouds is water vapor. Thus it is important for us to understand the general situation of water vapor in our atmosphere.

Water vapor is so pervasive in our daily life that many do not know that its concentration is actually quite small. Table 1.1 lists the five major gaseous constituents and their volume concentrations in the Earth's atmosphere. Water vapor ranks fourth after N_2, O_2, and Ar.

Water vapor is also different from the other four gases in another important aspect: whereas the concentrations of N_2, O_2, Ar, and CO_2 remain fairly constant from place to place, the concentration of H_2O is highly variable. The layer immediately above the warm tropical ocean surface is literally steaming with water vapor; the highest value is $\sim 4\%$ (tropical Indian Ocean), whereas the surface layer over the Sahara Desert in North Africa or the Taklimakan Desert in western China is close to 0%. Thus, even though the water vapor concentration in the whole atmosphere is more than that of CO_2, as shown in Table 1.1, there are places in the world where its concentration is less than that of CO_2.

Water vapor concentration also varies with height. In general, water vapor is most concentrated near the surface, and it becomes more and more dilute with height. This occurs naturally because the source of water vapor is at the Earth's surface, mainly the world's oceans, while rivers, lakes, polar ice sheets and glaciers also serve a minor role. Globally, the air in tropical regions generally contains more water vapor than in middle and higher latitudes. Fig. 1.1 shows the variation of zonal mean water vapor mixing ratio with latitude and altitude. The mixing ratio, in units of $g\ kg^{-1}$, is the number of grams of water vapor per kilogram of dry air and is a form of *absolute humidity*.

Table 1.1 *The five most concentrated gaseous constituents of the Earth's atmosphere.*

Name	Concentration (volume)
Nitrogen (N_2)	78.084%
Oxygen (O_2)	20.946%
Argon (Ar)	0.934%
Water vapor (H_2O)	~ 0.4%
Carbon dioxide (CO_2)	0.039%

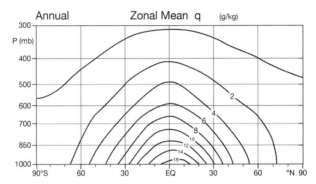

Fig. 1.1 Zonal mean water vapor mixing (g kg^{-1}) field as a function of latitude and altitude (in pressure units).

Another unusual characteristic of water vapor is that it can condense into either liquid or solid form in our atmosphere, a feat that no other gas in our natural environment can achieve. Indeed, water is the only natural substance that is present in all three phases in the Earth's surface environment – water vapor is of course ubiquitous in the atmosphere, but clouds consist of both liquid water droplets and ice particles. Liquid water oceans cover about 70% of the Earth's surface, along with many rivers and lakes. A large amount of ice exists in glaciers in high mountains and ice caps in polar regions.

This peculiar character of water makes the behavior of our atmosphere especially complicated and difficult to study. The phase change of water involves the release or consumption of latent heats that directly impacts the weather process, making accurate weather prediction difficult because it is not easy to predict precisely where, when, and how much condensation will occur. Condensation also makes the accurate estimate of moisture in the atmosphere a difficult task because it is difficult to measure accurately how much condensate there is in the atmosphere. The current estimates of cloud amounts and precipitation contain very large errors. The details of the cloud distribution (both horizontally and vertically) influence the radiation budget of the Earth–atmosphere system and should be considered carefully in any climate studies.

On longer time scales, the condensation of water also plays very important roles. In our atmosphere, clouds can produce precipitation, which represents one of the most powerful forces to shape the surface of the Earth. Precipitation forms rivers and glaciers that eliminate rugged high mountains and produce deep valleys over geological time scales. In fact, it was the formation of rain at the beginning of the current (secondary) atmosphere that dissolved most of the CO_2, SO_2, and other water-soluble gases that were present in the ancient atmosphere and led to our current nitrogen-dominated atmosphere. In contrast, the clouds in the atmosphere of our neighboring planets, Venus and Mars, do not produce rain, and consequently their atmospheres are very "dry" and dominated by CO_2 (see, for example, Walker, 1977).

1.2 Where do clouds occur in the atmosphere?

Although the mixing ratio of water vapor is an index of absolute humidity, the high value of mixing ratio alone does not necessarily mean lots of clouds. Rather, the occurrence of clouds depends more on the *relative humidity*, which measures the degree of saturation and hence the likelihood of condensation of the air.

Almost all (more than 99%) of the water vapor resides in the troposphere, with the stratosphere containing less than 1%. Thus, it is not surprising that almost all clouds that we encounter daily from a small fair-weather cumulus to very large cumulonimbus occur entirely within the troposphere. Even the clouds of very strong storm systems, such as hurricanes and tornadoes, are basically confined within the troposphere. Fig. 1.2 shows this situation graphically. Specific cloud types will be discussed later in the chapter.

Occasionally, we do see clouds in the stratosphere, but they are normally not water clouds. Rather, they are made mostly of sulfuric or nitric acid aerosol particles and are called *nacreous clouds* or *mother-of-pearl* clouds because they exhibit brilliant iridescent colors due to the diffraction of sunlight by the aerosol particles. Such clouds are particularly common in the stratosphere of polar regions, and hence are called polar stratospheric clouds (PSCs). They may play an important role in the depletion of stratospheric ozone in Antarctica and produce the "ozone hole" phenomenon.

Another kind of clouds, called *noctilucent* (night-glowing) clouds, occur even higher up – in the mesosphere. When they do appear, they can be seen as very high bright clouds even though it is already dark on the surface. They appear to be bright at night because the Sun is still shining on them due to their altitude. These clouds are not made of water either and are not the subject of this book.

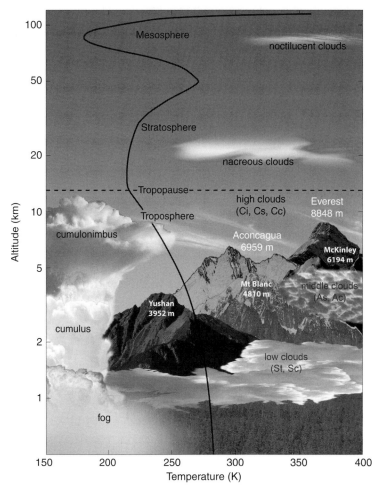

Fig. 1.2 A schematic guide of the height level of various cloud types. The solid curve represents the vertical temperature profile of the atmosphere. The elevations of a few mountains are indicated as a reference. Note that the height is in logarithmic scale.

1.3 Conventional classifications of clouds

The central point of scientific classification is to identify one or more suitable characteristics of the subject as the basis for classification. The two main characteristics upon which the conventional cloud classification scheme is based are (1) cloud visual texture and (2) cloud-base height. The visual texture of clouds is based on that seen by the observer on the ground. This scheme was originally introduced by the British meteorologist Luke Howard in 1803 and has since been modified and expanded; it was adopted by the World Meteorological Organization

(WMO) in 1956 after some modifications. Since the subject of cloud types is usually covered in more elementary meteorology textbooks, our discussion here will be brief.

The most commonly used scheme classifies clouds into four different height categories and three different basic textures. The three basic texture categories are the cirriform, stratiform, and cumuliform. Cirriform is only used to describe high clouds whose texture resembles curly hair or fibers. Stratiform clouds are clouds with more or less uniform, layer-like structure that tend to cover a large part or even the whole sky. Cumuliform clouds, on the other hand, have a patchy structure that looks like cells or fish scales. They can cover either a large part or just a small part of the sky. The reason they look like that has something to do with their dynamical environment. It is common to list the height groups and then to distinguish different types in that group according to their visual texture appearance. Each type may also have subtypes, which will not be discussed in this book, and sometimes it is not easy to determine exactly what designation a cloud should have. It is also possible that a cloud is undergoing transition from one type to another, and the designation may become ambiguous.

1.3.1 High clouds (base height greater than 6000 m)

This group consists of three types – cirrus (often abbreviated as Ci), cirrostratus (Cs), and cirrocumulus (Cc). Conventionally, all high clouds are thought to consist completely of ice crystals, although this notion has been called into question lately, as some observations have suggested that the liquid state in the form of aqueous solution droplets may exist in high clouds. Figs. 1.3–1.5 show the typical cases of the cirrus, cirrostratus, and cirrocumulus, respectively. Although "cirri" refers to the filament texture, in reality only cirrus has that characteristic when observed from the ground. Cirrostratus occasionally exhibits that characteristic but usually appears as a relatively uniform layer, but cirrocumulus never has that curly hair texture.

Fig. 1.3 shows a typical cirrus cloud with the clear texture of curly hair. This texture indicates the presence of wind shear at that level, and the falling ice crystals carried by the wind form the virga filaments. The filaments can be short, like the ones shown here, or very long, and may stretch across the whole sky.

Fig. 1.4 shows an example of a cirrostratus cloud. It has some recognizable curls but also a layer feature that is usually not easy to see if it is thin. The surest sign of the presence of Cs in the sky is the halo, a white circle around the Sun or Moon, as seen in this photograph. The sky inside the circle is slightly darker than outside. The Sun here is blocked by the satellite dish. But the presence of a Sun halo in this case positively identifies the cirrostratus layer in addition to the curls.

Fig. 1.5 shows an example of a cirrocumulus cloud. Both cirrocumulus and altocumulus (see below) have fish-scale-like texture, but cirrocumulus is usually

Fig. 1.3 Cirrus (Ci). Photo by P. K. Wang.

Fig. 1.4 Cirrostratus (Cs). Photo by P. K. Wang.

thinner and will not cast a shadow. The photograph shows the Sun's corona caused by the presence of the cirrocumulus.

High clouds are usually optically thin and allow sunlight (or moonlight) to shine through and they can produce spectacular optical phenomena called *halos* because of the ice crystals in them, as seen in Fig. 1.4. Halos are caused by the refraction of

Fig. 1.5 Cirrocumulus (Cc). Photo by P. K. Wang.

light by ice crystals and are different from the *coronae* that are caused by the interference of light diffracted by cloud particles, which can be either water droplets or ice particles.

Cirrus clouds also exist at the top of thunderclouds in the form of an anvil, but in that case they are considered as part of the thundercloud (cumulonimbus, see later in this section) unless they are blown away and become detached from the thundercloud.

Before the 1980s, high clouds were usually thought to be "harmless", that is, they did not seem to influence the atmospheric processes very much, although they have been used as an indicator of some weather processes. But since then it has been realized that the ice crystals in high clouds can interact strongly with solar and terrestrial radiation in both the visible and infrared bands, and can exert substantial impact on the global climate process. Cirrus clouds are now being studied more vigorously by the research community. We will have more discussions of the climatic impact of high clouds in Chapter 15.

1.3.2 Middle clouds (base height between 2000 and 6000 m)

There are two types of clouds that belong to this family: altostratus (As) and altocumulus (Ac). Examples of them are shown in Figs. 1.6 and 1.7. They consist of either water drops or a mixture of water drops and ice particles. Altostratus clouds

Fig. 1.6 Altostratus (As). Photo by P. K. Wang.

Fig. 1.7 Altocumulus (Ac). Photo by P. K. Wang.

are usually produced by large-scale lifting associated with a low-pressure system and hence may indicate the approach of bad weather. Altocumulus, on the other hand, is usually associated with good weather. Perhaps it is more useful to note that the weather is changing for the better if we observe that altostratus is gradually breaking up into patches and turning into altocumulus. This indicates that large-scale lifting, which is responsible for altostratus formation, is diminishing and is being replaced by smaller-scale weak convection indicated by the appearance of altocumulus, indicating better weather to come.

1.3.3 Low clouds (base height lower than 2000 m)

This group consists of three types: stratus (St), stratocumulus (Sc), and nimbostratus (Ns). Examples of them are shown in Figs. 1.8–1.10. They are often thick, dark, and cover a large part of the sky. Among the three, stratocumulus clouds show identifiable chunky structure, unlike stratus and nimbostratus, which are more uniform and continuous horizontally. All three can be associated with rain, although the rain intensity varies. Stratus and stratocumulus are usually associated with drizzle, light rain or no rain at all. The stratocumulus clouds sometimes form by the rain that falls from a layer of altostratus above it. Nimbostratus is always associated with rain (the prefix "nimbo" means rain) and have a more diffuse appearance at the cloud base. However, it is often difficult to distinguish nimbostratus from stratus by just looking at a photograph.

Low clouds are traditionally thought to consist of liquid droplets. However, more recent observations show that low clouds also contain ice particles when the environmental temperature is cold. Stratocumulus clouds are very common over the ocean and cover a wide area. They play an important role in the radiative balance of the Earth–atmosphere system.

Fig. 1.8 Stratus (St). Photo by P. K. Wang.

Fig. 1.9 Stratocumulus (Sc). Photo by P. K. Wang.

Fig. 1.10 Nimbostratus (Ns). Photo by P. K. Wang.

1.3.4 Clouds with vertical development

This last family of clouds has a common characteristic: they are formed in an environment of relatively strong updrafts. The normal synoptic-scale updraft is on the order of 1 cm s^{-1} (Holton, 2004). In contrast, the environment in which this family of clouds develops is often associated with updrafts greater than 1 m s^{-1}. This is to say that the environment has stronger convection. Hence this cloud family is sometimes called the *convective clouds*.

There are two members in this group: cumulus (Cu) and cumulonimbus (Cb). Examples of them are shown in Figs. 1.11–1.13. Just like other cloud types, there can be many subtypes. Cumulus, for example, can have subtypes such as fair-weather cumulus (small Cu seen on a fine day, Fig. 1.11) and cumulus congestus (fairly large and tall cumulus cloud but not yet a thundercloud, Fig. 1.12). Cumulus congestus can produce substantial precipitation and may develop into cumulonimbus when the top of the congestus starts to show filament structure, a sign that liquid drops are being transformed into ice crystals. Cumulonimbus clouds (Fig. 1.13) are also known as thunderclouds because they produce thunder and lightning.

The base of clouds in this family can vary from 0 m (if it is raining) to about 2000 m. The cloud-top height can vary from a few hundred meters for small fair-weather cumulus to greater than 10 km for cumulonimbus. It is known that the top of some strong cumulonimbus can be much higher than the local mean tropopause due to their strong upward momentum. This phenomenon is called *overshooting* and the part of the cloud that protrudes above the local tropopause is called the overshooting top.

The 10 cloud types mentioned above are of importance to weather forecasting and hence are still the mainstream cloud classification method. However, they do not exhaust all possible cloud types, as there are many other forms of clouds that are not included. The lenticular clouds (Fig. 1.14) that are associated with mountain waves are one example. They are sometimes called wave clouds because of that

Fig. 1.11 Fair-weather cumulus (Cu). Photo courtesy of Dr Martin Setvák.

Fig. 1.12 Cumulus congestus. The case here is also called towering cumulus because the vertical dimension is greater than the horizontal. Photo by P. K. Wang.

Fig. 1.13 Cumulonimbus (Cb). Photo by P. K. Wang.

Fig. 1.14 Lenticular clouds. Photo courtesy of NOAA/NWS.

Fig. 1.15 Wave-like clouds seen from an airplane flying over Wyoming. Photo by P. K. Wang.

association, not because of the appearance of the cloud itself. There are clouds that really look like waves in appearance (Fig. 1.15). Another peculiar cloud type is the so-called "morning glory" clouds that occur most often in Australia, which look like very long tubes often stretching tens or hundreds of kilometers. The list of "unlisted cloud types" can indeed be very long. A classical reference book for those interested in the wide variety of clouds is the one by Ludlam and Scorer (1957).

As mentioned before, the morphological appearance of clouds is mainly controlled by the dynamics of the cloud environment. The air is transparent and we cannot see the air in motion unless the winds move some visible objects. Clouds are such a visible tracer that they can disclose what the air is doing. By observing cloud appearance or dissipation, we can often discern the motions of air in that place. We will discuss in more detail the relation between clouds and air motion when we discuss cloud dynamics in Chapter 12.

1.4 Precipitation

When you are in the middle of rain or snow, you see the raindrops or snowflakes coming down. But how would rainfall or snowfall look like if observed from afar?

Fig. 1.16 shows an example of rainfall from a precipitating cloud. The dark fibrous feature that connects the cloud and the ground is the falling rain. If the rain is occurring between the light source and the observer (as in this case), the filaments appear to be dark because some light will be absorbed when it passes through the rain shaft. Darker regions represent heavier rain. If the light is coming from behind the observer, the rain filaments would appear as bright curtains extending from the cloud base to the ground because light will be scattered by the rain.

Fig. 1.16 The rain shaft coming down from a cumulonimbus near Denver, Colorado. Photo courtesy of Dr Martin Setvák.

Fig. 1.17 Virga clouds over San Francisco, California. Photo by P. K. Wang.

Sometimes raindrops fall from the cloud but evaporate before they reach the ground because the layer below the cloud is not humid enough for the survival of the drops. Then we would see that the filaments terminate in the air, forming the so-called "virga" as seen in Fig. 1.17. Since the virga usually consists of smaller droplets or ice crystals, it is easily influenced by the wind shear below the cloud and hence often looks like a hook, as seen here. If the wind shear is weak, the virga would be nearly vertical.

Although the examples shown here are about rain clouds, snow clouds would look more or less the same.

1.5 Observing clouds from an aircraft

The cloud characteristics we discussed in the previous section are based on observations made from ground level. It is of some interest to see how these clouds would look if observed from a flying aircraft. We will only examine some examples.

Fig. 1.18 shows a few cirrus unicus (against a background of an extensive altostratus cloud deck) seen from a commercial jet flying at about 12 000 m attitude, which is higher than the cloud level. The characteristic mare's tail feature of cirrus is still obvious seen from above. Like the one shown in Fig. 1.3, it appears that ice crystals formed first in the "head" region (the brighter and higher part of the cirrus) and some of them grow larger and fall to lower level and are stretched by the wind shear to form the long tail.

Fig. 1.19 shows another type of cirrus clouds as seen from the aircraft. It appears that the curls of the cirrus here were caused by the upward motion of the air rather than the fall of the ice crystals.

Fig. 1.18 Cirrus unicus as seen from an aircraft flying above it. Photo by P. K. Wang.

Fig. 1.19 Another cirrus seen from an aircraft flying above it. Photo by P. K. Wang.

Fig. 1.20 shows an optical phenomenon known as *subsun*, usually in the form of an oval-shaped bright white spot below the horizon as the Sun is above it. The appearance of subsun indicates the presence of ice cloud, often very thin to nearly invisible, that contains horizontally oriented crystals such as hexagonal ice plates.

Fig. 1.20 Subsun phenomenon caused by the reflection of sunlight by ice crystals, which implies that there is a layer of ice cloud below the aircraft. Photo by P. K. Wang.

These crystals behave like tiny mirrors that reflect the sunlight to form the subsun. The ice cloud that caused the subsun in this figure is most likely a cirrostratus. When the Sun's angle is high, the bright spot looks more circular. When the Sun's angle is low, the bright spot will look like an elongated beam, much like the reflection of the setting Sun on a lake.

Fig. 1.21 shows a field of fair-weather cumulus in the morning. The higher elevation of the aircraft affords a wide perspective of the spatial distribution of Cu, which appear to line up in rows in this case. Such an organization gives an indication of the mesoscale convection process that produces these clouds. Fig. 1.22 shows a more developed Cu field, which usually happens later in the day as the Earth's surface is heated by the Sun and convection becomes more vigorous.

In a favorable environment, cumulus congestus can grow rapidly into a towering cumulus, as in the example shown in Fig. 1.23. The top of this towering cumulus is higher than the aircraft level (flying at 10 km at the time). Observing a tall towering cumulus from an aircraft near it can be a stunning experience, similar to the sensation as one stands near the edge of a cliff and looks downward into the deep gorge below, except that a tall towering cumulus can be easily much taller than the highest peak on Earth, Mt Everest, which is "only" 8848 m in elevation! A cumulus congestus of this size can develop into a cumulonimbus rapidly.

Fig. 1.21 Fair-weather cumulus formed in the early morning as seen from an aircraft.
Photo by P. K. Wang.

Fig. 1.22 Medium-size cumulus formed around noon as seen from an aircraft.
Photo by P. K. Wang.

When a cumulus congestus develops to the tropopause level, its further upward
motion is suppressed by the stable stratification of the stratosphere and it is forced
to spread out horizontally, forming the anvil of a cumulonimbus. The view of a
cumulonimbus from an aircraft or a space shuttle differs from that of a towering

Fig. 1.23 Towering cumulus seen from an airplane flying at about 12 km altitude. Photo by P. K. Wang.

cumulus. Since the environment immediately around an active Cb is quite turbulent, an observer in an aircraft usually needs to observe the cloud from a distance. In a distant view, one is more impressed by the enormous size of the Cb rather than its tallness. The vastness of a large Cb system is well illustrated by the photo taken by the crew of a space shuttle (Fig. 1.24). Here the contrast of the size between a large Cb and an individual cumulus congestus surrounding it can be easily appreciated.

Fig. 1.25 shows a view of a lenticular cloud from above. The convex lens shape is clearly visible. There are also stripe-like textures around the cloud edge that must indicate certain dynamic processes related to cloud formation.

1.6 Cloud classification according to the phase of water substance

The cloud classification in Sec. 1.2 is not the only possible scheme. Another possibility for a cloud classification scheme can be formed based on the phase of

Fig. 1.24 Cumulonimbus clouds over the ocean as viewed from a space shuttle. Photo courtesy of NASA.

Fig. 1.25 A lenticular cloud as seen from an aircraft flying above it. Photo by P. K. Wang.

water substance in the cloud. As we know, clouds can consist of either liquid water drops (cloud droplets, raindrops) or ice particles (ice crystals, snowflakes, graupel, hail, etc.) or both. Based on the phase, we can divide clouds into the following three groups:

(1) *Liquid water clouds* – clouds with all of their constituent particles in the liquid state. Conventionally, it is taken that both middle and low clouds are all liquid water clouds, as are small cumuli. This statement may be correct most of the time, but it has to be remembered that some of these clouds may contain ice under cold environmental conditions. Sometimes the term "warm clouds" is used to represent these types of clouds. However, this can be misleading because some or all water clouds can have temperatures much colder than 0°C because water droplets can be supercooled easily in our atmosphere, and to call them "warm" tends to give the wrong impression.

(2) *Ice clouds* – clouds with all of their constituent particles in the form of ice. Conventionally, all high cloud types are considered as ice clouds, although there are certain probabilities that highly concentrated aqueous solution liquid drops may be present in some high clouds.

(3) *Mixed-phase clouds* – clouds with constituent particles in both liquid and ice phases. Perhaps the most prominent example of a mixed-phase cloud is the summertime cumulonimbus. Lower layers of cumulonimbus clouds are usually warm and therefore consist mainly of water drops (which can be both cloud droplets and raindrops). In the middle and higher layers, air temperatures are cold enough for ice particles to be present. Graupel and hailstones may exist in large quantities in a cumulonimbus, and they can be distributed from high levels in the cloud all the way down to the surface. Evidently, ice particles can be present even when the ambient air temperature is much higher than 0°C! The anvil of a cumulonimbus consists of nearly all ice particles, most of them probably being snow aggregates.

Other types of clouds may also be mixed-phase as well, as long as the environmental conditions allow. For example, precipitating altocumulus (Wang *et al.*, 2004) and some arctic clouds (for example, Turner, 2005) have been observed to consist of both liquid droplets and ice particles.

While this classification scheme may seem to be simplistic, it does have important utility, especially in the cloud–radiation interaction. The radiative properties of liquid water and ice in clouds are quite different, and in order to understand and assess the impact of clouds on the global climate, it is useful to distinguish clouds according to the phase of their constituent particles.

There may also be other possible classifications of clouds, for example, using cloud structures as observed by satellites and radar. Instead of individual cloud morphology, satellite cloud images show the large-scale structure of cloud ensembles, for example, the cellular clouds that span many thousand kilometers over the ocean and the narrow cloud bands that characterize the cold fronts (see Figs. 1.26 and 1.27).

On the other hand, radar and lidar cloud images show the internal structure of clouds instead of their exterior morphology. We will discuss the satellite and radar observations of clouds next.

1.7 Remote-sensing techniques of cloud observation

It is often useful to observe clouds from a distance using remote-sensing techniques. For instance, to understand the cloud distribution over a very large area, it is obviously impracticable to use *in situ* sampling techniques as described in the previous section. Rather, we can use a satellite orbiting the Earth to observe the clouds from high above. The data collected by satellites are very useful in understanding large-scale cloud structure and for long-term studies of cloud climatology because they cover a very wide area and can be continuous in time if necessary. For example, if we want to understand the global long-term variation of cirrus clouds (because they have a great impact on the global climate), only satellite techniques can provide data for this purpose.

Another instance is the observation of clouds of severe storms, which are dangerous for humans or aircraft to make direct *in situ* observations. In this case, radar can provide the internal structure of an approaching storm without actually dispatching humans or aircraft into the stormy area. Lidar is similar to radar except that it uses visible light instead of longer-wavelength electromagnetic radiation.

Remote-sensing techniques are highly complicated and hence we can only give a brief sketch in the following. We will mention just two subjects: radar/lidar and satellites. Previously, these two techniques were considered as distinct from each other, as radar/lidar was mainly ground-based whereas satellites are, of course, Earth orbiting. But, in reality, the two are not mutually exclusive because they merely refer to two different aspects of remote sensing: radars and lidars are about the instrument aspect, whereas satellites are about the observational platform aspect. There are radars and lidars on-board satellites to make cloud observations, as we will see in the next section.

1.7.1 Radar and lidar techniques

Meteorological radars can be used to observe the internal structure of a cloud system. Radar transmits electromagnetic waves in various bands but usually in the millimeter to centimeter wavelength range. When these waves intercept some particles in clouds, part of the wave energy will be reflected back – the backscattered waves, called "echo" – to the radar and received. Based on the time and direction in which the signal is received, the position of the scattering particle can be determined. Based on the strength of the signal, the size and the kind of the particle can often be inferred. While the principle seems to be simple, in reality it is very complicated and contains uncertainty. For example, the reflectivity of a water drop is usually higher than that of a snowflake, which has lower density. But if a snowflake melts slightly when it falls through a warm layer, it may have a thin layer of liquid water on the surface. Thus to a radar, the backscattered signal from this snowflake will be similar to that from a water drop, and an unsuspecting analyst is apt to misinterpret the signal as from a large drop because the dimension of a snowflake is often quite large, even though it contains very little water.

A recent development of radar technology is utilizing the polarization information of the backscattered waves in order to better distinguish the "shape" of the scattering particle in case ambiguity occurs. For example, a water drop is usually spherical whereas a snowflake is often like a flat plane. Thus a spherical drop will reflect evenly the horizontally and vertically polarized waves, whereas a flat snowflake will reflect more strongly the horizontally polarized waves. Hence a radar set equipped with *dual polarization* capability can detect the difference between horizontally and vertically polarized waves and thus can distinguish a spherical drop from a flat snowflake.

Those radars using longer wavelengths (e.g. 10 cm) can see through a storm cloud easily and reveal the overall structure of the storm.

Lidar is essentially the same as radar except that it utilizes visible light instead of microwaves. The light source is a laser. Because visible light cannot penetrate very thick clouds, lidar is usually used for thin cloud (such as cirrus) and aerosol studies.

Both radar and lidar can be either ground-based or carried by aircraft or spacecraft (satellites, space shuttles). The recent CloudSat and TRMM (Tropical Rainfall Measuring Mission) satellites carry on-board radars that provide the vertical cross-sectional structure of clouds and precipitation. CALIPSO (Cloud-Aerosol Lidar and Infrared Pathfinder Satellite Observation) combines an active lidar instrument with passive infrared and visible imagers to probe the vertical structure and properties of thin clouds and aerosols over the globe. CALIPSO and CloudSat are highly complementary and, if combined with conventional satellite images, they can provide 3D perspectives of how clouds and aerosols form, evolve, and affect weather and climate.

1.7.2 Satellite techniques

Satellite techniques refer to observational methods that use a satellite as the platform that carries the instruments to observe clouds. Meteorological satellites can be operated in either passive or active mode. For example, the CloudSat mentioned in the previous section operates in active mode. But there are also satellites that carry cameras or radiometers that measure the light or other electromagnetic radiation emitted by clouds, and hence operate in passive mode.

There are two broad categories of meteorological satellites in terms of their orbit characteristics: geostationary and polar orbiting. Geostationary satellites are those orbiting with the same angular speed as the Earth, and hence they are stationary with respect to the Earth. To an observer on the surface, a geostationary satellite would appear to be "parking" on the same spot in the sky all the time. This is only possible if the satellite is orbiting at an altitude of 22 236 miles (35 786 km) right above the equator. The main benefit of a geostationary satellite is that it can provide weather-related images (clouds, water vapor, volcanoes, etc.) of an area continuously. The main disadvantage is that it is orbiting at a fairly high altitude and the resolution of its images is usually lower than that of polar orbiting satellites of the same genera-tion. The GOES-East and GOES-West satellites operated by NOAA of the USA, the Meteosat operated by the EUMETSAT of the European Union, and GMS operated by the Meteorological Agency of Japan are all geostationary meteorological satel-lites. The best resolutions of the current-generation GOES satellites are 1 km for visible channels and 4 km for infrared channels. Fig. 1.26 shows an example of the visible image of cloud systems obtained from a geostationary satellite.

In contrast to geostationary satellites, polar orbiting satellites do not stay in the same spot relative to the Earth's surface. They usually orbit the Earth in a north–south (meridional) sense and nearly pass above both poles. Many of them assume the so-called Sun-synchronous orbit. This makes the satellite always pass over a given point on the surface at the same local mean solar time, for example, 6:00 pm every day. The orbit is a low-altitude one compared to the geostationary orbit, usually a few hundred kilometers above the ground. Since they are not stationary relative to the surface, they cannot provide images of a fixed location continuously. Rather, they provide snapshots only for that location. But because of their low orbits, they can provide much higher-resolution images than geostationary satel-lites. Examples of polar orbiting meteorological satellites are the Aqua and Terra satellites that carry the MODIS instrument on-board. The resolution for visible images is 250 m and for infrared is 1 km. This high resolution can reveal many details of cloud-top structures, which is very useful in many studies such as thunderstorm dynamics. Fig. 1.27 shows an example of a MODIS visible image of a thunderstorm system. The thunderstorm shows vigorous dynamic activity at the cloud top and an ice plume system is seen lying over the storm anvil (see Chapter 15 for further discussion of the above anvil–plume phenomenon).

Most meteorological satellites carry instruments for detecting radiation in the visible and infrared channels but some also carry microwave detectors. The visible and infrared channels are usually for detecting clouds and water vapor, whereas the microwave channels are mostly for detecting precipitation.

Fig. 1.26 The first GOES-15 full disk visible image on 6 April 2010 at 1732 UTC.
Photo courtesy of NOAA/NASA.

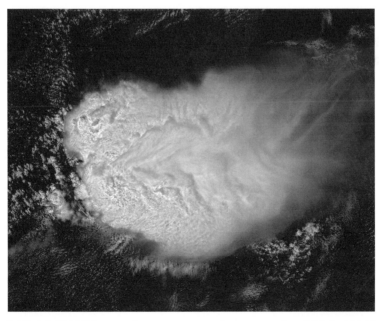

Fig. 1.27 A MODIS image of a storm system that occurred on 24 July 2004 in the
Wisconsin–Illinois area of the USA, showing extensive above-anvil ice plumes.
Photo courtesy of NOAA.

Satellite cloud images provide additional possibilities for cloud classification. There is a wide variety of cloud textures that one can identify from inspecting satellite cloud images, for example, the whirlpool clouds of hurricanes, the open and closed cellular stratocumulus over the vast ocean surface, and the well-known frontal band of extratropical cyclones, etc.

Problem

1.1. In this chapter, we presented two cloud classification schemes. The first one is based on the cloud-base height and appearance (cloud texture), and the second one is based on the phase of particles in the cloud. Can you come up with other cloud classification schemes? The importance is to identify a property (physical or chemical) that can serve as the basis for the classification.

2

The shape and size of cloud
and precipitation particles

The cloud observations that we described in the last chapter are concerned mainly with the exterior appearance of clouds – except for radar and lidar, which also probe the cloud interior – but we did not discuss them in any depth. In this chapter, we want to look into the internal structure of clouds and precipitation. In particular, we want to examine the microstructures, such as the shape and size of the particles that constitute the clouds and precipitation, in some detail.

2.1 Clouds as a colloidal system

Clouds are a system of condensed particles of water substance (water drops, ice particles) suspended in a fluid medium (air). Such a system is called a *colloid*. If the carrier fluid of a colloid is a gas (such as air), the colloidal system is called an *aerosol*. If the carrier fluid is a liquid (such as water or oil), the colloid is called a *hydrosol*. You will occasionally find in the literature that clouds are referred to as an aerosol. Thus, for example, the term "Venusian aerosol" sometimes refers to the clouds in the atmosphere of the planet Venus. In the case of the Earth's atmosphere, however, the term "aerosol" is usually reserved to mean those particles consisting mainly of chemicals other than water (though they may contain a small amount of water too).

One of the most important characteristics of a colloid system is the *size distribution* of its constituent particles. This is because the particle size distribution is closely related to the stability of the colloidal system. If a colloid is stable, it tends to remain in the colloidal state, i.e. particles remain suspended and dispersed in the fluid medium. If the colloidal system becomes unstable, it tends to conglomerate and precipitate – particles will settle out of the carrier fluid medium. Consequently, the colloid disappears.

The same stability description applies to an atmospheric cloud system. If a cloud system is stable, it tends to stay as a cloud, that is, cloud particles remain dispersed.

The particles in this cloud would not easily coagulate and precipitate. If it becomes unstable, then it tends to precipitate to produce rain or snow. Thus, the particle size distribution in a cloud system tells us about the stability of the cloud and its tendency to precipitate.

Particle size distributions can be very different in different clouds. This is especially so if we compare small clouds, such as fair-weather cumulus, that contain only small cloud droplets, with precipitating clouds, such as cumulonimbus, which contain water drops and ice particles of both small and large sizes.

2.2 Frequency of liquid water and ice clouds in subfreezing environment

We already know that clouds can be all liquid water, all ice or mixed-phase. Before we examine the size distribution issue, let us first look at the frequency of particle phase in clouds. For clouds formed in an environment warmer than 0°C, it is generally assumed that they consist of virtually all liquid water. It is quite possible that occasionally they may contain partially melted snow or hail, but this is usually ignored, as the amount is usually small.

But what types of clouds will we expect to see in an environment colder than 0°C? Conventional wisdom would say that liquid water freezes into ice at 0°C and therefore all clouds should consist completely of ice once its temperature drops to below freezing. But actual cloud observations show otherwise. Fig. 2.1 shows the observed frequency of liquid water clouds (i.e. clouds that do not contain ice) as a function of temperature. This chart shows that, as the cloud temperature becomes colder, the percentage of liquid water clouds indeed decreases (i.e. more and more clouds contain ice) but does not immediately drop to zero. In fact, at a temperature as cold as −12°C, there is still about 40% of clouds that consist of no ice but just supercooled water drops. Only when the environmental temperature is close to −40°C would we find that all clouds are ice clouds. When water remains liquid at a temperature colder than 0°C, it is called *supercooled water*. Fig. 2.1 shows that clouds are often supercooled even at very cold temperatures.

The average frequency curve in the chart can be fitted by the following empirical equation:

$$F = a_0 + a_1 T + a_2 T^2 + a_3 T^3 + a_4 T^4 + a_5 T^5, \tag{2.1}$$

where F is the frequency in percent and

$$a_0 = 98.90, \quad a_1 = 4.291, \quad a_2 = -0.318, \quad a_3 = -2.623 \times 10^{-2},$$
$$a_4 = -6.398 \times 10^{-4}, \quad a_5 = -5.279 \times 10^{-6}, \tag{2.2}$$

are the coefficients. The unit of the temperature T must be in °C. Obviously, there will be large variations of this frequency in different geographical locations, so the

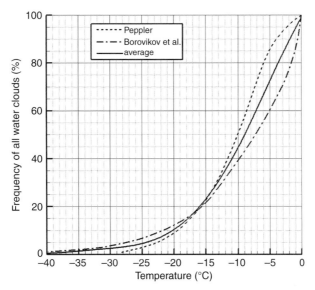

Fig. 2.1 Frequency of clouds that consist of completely liquid water drops. The dashed curve is a fit of the observations by Peppler (1940), while the dashed-dotted curve is a fit of the observations by Borovikov *et al.* (1963). Data adapted from Pruppacher and Klett (1997). The solid curve represents Eq. (2.1), which is a fit from the average values of these two sets of observations.

above curve can only be regarded as a rough guide for the probability of an all-supercooled-water cloud that one can expect at a certain subfreezing temperature. Nevertheless, such a guide is useful for various purposes. For example, a numerical global circulation model may need to generate cloud fields and it will need to decide whether, at given pressure, temperature, and humidity, the cloud should be a supercooled liquid water cloud or an ice cloud or a mixed-phase cloud. Then (2.1) can be used to produce a probability estimate.

2.3 Types of particles in clouds and precipitation

The collective name for the condensed particles of water substance in clouds is *hydrometeor*. When the updraft in clouds is strong enough to support hydrometeors from falling, the hydrometeors are usually considered as cloud particles. If the updraft is not strong enough to keep the hydrometeors suspended, then the hydrometeors become precipitation particles. Thus the definition depends on the strength of the updraft and there is no clear-cut size boundary between cloud and precipitation particles. People loosely call smaller particles "cloud particles" and larger particles "precipitation" without specifying their sizes clearly. Nevertheless, we can still speak of the typical size of these particles.

We usually distinguish the following six kinds of particles in clouds. The size indicated refers to the radius of the particle:

- Cloud drops – water drops that are suspended by updraft. Typical size range is from a few micrometers to ~400 μm, and characteristic size is ~10 μm.
- Raindrops – water drops that fall against the updraft and may eventually reach the ground. Typical size range is from a few hundred micrometers to ~3 mm, and characteristic size is ~1 mm. Drizzle drops are a subcategory of raindrops with radius smaller than 250 μm.
- Ice crystals – these refer to the relatively clean crystalline ice particles of predominantly hexagonal shape. Typical ice crystals in clouds range from a few tens of micrometers to a few hundred micrometers. They are sometimes called cloud ice.
- Snowflakes – crystalline ice particles of relatively large size that fall against the updraft. They can be single crystals or aggregates of crystals, the latter often being referred to as snowflakes or snow aggregates, and the former as snow crystals. The typical size range is from a few hundred micrometers for clean snow crystals to a few centimeters for large snowflakes. Some snowflakes can consist of hundreds of individual crystals.
- Graupel – when an ice or snow crystal collides with supercooled droplets, the droplets will freeze on the surface of the crystal to form rime. If the rimes are not so thick as to cover up the crystal, then the particle is referred to as a rimed crystal. But continued riming will eventually produce a particle that is completely covered by rimes such that the original crystal shape can no longer be recognized. The particle now becomes a graupel (sometimes also called a soft hail or snow pellet). Graupel often have conical shape akin to that of a pear or a wax apple and they represent the precursor of hail. By convention, a graupel should be smaller than 5 mm in diameter. If it becomes greater than 5 mm in diameter (2.5 mm in radius) it will be called hail.
- Hail – if present, hailstones are usually the largest particles in a precipitation system. By convention, a hailstone should be larger than 5 mm in diameter (about pea size) but they can reach more than 15 cm in size (similar to a grapefruit).

We list these particles here not only for the purpose of knowing what may exist in clouds but to study their shape and size distributions, densities, and surface features. The study of these properties is important not only to determine the colloidal state of the cloud but to provide a microstructural description of particles upon which many other studies of clouds can be based. For example, the radar echo depends on the size, shape, surface features, and number concentration of the particles that the radar waves encounter. Before we can design a radar technique to remotely sense the internal structure of a cloud, we would need to know what kinds of particles are present and how many of them the transmitted radar beam may expect to intercept in the cloud. We can meaningfully interpret the observed echo only with such knowledge as a foundation. In principle, the more accurately we know the details about the properties of these particles beforehand, the better we can correctly interpret the observed echo afterwards.

2.4 Sampling of cloud and precipitation particles

In order to understand the properties of these particles, we need to collect physical samples of them. Although remote-sensing techniques have been used to obtain the properties of cloud and precipitation particles, they do not collect physical samples of particles but retrieve the information based on the interpretation of the remote data. Having said this, the gap between the physical sampling and remote-sensing techniques seems to be narrowing, and sometimes it is becoming difficult to clearly distinguish them.

Some of the classical methods of collecting *in situ* cloud droplet samples have been described by Mason (1971), most of which involve impact collection. For example, a glass slide coated with a layer of oily mixture or magnesium oxide is carried on-board an aircraft and exposed to the cloud at a suitable time. Since the aircraft flies at substantial speed, this will cause some cloud droplets to impact on and make smudges on the slide. The size of the smudge, in principle, is correlated with the original drop size, which can be established beforehand by laboratory calibration. Thus by measuring the smudges on the slide we can determine the drop size distributions. There are many caveats to using such direct collection methods, including the following, for example: it requires a large number of slides to obtain a representative sample of cloud droplets; there is a collection efficiency concern because small droplets tend to miss the slide more than large ones; droplets may impact on a pre-existing smudge; large drops may shatter upon impact and create many small droplets. Later improvements included applying the coating on glass rods that can be retrieved quickly or using Formvar solution to coat 16 or 35 mm films, which resulted in more efficient sampling. However, one common problem of this kind of technique is the labor that it takes to measure the size.

It is also possible to take direct in-cloud photographs of cloud droplets. This has the advantage of conveying directly how droplets are distributed in the cloud. But here the problem is that the sampling volume is very small – only a few cubic millimeters – and one needs a large number of photographs to obtain a representative sample. Borrmann *et al.* (1993) developed a holographic technique that can show the 3D droplet distribution and tested it on a stratus cloud. Fig. 2.2 shows one example of their samples. Naturally, it is impossible to show the holograph, but the 2D projection here gives a sense of the dispersive state of the droplets in the cloud.

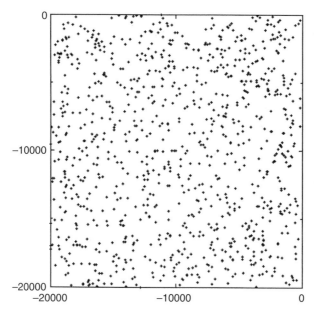

Fig. 2.2 Projection of the droplet positions onto the bottom area of the scanned holographic image volume. The numbers indicate the lengths in micrometers along the *x* and *y* coordinates. From Borrmann *et al.* (1993), courtesy of Prof. S. Borrmann.

Presently, the most popular cloud particle sampling technique is probably the one utilizing the *in situ* laser imaging of them. Two examples to be mentioned are the so-called Knollenberg probes (also called PMS probes) and the Cloud Particle Imager (CPI), both using lasers to illuminate the cloud particles. The former operates on the same principle as a shadowgraph, in which a monochromatic laser beam shines on cloud particles flying through a sampling gap of the probe. The opposite side of the sampling gap is an array of photodiodes. If the laser beam is blocked by the cloud particle, a shadow is registered by the photodiode. By analyzing the output of the photodiode array and the time information, the shadow of the particle can be retrieved. There are different models for different purposes. The PMS two-dimensional cloud (PMS 2D-C) probe (Fig. 2.3) is usually mounted on a research aircraft, for example, under the wing, to sample cloud particles. Its measuring range is 25–800 μm. Fig. 2.4 shows an example of the sampled data. Computer algorithms have been developed to analyze such images, so the method is much more efficient than old impaction methods.

The CPI, on the other hand, employs three lasers together with a charge-coupled device (CCD) camera and is able to obtain a better quality of particle images, not just shadowgraphs like the PMS 2D-C type. It is especially suitable for recording larger ice particles and those with more 3D structures. Fig. 2.5 shows an actual in-cloud sample of ice bullet rosettes. The measurement range is 15–2500 μm.

Fig. 2.3 The PMS 2D probe (right). Photo courtesy of NCAR/UCAR.

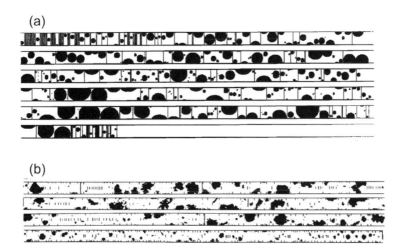

Fig. 2.4 Example images of PMS 2D probe. (a) Images of drops from a low-level penetration (330 m) of a small precipitation shaft below a band cloud. From Beard *et al.* (1986). (b) Images of ice particles taken in a snow cloud. From Braham (1990). Reproduced by permission of the American Meteorological Society.

Older precipitation sampling methods are also summarized by Mason (1971). They were mainly impaction-type measurements like that in cloud drop sampling. These were usually performed on the ground. Filter papers were used earlier but with the problem of splashing when drops hit the surface with high velocity. Later wire meshes coated with soot and nylon meshes treated with a benzene–lanolin

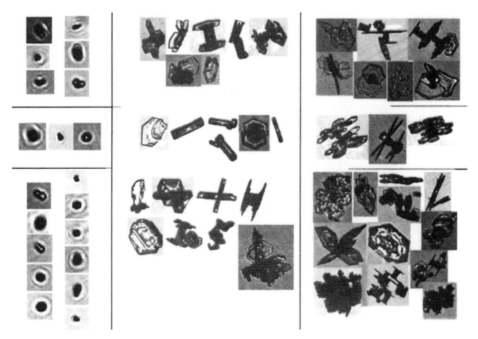

Fig. 2.5 An example of a CPI image. From Heymsfield *et al.* (2002). Reproduced by permission of the American Meteorological Society.

solution and covered with sugar powders were used to solve the splashing problem because drops will pass through these meshes without splashing. Since then, there have been new developments using either acoustic or optical techniques to measure the raindrop size distributions and they are called disdrometers. The optical methods operate by the same principle as the PMS cloud probes. The acoustic disdrometers operate on the principle that drops of different sizes hitting a surface would produce different acoustic signatures. By analyzing the spectrum of the acoustic signals, one can (in principle) resolve the drop size distribution.

Use of a video camera to take photographs of hydrometeors has also been developed. The device is called a hydrometeor videosonde and is attached to balloons ascending into the cloud to obtain hydrometeor images. Murakami and Matsuo (1990) provide the details.

Snowflakes on the ground can be directly sampled or allowed to fall into chilled hexane to preserve them and measure their sizes. There are also electronic techniques for measuring the snow size distributions (see e.g. Schmidt, 1984). Hail size distributions are commonly measured using the hail pad – a Styrofoam board about 1 inch thick covered with a layer of heavy-duty aluminum foil. When hailstones impact on the pad surface, they leave dents much like impact craters on the Moon, with the dent size related to the hail size. Fig. 2.6 shows such an example.

Fig. 2.6 A hail pad damaged by hailstones. The largest dent is about the size of a table tennis (ping-pong) ball. Photo courtesy of Mr Henry Reges.

Using the sampling methods described above and making the necessary measurements, we can form the size distributions of cloud and precipitation particles and study their characteristics. In the following, we will study the properties of hydrometeor size distributions.

2.5 Cloud droplet size distributions

Being small, cloud drops are nearly perfect spheres, as the hydrodynamic force of the air acting on them is not big enough to cause noticeable distortion. Since they are spherically symmetric, a single parameter, either radius or diameter, is enough to characterize their size completely. Hence there is no ambiguity in the term "drop size distribution" here.

The drop size distribution is most commonly expressed graphically as a chart whose vertical axis is the drop number concentration and horizontal axis is the particle size. Note that the number concentration is expressed as the number of drops per unit volume *per size interval*, so the unit is, for example, $cm^{-3} \, \mu m^{-1}$ instead of just cm^{-3}.

The discussions in this section will be mainly on the non-precipitating clouds, as deep convective clouds are too complex. Fig. 2.7 shows the drop size distributions of four different non-precipitating clouds; three of them are maritime clouds from Hawaii, and one is a continental cumulus from Australia. While the appearance of these clouds would look different (hence the different names), their liquid water contents W_L differ little, being from 0.35 g m^{-3} to 0.5 g m^{-3}. The orographic cloud in Hawaii has the widest spectrum (5–160 μm) but the lowest number

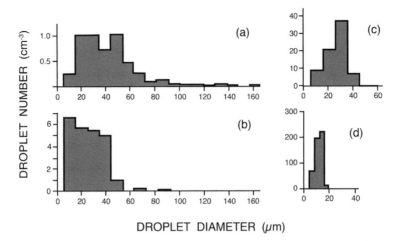

Fig. 2.7 Examples of cloud drop size distributions taken 2000 ft (~ 600 m) above cloud base: (a) orographic cloud over Hawaii, W_L = 0040 g m^{-3}; (b) dark stratus over Hilo, Hawaii, W_L = 0.34 g m^{-3}; (c) trade wind cumulus over Pacific off the coast of Hawaii, W_L = 0.50 g m^{-3}; (d) continental cumulus over Blue Mountains near Sydney, Australia, W_L = 0.35 g m^{-3}. The orographic and dark stratus values are average. Adapted from Fletcher (1962).

concentration (peak number about 1 cm^{-3}). The Hawaiian dark stratus and trade wind cumuli have narrower spectra (~5–90 μm and 5–45 μm, respectively) but higher concentration (peak at ~6 and 36 cm^{-3}, respectively). The Australian continental cumulus has the narrowest spectrum (~5–20 μm) but highest concentration (peaking at > 200 cm^{-3}).

Several conclusions can be drawn from these spectra. First of all, larger mean drop size tends to correspond to lower number concentration. This is not surprising, since W_L are similar and so larger drop size must result in lower number concentration. Secondly, those clouds with larger mean drop size also tend to have broader spectra – having both small and large drops – whereas those with smaller mean size, such as the two cumulus cases, tend to have narrow spectra. Finally, for the same type of clouds, the continental cumulus has a narrower spectrum and much higher concentration than the maritime cumulus.

Fig. 2.8 shows the statistics of putting all four cloud data together. We see that there is a power function relation between the mean drop concentration and the mode drop diameter in these clouds. This power-law type of relation appears to be fairly common for clouds and, as we see later, precipitation.

In a colloidal system, a narrow spectrum often indicates higher stability and hence less likelihood of producing precipitation. In the above examples, we therefore would expect that the continental cumulus in Australia is the least likely to produce rain, whereas the Hawaiian orographic cloud has the highest potential for rain

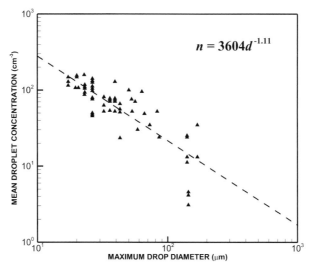

$$n = 3604d^{-1.11}$$

Fig. 2.8 Mean drop concentration as a function of drop size from the data presented in Fig. 2.7.

among the four clouds. How these clouds produce different drop size spectra is naturally due to the physicochemical processes and the dynamical environment in which they develop. For example, the high concentration of drops in the Australian cumulus is thought to be due to the abundance of cloud condensation nuclei over the continent, which produces such a high number of drops. In contrast, the maritime air mass of Hawaii contains many fewer available nuclei and each drop can grow to a larger size due to less competition from others. We will discuss the condensation nuclei in Chapter 6 and nucleation in Chapter 7.

Clouds developed in the same general region tend to have similar spectra if sampled at a similar level. Fig. 2.9 shows the drop size spectra of two adjacent samples spaced 100 m apart but at the same level near the top of a Cu. The two spectra are indeed very similar not only in the concentration but in the width and shape as well. Both show a clear bimodal distribution. The first mode occurs at $d \sim 10\,\mu m$ and the second at $d \sim 35\,\mu m$. The bimodal distribution is most likely the result of two physical processes whereas a unimodal distribution usually indicates a single physical process in forming the spectrum. More discussion of this will be given in Chapter 12.

On the other hand, drop size spectra may be quite different in different parts of the same cloud. Fig. 2.10 shows the comparison of size spectra sampled at three different height levels of two clouds. In both cases, the mean drop size becomes larger and the spectrum broader and more pronouncedly multimodal as the height increases, indicating that the colloidal instability increases with height.

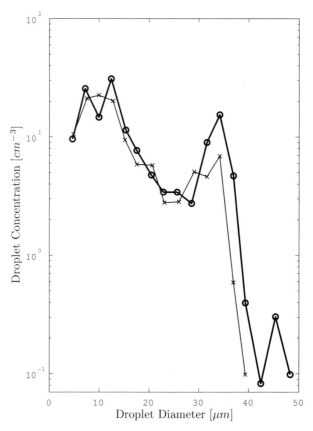

Fig. 2.9 Cloud drop size spectra of adjacent samples taken 100 m apart near the top of a cloud 1400 m deep. Both show a strongly bimodal distribution. Adapted from Warner (1969).

2.5.1 Mathematical expressions of cloud drop size distributions

For statistical and calculation purposes, it is convenient to have mathematical equations to describe the observed cloud drop size spectrum. Two expressions are often mentioned: (1) Khrgian–Mazin distribution, and (2) log-normal distribution.

The Khrgian–Mazin (KM) distribution is a special case of the so-called gamma distribution that has been studied extensively by statisticians and hence many of its mathematical properties are well established. The KM distribution is

$$n(a) = Aa^2 \exp(-Ba), \tag{2.3}$$

where $n(a)da$ represents the number concentration of drops of radius between a and $a + da$, and A and B are adjustable constants that are used to fit the observed spectrum. This is a special case of the gamma distribution given by

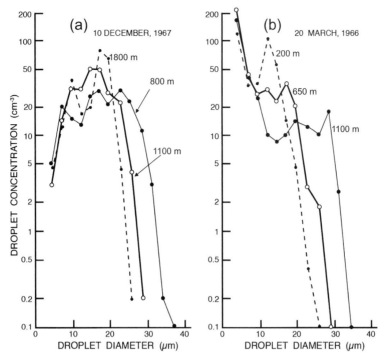

Fig. 2.10 The variation with height of the droplet spectrum in two clouds, showing also the height above cloud base at which the samples were taken; the average total drop concentration at that height. The height value is the height above cloud base. Adapted from Warner (1969).

$$n(a) = Aa^{\beta}\exp(-Ba^{\gamma}), \tag{2.4}$$

where β and γ are adjustable parameters to fit the data. Fig. 2.11 shows an example of the KM distribution.

It is easy to obtain the following primary statistical properties of the cloud drop size spectrum represented by (2.3):

- mean drop size

$$\bar{a} = \frac{\int_0^{\infty} an(a)da}{\int_0^{\infty} n(a)da} = \frac{3}{B}, \tag{2.5}$$

- liquid water content

$$W_{\mathrm{L}} = \frac{4\pi}{3}\rho_{\mathrm{w}}\int_0^{\infty} a^3 n(a)da, \tag{2.6}$$

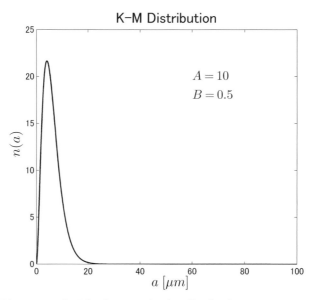

Fig. 2.11 The curve of a Khrgian–Mazin size distribution.

- total drop concentration

$$N = \int n(a)da = \frac{2A}{B^3},\qquad(2.7)$$

where ρ_w represents the density of water. The KM distribution is a unimodal distribution, as can be easily deduced from Eq. (2.3). We see in this equation that the factor a^2 causes $n(a)$ to increase monotonically with the drop radius a, whereas the factor $\exp(-Ba)$ causes a monotonic decrease. Consequently, there exists a single maximum $n(a)$ in the middle corresponding to a modal radius a_m :

$$a_m = \frac{2}{B}.\qquad(2.8)$$

The KM distribution thus predicts that the modal drop size is smaller than the mean drop size, and hence the "tail" of the distribution curve is on the larger drop side (called positive skew) as shown in Fig. 2.11. We have seen before that many cloud drop size spectra are actually bimodal and hence they cannot be fitted realistically by the KM distribution. Theoretical studies (such as in some cloud models) that assume the KM distribution for the drop size spectra may therefore introduce certain errors in the results.

The log-normal distribution is given by

$$n(a) = \frac{N}{\sqrt{2\pi}a\sigma}\exp\left[-\frac{1}{2}\left(\frac{\ln a - \ln a_m}{\sigma}\right)^2\right],\qquad(2.9)$$

where a_m is again the modal radius and σ is the logarithmic standard deviation (i.e. the spectrum width). It can be proven that the log-normal distribution is also a

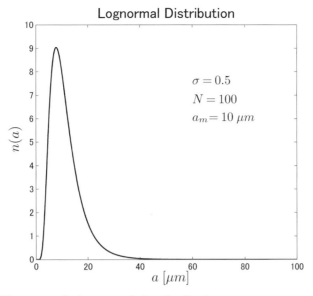

Fig. 2.12 The curve of a log-normal size distribution.

unimodal distribution and hence suffers from the same kind of problem as the KM distribution when dealing with bimodal or multimodal spectra. Fig. 2.12 shows an example of a log-normal drop size distribution. We see that the distribution looks similar to the KM distribution in that it is also positively skewed. One important difference, though, is that the log-normal distribution will not extend to size zero, which corresponds better to the observed drop spectra, where the size never goes to zero.

Both kinds of distributions have been applied widely and the choice may be just a matter of taste. Sometimes, though, one kind may fit better than the other.

2.6 Raindrop size distributions

Like cloud drops, the raindrop size spectrum is an important characteristic of the precipitation process. It is especially important to radar observation of precipitation, as the scattering of radar waves depends strongly on the concentration and size of the precipitation particles. Consequently, the interpretation of the radar echo depends strongly on the particle size distributions assumed.

Unlike the cloud droplets, which are largely spherical, the shape of raindrops deviates significantly from a sphere. Raindrops are typically about 1 mm in radius but can range from about 500 μm to more than 3 mm for large raindrops. Fig. 2.13 shows photographs of raindrops suspended in a vertical wind tunnel simulating the natural rainfall situation (but without the turbulence in the natural rain environment, of course). We see that the drop has a relatively flat bottom and relatively round top similar to that of a hamburger

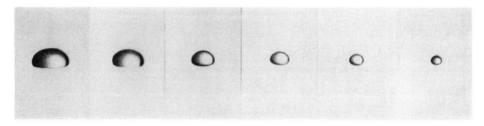

Fig. 2.13 Shape of large drops falling at equivalent spherical radius of $D = 8.00, 7.35,$ 5.80, 5.30, 3.45 and 2.70 mm. From Pruppacher and Beard (1970). Reproduced by permission of Royal Meteorological Society (UK).

bun, making it asymmetric in the vertical direction. Its horizontal and vertical dimensions are different, and the term "size" becomes ambiguous.

The conventional way of dealing with raindrop size is to use the "equivalent size", that is, the size of a spherical drop that has the same volume as the deformed drop. Raindrop size spectra have been observed extensively and they tend to be approximately exponentially distributed with respect to size, as pointed out by Ohtake (1970). A well-known empirical formula to represent the spectra is the Marshall–Palmer (MP) distribution, which was based on observations using a dyed filter technique to correlate with the radar echo of rain (Marshall and Palmer, 1948):

$$n(D_0) = n_0 \exp(-\Lambda D_0), \tag{2.10}$$

where D_0 (in mm) is the equivalent diameter of the drop, and $n(D_0)$ (in $\text{m}^{-3}\,\text{mm}^{-1}$) is the drop concentration of equivalent diameter D_0. The factor Λ (in mm^{-1}) is a function of the rainfall rate R (in mm h^{-1}):

$$\Lambda = 4.1 R^{-0.21}. \tag{2.11}$$

Note that, since this is an empirical formula, the units of the quantities must be exactly as specified above; otherwise the values of the empirical constants may change. The quantity n_0 is the intercept of the curve with the y axis and has the value 8×10^3 $\text{m}^{-3}\,\text{mm}^{-1}$. Fig. 2.14 shows the MP distribution along with the observational data.

Note that the MP distributions appear as straight lines when plotted on a semi-logarithmic chart such as Fig. 2.14. For a fixed rainfall rate R, the MP distribution predicts a monotonic decrease of raindrop concentration with increasing size. Secondly, the slope of the line becomes less as the rainfall rate increases (corresponding to the increase of the area under the line). This also indicates that, for a fixed size, the number concentration of raindrops increases as R increases. All these lines converge to the left to a point where the value of $n(D_0)$ is 8000 corresponding to $D_0 = 0$.

While the MP distribution has been used extensively in representing the raindrop size spectra, there are also discrepancies between this fit and the observed spectra. First, (2.10) seems to give a good description of the distribution for larger drops but

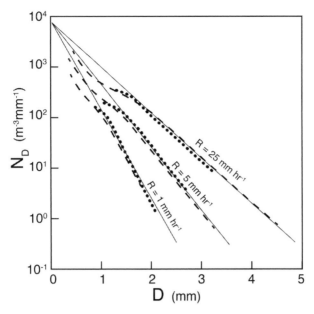

Fig. 2.14 The Marshall–Palmer distribution of raindrops (solid lines). Also plotted are the measurements by Laws and Parsons (1943) (dashed curves) and Ottawa observations by Marshall and Palmer (1948) (dotted curves). Adapted from Marshall and Palmer (1948).

tends to overpredict the number concentrations of smaller drops. In particular, there is a shoulder-like kink in the slope at D_0 between 1 and 2 mm that deviates significantly from the MP distribution. Secondly, it has to be noted that the MP distribution is the result of averaging over many raindrop spectra. At any instant during a rainfall event, however, the raindrop size spectrum may not look like the MP distribution. Fig. 2.15 shows two short-time raindrop spectra observed in Switzerland during a rainfall event. The rainfall rate was fairly constant during the whole period, yet the raindrop size spectra changed markedly.

If we insist on using an exponential formula such as (2.10) to fit these spectra, then the intercept factor n_0 cannot be fixed at 8000 but has to change to 25 000 and 2790, respectively. During the rain, the value of n_0 changed from one value to the other rapidly and has been called the " n_0 jump" (Waldvogel, 1974).

The original spectra obtained by Marshall and Palmer lack the drop size data for $D < 1$ mm because of the instrument limit. Later observations showed that there is usually a maximum at the small size end of the spectrum which cannot be represented by an exponential distribution such as (2.10). For these spectra, the gamma and log-normal distributions, given by (2.4) and (2.9) respectively, can be used. Figs. 2.16 and 2.17 show examples of such fits. For the details of the fitting process

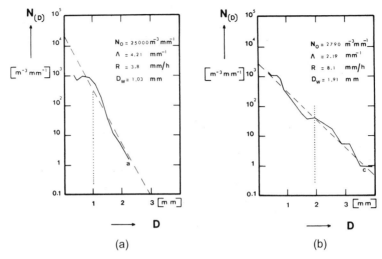

(a)   (b)

Fig. 2.15 Raindrop size distributions observed (solid) and fitted (dashed) during a
rainfall event in southern Switzerland. From Waldvogel (1974). Reproduced by
permission of the American Meteorological Society.

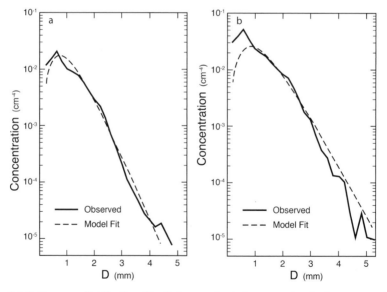

Fig. 2.16 Gamma distribution fit of observed raindrop spectra during convective
storms. Mean drop size distributions from observations at 3000 m altitude, and
gamma function fit, for rain rates: (a) 25–62.5 mm h^{-1} and (b) 62.5–125 mm h^{-1}.
Adapted from Willis and Tuttleman (1989).

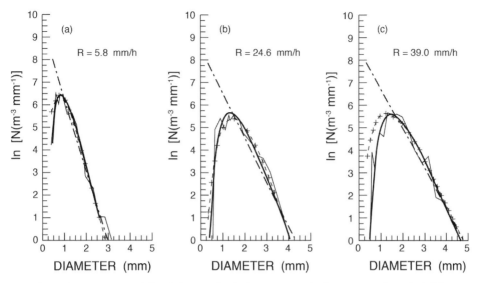

Fig. 2.17 Log-normal fit of observed raindrop spectra during convective rainfall in Israel. Adapted from Feingold and Levin (1986).

and the equations, the reader is referred to Willis and Tuttleman (1989) and Feingold and Levin (1986), respectively.

The MP distribution has been widely applied in the rainfall estimate using conventional radar techniques. The radar-measured reflectivity is interpreted assuming that the raindrop size distribution is MP and the total rainfall is estimated based on this assumption. Brandes *et al.* (2006) compared the radar estimate using the MP distribution assumption and the disdrometer observation and concluded that the MP assumption tended to underestimate the total drop concentration in storm cores and to overestimate significantly the concentration in stratiform regions. Rain water contents in strong convection were underestimated by a factor of 2–3, and drop median volume diameters in stratiform rain were underestimated by 0.5 mm. The more general gamma distribution fit (Ulbrich, 1983)

$$N(D) = N_0 D^{\mu} \exp(-\Lambda D), \qquad (2.12)$$

where μ is the distribution shape parameter, performs better than the single moment (size only) MP model.

2.6.1 Double-gamma distribution

However, even the gamma and log-normal fits fail to deal with the shoulder kink and bimodal and multimodal features that show up in many spectra, and they certainly represent some cloud microphysical processes that eventually need to be addressed.

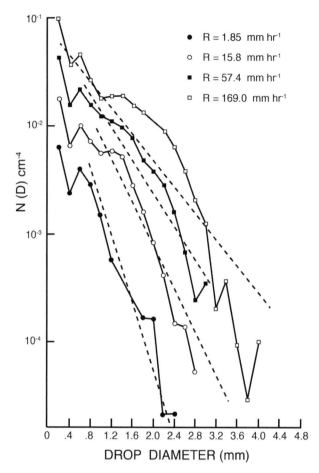

Fig. 2.18 Comparison of observed raindrop size distributions to fitted MP distributions. Adapted from Willis (1984).

Fig. 2.18 shows a few such spectra that will not be described either by the MP distribution or the gamma/log-normal distributions.

There are many ways to fit an observed bimodal spectrum and one example is by the double-gamma distribution method. Fig. 2.19 shows such a fit to an observed raindrop spectrum. The double-gamma distribution is given by (see Sun, 1993):

$$n(d_0) = 0.33d_0^{1.69} \exp(-3.20d_0) + 0.013d_0^{12.23} \exp(-4.94d_0), \qquad (2.13)$$

which is obtained by adding two gamma distributions together. We see that the double-gamma can indeed fit the observed bimodal spectrum rather well. The trade-off is that the mathematics is much more complex than a single-moment distribution. With the aid of a digital computer, however, this complexity may not be much of a problem. Obviously, a double-log-normal fit can also achieve similar results. Again, the choice

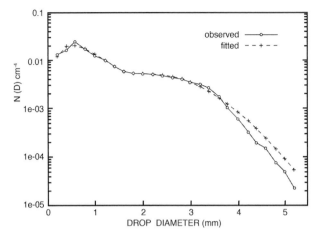

Fig. 2.19 A double-gamma fit to a raindrop size distribution observed by Willis and Tuttleman (1986). From Sun (1993).

is just a matter of taste, but sometimes it appears that one fits better than the other. Kogan *et al.* (2009) used a drop size distribution dataset generated by an explicit microphysics large-eddy model for stratocumulus cases observed during a field project. The fidelity of analytic log-normal- and gamma-type functions is evaluated according to how well they represent the higher-order moments of the drop spectra, such as precipitation flux and radar reflectivity. They concluded that, for boundary-layer marine drizzling strato-cumuli, a distribution based on the two-mode double-gamma distribution provides a more accurate estimate of precipitation flux and radar reflectivity than that based on the log-normal distribution. The gamma distribution also provides a more accurate radar reflectivity field in two- and three-moment bulk microphysical models compared to the conventional reflectivity–rainfall (Z–R) relationship.

2.7 Raindrop shape problem

All the above discussions assume that the size of raindrops can be addressed by a single parameter, the equivalent diameter D_0. But the shape of raindrops provides another parameter that can be utilized to probe the cloud microphysical structure, and hence it is useful to discuss the shape problem. Presently there are two ways to address this problem.

2.7.1 *Quasi-spheroid approach*

This is one step further in the improvement of representing raindrop shapes as raindrops are indeed quasi-oblate spheroids. An oblate spheroid has separate hor-izontal and vertical symmetry and the shape factor can be represented by the *axis*

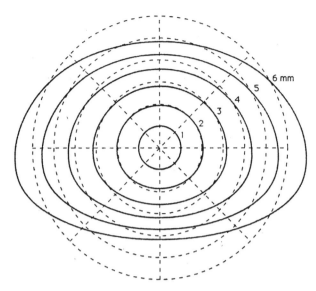

Fig. 2.20 Computed shapes of diameter $D_0 = 1, 2, 3, 4, 5, 6$ mm with origin at center of mass. Shown for comparison are dashed circles of D_0 divided into $45°$ sectors. From Beard and Chuang (1987). Reproduced by permission of the American Meteorological Society.

ratio (b/a), where a and b are the length of the semi-major (horizontal) and semi-minor (vertical) axis, respectively.

The axis ratio of large water drops has been measured in vertical wind tunnel experiments and also inferred from radar observations. Note that a raindrop may oscillate during the fall, which may change its axis ratio substantially, and what we are discussing here is the "equilibrium" axis ratio that does not take the oscillation into account. Nor do we consider the influence of a strong external electric field on the drop shape either, as would happen during a thunderstorm.

Beard and Chuang (1987) solved the equilibrium shape of raindrops based on Laplace's equation, and Fig. 2.20 shows their results. Their model shapes have been fitted by cosine series. But here we want to focus on the axis ratio of these shapes because that is what the spheroidal approach is based upon.

The axis ratio (b/a) of these shapes can be fitted by the following equation (Andsager *et al.*, 1999):

$$a_N = 1.0048 + 0.0057D_0 - 2.628D_0^2 + 3.682D_0^3 - 1.677D_0^4, \qquad (2.14)$$

where a_N is the axis ratio (b/a) and the diameter D_0 should be in cm. Fig. 2.21 shows the curve for this fit. We see that the axis ratio in general decreases with drop size, indicating that drops become flatter as they become larger. Using (2.14), one can construct an ensemble of oblate spheroidal raindrops, using, for example, the MP

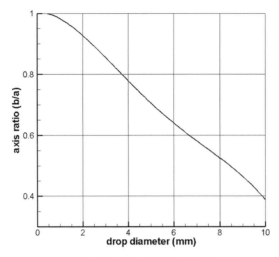

Fig. 2.21 The curve generated by the polynomial Eq. (2.14).

distribution, that contains not only size but, to a first approximation, also shape information. It is then possible to perform various calculations, e.g. radar backscattering cross-section, of such a drop ensemble. Indeed current dual-polarization radar techniques for detecting precipitation, especially the one based on the linear polarization technique, assume that precipitation particles are spheroids and essentially measure the axis ratio of these particles. Details of such techniques can be found in Bringi and Chandrasekar (2001).

2.7.2 *Conical particle approach*

While the oblate spheroid is a useful practical approximation, it is clear that it has its limitations, especially in representing the vertical asymmetry of raindrops. Another approach that goes one step further than the oblate spheroid is the conical particle approach that takes the vertical asymmetry into account. Wang (1982) proposed the following expression to represent such a conical particle:

$$ x = \pm a \sqrt{1 - \frac{z^2}{c^2}} \cos^{-1}\left(\frac{z}{\lambda c}\right), \tag{2.15} $$

where x and z are the horizontal and vertical coordinates, and a and c are the horizontal and vertical semi axis lengths, respectively, of an ellipse. The portion of (2.15) without the arccosine function is the equation of an ellipse (called the generating ellipse as indicated by the dashed ellipse (1) in Fig. 2.22). When this ellipse is modified by the arccosine function, it transforms into a conical particle. The parameter λ, whose value should vary between 1 and ∞, determines the sharpness of the apex of the cone: small λ

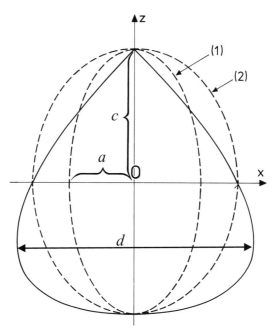

Fig. 2.22 The conical curve (solid) generated by Eq. (2.15). The dashed ellipse (1) is the generating ellipse and dashed ellipse (2) is the limiting ellipse.

produces sharp apex whereas large λ produces a round apex. When $\lambda \to \infty$, Eq. (2.15) produces another ellipse, called the limiting ellipse (the dashed ellipse (2) in Fig. 2.22). While this equation can fit some raindrops of smaller size rather well, it does not produce flat enough bases for larger raindrops. Instead, a modified formula proposed by Thurai *et al.* (2007, 2009) based on (2.15) fits very well for all raindrop sizes. This modified equation is

$$x = \pm a \sqrt{1 - \frac{z^2}{c^2}} \cos^{-1}\left(\frac{z}{\lambda c}\right)\left(1 + k\frac{z^2}{c^2}\right), \qquad (2.16)$$

where k is an additional parameter to better fit the shape of large raindrops with diameter > 4 mm. The size dependence of these parameters is given by:

$$a = \pm\frac{1}{\pi}(0.02914D_0^2 + 0.9263D_0 + 0.07791),$$

$$c = -0.01938D_0^2 + 0.4698D_0 + 0.09538,$$

$$\lambda = -0.06123D_0^3 + 1.3880D_0^2 - 10.41D_0 + 28.34, \qquad (2.17)$$

$$k = -0.01352D_0^3 + 0.2014D_0^2 - 0.8964D_0 + 1.226 \text{ for } D_0 > 4\,\text{mm},$$

$$k = 0 \text{ for } 1.5\,\text{mm} \le D_0 \le 4\,\text{mm}.$$

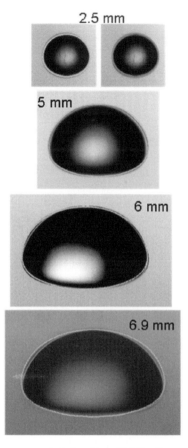

Fig. 2.23 Drop shapes generated by Eq. (2.16). From Thurai *et al.* (2009). Reproduced by permission of the American Meteorological Society.

Fig. 2.23 shows a few examples of such fits using (2.15) and (2.16), and the results are indeed remarkably close to the observed shapes. This shape-fitting scheme has been used to generate a drop shape–size distribution for the purpose of radar studies of precipitation.

2.8 Size and shape of graupel and hail

As mentioned in Sec. 2.1, graupel are rimed particles smaller than 5 mm in diameter by definition. Fig. 2.24 shows an example of graupel. Nearly all of them are conical in shape and many studies have pointed out that this is probably the most prevalent shape. The apex angle of the cone varies from graupel to graupel. Some non-conical ones also appear to be initially conical but with the cone tip having fallen off later. The surface of a graupel is usually not smooth but grainy due to the rime formed by

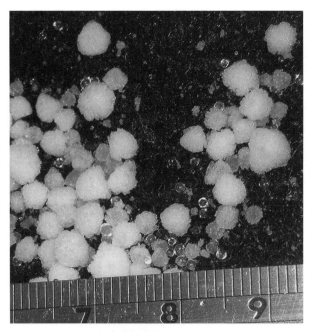

Fig. 2.24 Graupel that fell in Madison, Wisconsin, on 28 January 2008. Photo by
P. K. Wang.

the frozen droplets. Both graupel and hail are ice particles, but their bulk density is
generally smaller than that of solid ice because they are not tightly packed. There are
air bubbles and open spaces inside these particles that reduce the bulk density.
Graupel density can vary anywhere from 0.05 to 0.89 g cm^{-3} depending on their
stage of growth. Sometimes graupel can have a liquid water coating on them.

Hailstones are rimed ice particles larger than 5 mm in diameter by definition.
They vary greatly in size, from blackberry to golf ball (Fig. 2.25) and baseball, to as
large as volleyball size. The record hailstone to date measures about 20 cm (8 in) in
diameter and weighs nearly 0.88 kg (1.9375 lb) (Fig. 2.26). Hailstone density can
vary from 0.7 to 0.9 g cm^{-3}.

Like raindrops, hailstones are not spherical. They are often quasi-spheroidal or
conical, sometimes with lobes protruding from the surface, and sometimes are
irregularly shaped. Because of the non-spherical shape, in principle their size cannot
be characterized by a single parameter. Nevertheless, it is customary to take the
largest dimension of the particle as its "size" and use it to form the size distribution.
In many studies, graupel and hail are considered to belong to one and the same
category because both are rimed ice particles.

Fig. 2.27 shows the hail size distribution reported by Auer and Marwitz (1972),
who found that the curve can be fitted by a power-law distribution

Fig. 2.25 Hail that fell in Madison, Wisconsin, on 12 April 2006. Photo by P. K. Wang.

Fig. 2.26 The largest hailstone recorded to date. The stone fell in a thunderstorm that occurred on 23 July 2010 at Vivian, South Dakota, USA. It is about volleyball size with a diameter about 20 cm and weighs 0.88 kg. Courtesy of NOAA/NWS/ photo by David Hintz.

$$n(d) = Ad^B, \qquad (2.18)$$

where d is the particle size and A and B are empirical constants. There are others who propose that the exponential fit is more appropriate (e.g. Cheng and English, 1983)

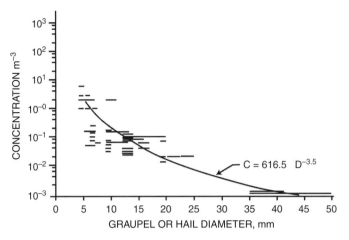

Fig. 2.27 Hail size distribution and the power-law fit (solid curve). Adapted from Auer and Marwitz (1972).

$$n(d) = n_0 \exp(-\Lambda d), \tag{2.19}$$

where n_0 and Λ are the intercept and slope parameters to be determined from the observational data and are not assumed to be constant.

As mentioned before, Eq. (2.15) can be used to simulate the shape of conical graupel and hailstones. Fig. 2.28 provides examples of actual fits, which appear rather satisfactory.

Using this method, Wang *et al.* (1987) analyzed the shape and size of 679 hailstones collected on 22 June 1976 in Colorado during a hailstorm. Most of these stones are conical and fit (2.15) well. Fig. 2.29 shows the distribution of the parameters a, c and λ. Using these parameters, the volume V, axial cross-sectional area A_r and surface area A_s of the stones can be easily calculated using the formulas given in Wang *et al.* (1987). Their distributions are shown in Fig. 2.30. Obviously the λ distribution is exponential, whereas all the others are of gamma function type:

$$\begin{aligned} n(\lambda) &= A \exp[-\beta(\lambda - 1)], \\ n(X) &= A(X - \gamma)^{\alpha} \exp[-\beta(X - \gamma)]. \end{aligned} \tag{2.20}$$

Here X represents any of the geometrical quantities discussed above (except λ) and A, B, a, b, and γ are fitting constants. The values of these constants for this sample of stones are given in Table 2.1.

Unlike the simple size distribution, the method described in this section can describe both the shape and size of the observed particles, thus preserve more complete observational information. Because the shape factor influences the scattering properties of precipitation particles, as has been demonstrated by current dual-polarization

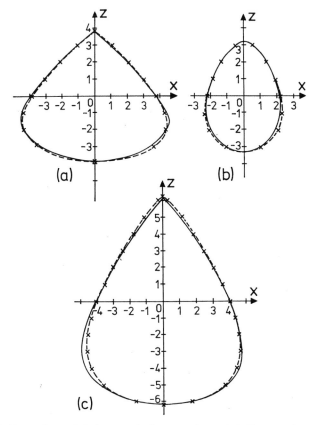

Fig. 2.28 Examples of fitting conical graupels and hailstones by Eq. (2.15): (a) $a = 2.42$, $c = 3.80$, $\lambda = 1.0$; (b) $a = 1.40$, $c = 3.20$, $\lambda = 1.72$; and (c) $a = 2.55$, $c = 6.20$, $\lambda = 1.045$. (a) and (b) are from Mason (1971, Figs. 6.20 and 6.21), and (c) is from Iribarne and Cho (1980, Fig. V-21). The actual dimension (mm) of a and c can be obtained by dividing the numbers by scale factors. The scale factor is 1.639 for (a) and (b), and 1.575 for (c). From Wang (1982). Reproduced by permission of the American Meteorological Society.

radar techniques, it is important to have such information. But the conical scheme goes one step further than the quasi-spheroid scheme in providing information about the asymmetry in the vertical (or along the fall path) direction. Indeed, Sturniolo *et al.* (1995) calculated the backscattering cross-section of conical particles by 35.8 GHz microwaves and demonstrated that the cross-section is sensitive to the shape factor λ.

Such sensitivity can potentially be exploited for further improvements in the radar detection technology. Current dual-polarization techniques do not utilize this sensitivity and that may sometimes result in ambiguities in the interpretation of radar echo. For example, during a thunderstorm under the high vertical external electric

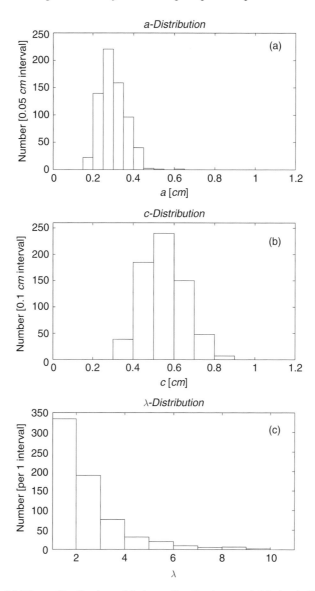

Fig. 2.29 (a) The *a* distribution, (b) the *c* distribution, and (c) the λ distribution. Adapted from Wang *et al.* (1987).

field condition, a raindrop may be elongated vertically, resembling a conical hailstone or graupel with a thin layer of liquid on its surface (see Fig. 14.19). The present dual-polarization radar will have difficulty in distinguishing these particles. But if sufficient is known about the λ sensitivity, it is possible to develop a new radar technique to clearly distinguish them.

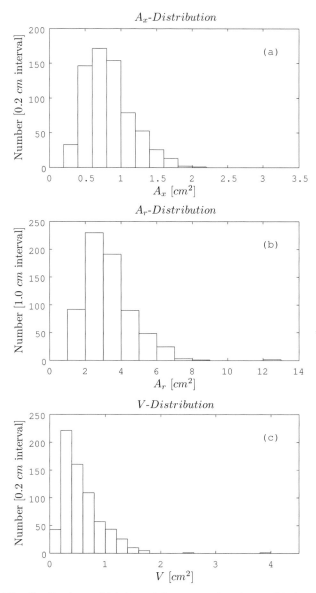

Fig. 2.30 The distributions of (a) the axial cross-sectional area, (b) the total surface area, and (c) the volume of a hail ensemble. Adapted from Wang *et al.* (1987).

2.9 Shape and size of ice crystals and snowflakes

Down to the molecular scale, the common ice particles are all hexagonal in structure except those near the tropical tropopause (see Chapter 3) where they may be cubic. But in macroscopic scale their shapes are much more complicated. In this section we focus on the shape and size of ice crystals and snowflakes.

Table 2.1 *Values of the constants in Eq. (2.20) for various distributions;*
R is the correlation coefficient between the original and fitted curves.

Distribution	Interval size	A	α	β	γ	R
λ	1.0	516.97	–	0.74	–	0.9986
a	0.05 cm	1.02×10^7	3.472	28.54	0.150	0.9961
C	0.1 cm	1.94×10^7	4.612	19.523	0.277	0.9930
A_x	0.2 cm^2	6598.8	2.012	4.531	0.201	0.9972
A_r	1.0 cm^2	552.26	2.194	1.226	0.854	0.9964
V	0.2 cm^3	9496.9	1.551	6.192	0.065	0.9923

2.9.1 Habit of ice crystals

As will be explained in Chapter 3, hexagonal ice has two principal axes, the c and a axes. Those crystals growing in the c axis direction will become columnar and needles, while those growing in the a axis direction will become planar crystals such as plates and dendrites. These are the two fundamental *habits* of ice crystals. But an amazing amount of variations can occur between them, such as hollow columns, capped columns, stellar, sectors, dendrites, etc., and then there are also complex combinations, such as spatial dendrites, bullet rosettes, etc., which was probably the rationale behind the well-known folklore that "no two snowflakes look exactly alike". Fig. 2.31 shows some examples of the milliard variety of ice crystals.

In addition to the already complex patterns, there is one more complexity that most casual observers of snow crystals tend to ignore, which is that some seemingly two-dimensional crystals are actually three-dimensional when one inspects carefully. It is often difficult to see the 3D structure from regular 2D photographs of snow crystals. This is clearly demonstrated by Iwai (1983), who showed that a common dendrite with a hexagonal plate at the center as shown in Fig. 2.32(a) actually consists of a hexagonal plate separated from the dendrite below by a frozen drop in between, as shown in Fig. 2.32(b). This indicates that the plate and the dendrite are not on the same plane and the crystal is indeed 3D in nature. This is often the case for many snow crystals that have such seemingly different "center piece" design.

The 3D structure of snow crystals can be revealed better by stereographs such as the examples in Figs. 2.33 and 2.34, which are taken from Iwai (1989). The 12-branched dendrite in Fig. 2.33 actually consists of crystals combined in four different planes. Fig. 2.34 shows a seven-branch bullet rosette crystal, which of course is very 3D in nature.

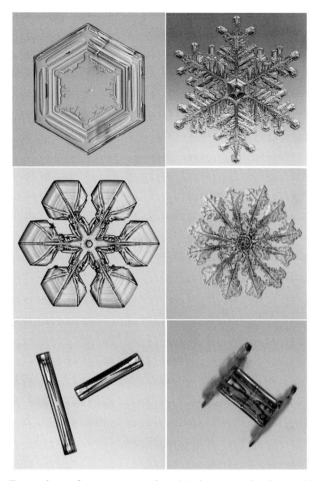

Fig. 2.31 Examples of snow crystals: (a) hexagonal plate; (b) dendrite; (c) dendrite with plates at the ends; (d) dendrite with 12 branches; (e) hollow long columns; and (f) short column with plates. Photos courtesy of Prof. Kenneth Libbrecht.

2.9.2 *Magono–Lee classification*

The fact that ice crystal shapes are so variable and complex causes a problem of how to describe them in a systematic way. In order to standardize descriptions of different ice and snow particle habits, Magono and Lee (1966) designed a classification scheme for this purpose (Fig. 2.35). Using the code specified in this chart will help to reduce the ambiguities in colloquial descriptions. Note that this classification is not limited to snow crystals only but applies to other ice particles as well.

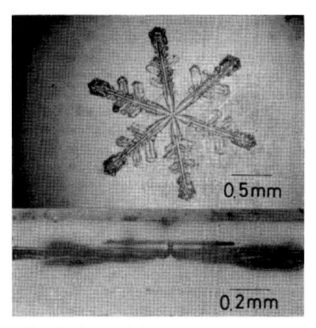

Fig. 2.32 Dendrite with a hexagonal plate at the center: (a) normal view and (b) side view. From Iwai (1983). Reproduced by permission of the Meteorological Society of Japan.

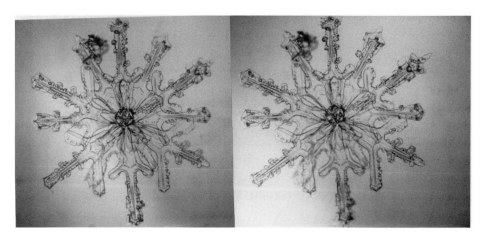

Fig. 2.33 Stereographic pair of a 12-branch dendrite with four different planes. From Iwai (1989). Photos courtesy of Prof. Kunimoto Iwai. Reproduced by permission of Elsevier B.V.

2.9.3 Dimensional relations

Observational studies showed that the dimensions of ice and snow crystals, for example, the length and diameter of ice columns, are not arbitrary but are related by certain relations. The relation is generally expressed as a power-law-type equation:

Fig. 2.34 Stereographic pair of a seven-branch bullet rosette crystal. From Iwai (1999). Photos courtesy of Prof. Kunimoto Iwai. Reproduced by permission of the Japanese Society of Snow and Ice.

Code	Name	Code	Name	Code	Name
N1a	Elementary needle	C1f	Hollow column	P2b	Stellar with sectorlike ends
N1b	Bundle of elementary needles	C1g	Solid thick plate	P2c	Dendrite with plates at ends
N1c	Elementary sheath	C1h	Thick plate of skeletal form	P2d	Dendrite with sectorlike ends
N1d	Bundle of elementary sheaths	C1i	Scroll	P2e	Plate with simple extensions
N1e	Long solid column	C2a	Combination of bullets	P2f	Plate with sector extensions
N2a	Combination of needles	C2b	Combination of columns	P2g	Plate with dendrite extensions
N2b	Combination of sheaths	P1a	Hexagonal plate	P3a	Two branches
N2c	Combination of long solid columns	P1b	Sector plate	P3b	Three branches
C1a	Pyramid	P1c	Broad branch	P3c	Four branches
C1b	Cup	P1d	Stellar	P4a	Broad branch with 12 branches
C1c	Solid bullet	P1e	Ordinary dendrite	P4b	Dendrite with 12 branches
C1d	Hollow bullet	P1f	Fernlike dendrite	P5	Malformed crystal
C1e	Solid column	P2a	Stellar with plates at ends	P6a	Plate with spatial branches

Code	Name	Code	Name	Code	Name
P6b	Plate with spatial dendrites	CP3d	Plate with scrolls at ends	R3c	Graupel-like with nonrimed extensions
P6c	Stellar with spatial plates	S1	Side planes	R4a	Hexagonal graupel
P6d	Stellar with spatial dendrites	S2	Scalelike side planes	R4b	Lump graupel
P7a	Radiating assemblage of plates	S3	Side planes with bullets and columns	R4c	Conelike graupel
P7b	Radiating assemblage of dendrites	R1a	Rimed needle	I1	Ice particle
CP1a	Column with plates	R1b	Rimed columnar	I2	Rimed particle
CP1b	Column with dendrites	R1c	Rimed plate or sector	I3a	Broken branch
CP1c	Multiple capped column	R1d	Rimed stellar	I3b	Rimed broken branch
CP2a	Bullet with plates	R2a	Densely rimed plate or sector	I4	Miscellaneous
CP2b	Bullet with dendrites	R2b	Densely rimed stellar	G1	Minute column
CP3a	Stellar with needles	R2c	Stellar with rimed spatial branches	G2	Germ of skeletal form
CP3b	Stellar with columns	R3a	Graupel-like snow of hexagonal type	G3	Minute hexagonal plate
CP3c	Stellar with scrolls at ends	R3b	Graupel-like snow of lump type	G4	Minute stellar
				G5	Minute assemblage of plates
				G6	Irregular germ

Fig. 2.35 Magono–Lee diagram (Magono and Lee, 1966). Reproduced by permission of Hokkaido University.

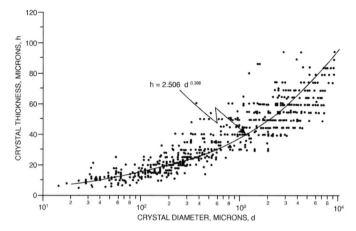

Fig. 2.36 Ice crystal thickness and dimension relation. Adapted from Auer and Veal (1970).

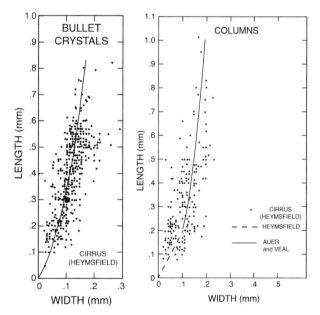

Fig. 2.37 Length vs. width for rosettes (left) and columns (right). From Heymsfield (1972).

$$h = ad^b \quad \text{for plates,}$$
$$d = aL^b \quad \text{for columns,}$$
$$(2.21)$$

where h, d and L represent the thickness (of plates), the diameter, and the length (of columns). Figs. 2.36 and 2.37 show some examples of such relationships.

2.9.4 Ice crystal and snowflake size and shape distribution

The typical concentration of ice crystals in cirrus clouds is in the range of 5×10^4 to 5×10^5 m^{-3} and the size distribution can be described by a power-law-type function

$$N = A w_i d^B, \tag{2.22}$$

where N is the total number concentration, A and B are fitting constants, d is the largest dimension of the ice crystal, and w_i is the ice water content of the cirrus cloud (Heymsfield, 1975).

For snowflakes, the often used distribution is the Gunn–Marshall distribution (Gunn and Marshall, 1958):

$$n(D_0) = n_0 \exp(-\Lambda D_0), \tag{2.23}$$

which is analogous to the Marshall–Palmer distribution of raindrops. Here D_0 is the diameter of the water drop formed by melting the snow crystal, and

$$\begin{aligned} n_0 &= 3.8 \times 10^3 \, R^{-0.87} \text{m}^{-3} \text{mm}^{-1}, \\ \Lambda &= 25.5 \, R^{-0.48} \text{cm}^{-1}, \end{aligned} \tag{2.24}$$

where R is the precipitation rate in millimeters of water per hour. This is unlike the MP distribution where the intercept n_0 is constant. The Gunn–Marshall distribution is shown in Fig. 2.38.

Sekhon and Srivastava (1970) proposed the same expression as (2.23) but with the following modifications:

$$\begin{aligned} n_0 &= 2.50 \times 10^3 R^{-0.94} \text{m}^{-3} \text{mm}^{-1}, \\ \Lambda &= 22.9 \, R^{-0.45} \text{cm}^{-1}. \end{aligned} \tag{2.25}$$

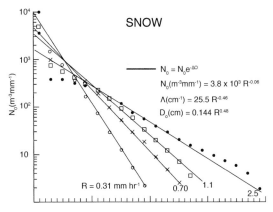

Fig. 2.38 The Gunn–Marshall size distribution. Adapted from Gunn and Marshall (1958).

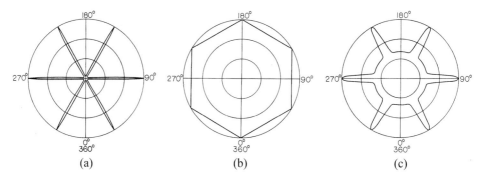

Fig. 2.39 Ice crystal shapes fitted by Eq. (2.26): (a) $a = 7.37$, $b = 50$, $c = 0.25$; (b) $a = 1.02$, $b = 0.35$, $c = 6.6$; and (c) $a = 4.22$, $b = 12$, $c = 3.4$. From Wang and Denzer (1983). Reproduced by permission of the American Meteorological Society.

Other investigators formed snow size distributions using the actual dimensions of the crystal instead of the melted diameter D_0.

2.9.5 *Mathematical representations of ice and snow crystal shapes*

The shapes of ice and snow crystals are even further away from spherical and hence a single "size" parameter is grossly inadequate to really describe their geometrical properties. To include the description of the basic crystal shape in addition to size, Wang and Denzer (1983) introduced the following simple expression in polar coordinates:

$$r = a\left[\sin^2(n\theta)\right]^b + c, \tag{2.26}$$

where r and θ are the radial and angular coordinates, and a, b, and c are fitting constants. The parameter n determines the number of symmetric sides of the crystal, which is simply $2n$. Hence for hexagonal crystals, $n = 3$. The two-dimensional shape and size of the crystal are specified once a, b, c, and n are fixed. One can also use (2.26) to generate various snow crystal shapes. Fig. 2.39 shows some examples. Both very skeletal crystals as well as hexagonal plates can be reproduced fairly closely.

Using this method, we can also form an ensemble of simulated snow crystals with certain shape and size distributions as shown in Figs. 2.40 and 2.41.

The expression (2.26) characterizes only 2D shape and size. The following expression can produce simulated 3D hexagonal columns and plates:

$$\frac{(x^2 + y^2)}{\left\{a - A\left[\frac{y^2(3x^2 - y^2)}{(x^2 + y^2)^3}\right]^B\right\}^2}\left(1 + \varepsilon - \frac{z^2}{c^2}\right) + \frac{z^2}{c^2} = 1. \tag{2.27}$$

Here ε is a very small number that one can set to make the edges of the crystal very sharp, and a, c, A, and B are fitting constants. The derivation and other details of the equation are

Fig. 2.40 A simulated sample of 15 ice crystals generated by Eq. (2.26) with the distributions of a, b, and c as given in Fig. 2.41. From Wang (1997). Reproduced by permission of the American Meteorological Society.

given in Wang (1997, 1999). Other equations that can generate simulated crystal shapes such as bullet rosettes and spatial dendrites are also given in these two references.

Aside from quantitative characterization of observed ice and snow crystals, these equations can also be used to generate simulated crystals for the purpose of theoretical studies such as the scattering of electromagnetic radiation from ice crystals, as they provide a convenient way to specify the inner boundary conditions for such problems.

There are currently no analytical expressions to simulate complicated snow aggregates that consist of many overlapping or interlocking single crystals.

Problems

2.1. This problem is designed to familiarize you with the equation. You do not need to worry about the units.

    (a) Use the theorems in calculus to prove that the Khrgian–Mazin (KM) size distribution equation (2.3) for cloud droplets is a unimodal (i.e. one-peak) distribution. Determine the modal drop size (the drop size with highest concentration).

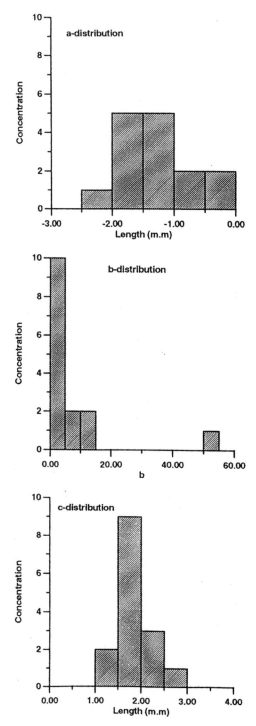

Fig. 2.41 The *a*, *b* and *c* distribution of the 15 ice crystals shown in Fig. 2.40. From Wang (1997). Reproduced by permission of the American Meteorological Society.

(b) Plot the KM distribution for $A = 1000$, 500, and $B = 0.4, 0.5, 0.6$ (so you should have six plots) for drop radius 0 to 20. Then discuss the effect of the parameters A and B on the distribution.

(c) Determine the mean drop radii, liquid water contents, the modal drop sizes, and the total number of drops for the distributions in (b).

2.2. This problem is for the log-normal distribution.

(a) Derive expressions for the mean and modal drop sizes, liquid water content and total drop concentration for the log-normal distribution in Eq. (2.9).

(b) Plot $n(a)/N$ vs. a for $a = 0$ to 50 and $\sigma = 0.25, 0.5$, and 1.0, and discuss the impact of σ on the distribution.

2.3. Plot the following curves using the function

$$r = a\left[\sin^2(n\theta)\right]^b + c$$

in **polar coordinates**. (It would be best if you could use computer software to do the plots. Make sure the x and y scales are the same. Otherwise, your plots may look distorted.)

(a) $n = 3$, $a = 0.5$, $b = 20$, $c = 0.0$

(b) $n = 3$, $a = -0.067$, $b = 0.397$, $c = 0.5$

(c) $n = 3.5$, $a = 0.5$, $b = 1.0$, $c = 0.1$.

(d) Discuss the effects of changing a, b, c, and n.

2.4. Plot the following function (in x, z coordinates):

$$x = \pm a\sqrt{1 - \left(\frac{z}{c}\right)^2}\cos^{-1}\left(\frac{z}{\lambda c}\right)$$

with

(a) $a = 2.4$, $c = 3.8$, $\lambda = 1.0$

(b) $a = 2.4$, $c = 3.8$, $\lambda = 3.0$

(c) $a = 2.4$, $c = 5.0$, $\lambda = 1.0$

(d) $a = 4.5$, $c = 5.0$, $\lambda = 15.0$.

(e) Discuss the effects of changing a, c, and λ.

2.5. Determine the mean distance between two drops for a cloud consisting of uniform droplets of 10, 20 and 30 μm radii for (a) $W_L = 0.1$ g m^{-3}, (b) $W_L = 0.5$ g m^{-3} and (c) $W_L = 1.0$ g m^{-3}.

2.6. (a) Normalize the size of ice crystals shown in Fig. 2.39 so that they all have a new radius 1. You will get a new set of (a, b, c) values for each crystal.

(b) Calculate the volume of these normalized ice crystals by assuming that they all have thickness $= 0.1$. Compare the volume of these ice crystals with that of a circular disk of the same radius and thickness and a sphere of the same radius. This will give you some idea about the water content of such crystals.

3

Molecular structures of water substance

Clouds are made of water in all three phases – vapor, liquid, and solid – and they can turn into each other via phase change under specific environmental conditions. We are visualizing these phase changes when we see clouds grow or dissipate, and when we see rain or snow coming out of precipitating clouds. To fully appreciate how these phase changes occur and what the end products are in such processes, we need to understand the molecular structure of water substance in different phases. We shall start with that of a single water molecule. Then we will discuss water vapor, which is essentially an ensemble of weakly interacting water molecules. Next we will look at the solid phase – ice – which is an ensemble of strongly interacting water molecules. We will treat liquid water last because, surprisingly, it is the most complicated of the three, even though it is probably the most familiar to us in our daily life.

3.1 Single water molecule

A water molecule is composed of two hydrogen atoms and one oxygen atom, H_2O, and hence is a triatomic molecule. But how do these atoms arrange themselves geometrically in a molecule? We call such structure the *molecular structure*. To fully explain it, we would need to invoke quantum mechanics, which is beyond the scope of this book. The discipline of quantum chemistry is the proper forum for that type of discussion. Here we will use the somewhat simplified "chemical bond" model to describe such structures.

A triatomic molecule may be arranged into two major types of structure: linear and nonlinear. An example of a linear triatomic molecule is CO_2, which has the structure $O{=}C{=}O$ (Fig. 3.1). A linear molecule has two rotational and four vibrational degrees of freedom. Water, on the other hand, is a nonlinear molecule, which has three rotational and three vibrational degrees of freedom. The importance

Fig. 3.1 CO_2 – a linear molecule.

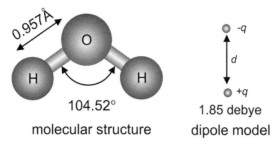

molecular structure   dipole model

Fig. 3.2 (a) The molecular structure of H_2O, showing the angle between the two covalent OH bonds. (b) The dipole model of the electric structure of the H_2O molecule.

of these degrees of freedom in this book is that different degrees of freedom result in different specific heats, which have a profound influence on the physical and chemical properties of matter. The molecular structure of water is shown in Fig. 3.2.

The length unit suitable for the molecular scale is the angstrom (Å), where $1\,\text{Å} = 10^{-10}$ m.

3.1.1 Electronic structure of the water molecule

The O—H bond is a covalent bond formed by a covalent electron pair. A somewhat simplified view of the electron pairs is that a water molecule has sp^3-hybridized electron pairs, arranged approximately in a tetrahedron. Two of the pairs are associated with hydrogen atoms to form the O—H bonds, leaving the two remaining lone pairs. Strict tetrahedral arrangement would result in an angle of 109.47° between bonds, which is not observed. Instead, the angle is about 104.52° when the atom is in equilibrium. This is because the atoms in the molecule are not static but are constantly in motion, which results in the different angular arrangement.

3.1.2 Electric dipole moment

The water molecule has a substantial electric dipole moment, with a value (Clough *et al.*, 1973) of about 1.85 D, where D = debye is a unit for dipole moment and 1 debye = 10^{-18} C cm = $3.335\,364 \times 10^{-30}$ C m. The dipole moment arises from the non-uniform distribution of electron density in the molecule. The electron pairs that form the two O—H bonds are actually closer to the oxygen atom, making the oxygen atom region more electronegative, whereas the two hydrogen atom regions are more electropositive, even though as a whole the molecule is electrically neutral. Consequently, to the first order of approximation, the electrical property of a water molecule can be represented as an electric dipole of 1.85 D. The large electric dipole moment of water implies that water molecules react easily with polar materials, such as salts. They will also respond strongly to an external electric field. Putting water substance under a strong external electric field may change its physical and chemical behavior substantially.

The idea that a water molecule can be considered as an electric dipole with point charges of equal and opposite signs separated by a distance d is called the point charge model of the water molecule. This model is useful in understanding and predicting many physical and chemical (especially electrochemical) properties of water substance.

3.1.3 Water isotopes

Since both hydrogen and oxygen atoms have three isotopes – namely, ^{1}H, ^{2}H (or D, deuterium), ^{3}H (or T, tritium), and ^{16}O, ^{17}O, and ^{18}O – their combinations form 18 possible kinds of isotopic water. Most of them (aside from the normal ^{1}H$_2$^{16}O, for which we will simply use H$_2$^{16}O) are, however, present in extremely small amounts or are not observed at all in our atmosphere. Only two of them, namely, HDO (^{1}H^{2}H^{16}O) and H$_2$^{18}O, are considered as stable isotopes in the atmosphere and hence have been better studied. Both isotopic species are heavier and tend to condense faster than the regular H$_2$O in the same environment. This property has been exploited by researchers, who use them as chemical tracers in the atmosphere. HDO has been utilized to understand the transport processes in the atmosphere (such as the stratosphere–troposphere exchange), whereas H$_2$^{18}O has been utilized to understand past climatic changes, especially in association with the temperature fluctuation over thousands of years.

3.2 Hydrogen bonds

Hydrogen bonds play a central role in the structure of water substance. The O—H bond discussed in the previous section is a covalent bond formed by the pairing of two valence electrons. The hydrogen bond, on the other hand, is not a covalent bond but a bond formed by the attractive force between a hydrogen atom and a more electronegative atom such as oxygen or nitrogen. In water substance, the hydrogen bond is an intermolecular interaction between an H atom of a water molecule and the

O atom of another molecule, and is commonly expressed as H$\cdots$O. The H bond is weaker than the O—H bond but stronger than the van der Waals force.

Note that a hydrogen bond can also be intramolecular, where an H atom in the molecule interacts with a different part of the same molecule, for example, in some organic molecules such as acetylacetone, but this is not the case for H_2O.

Hydrogen bonds are not considered important in water vapor (except maybe near saturation) but certainly are very important in liquid water and ice. The strong hydrogen bonding in water is responsible for the fairly high boiling point (100°C at 1 atm pressure) of water because it takes a lot of energy to break these bonds.

3.3 Structure of water vapor

Water vapor is the gaseous state of water substance and most of the time it is treated as an ideal gas. According to classical thermodynamics, an ideal gas consists of single molecules that do not interact except during collisions and are distributed and moving in a perfectly random fashion. Such an ensemble of molecules obeys the well-known ideal gas law that relates the pressure, temperature, and density (or volume) of that gas. The thermodynamic conditions that seem to fit the above description most closely will be gases with very weak dipole moment in a high-temperature (so very random) but low-pressure (very few in number concentration, hence very small chance of interacting with others) system.

But water vapor cannot be a true ideal gas because it is known that the water molecules in the vapor form *do* interact considerably before they collide. In fact, the substantial electric dipole moment in the H_2O molecule mentioned in the previous section already implies that there will be long-range interactions between molecules because the electromagnetic force is a long-range force. Experimentally, it has been found that water vapor consists of not only single individual molecules but also clusters of molecules. Dimers (two-molecule clusters), trimers (three-molecule clusters) and higher-order polymers are present in the vapor, and their concentrations depend on the degree of saturation; the closer to saturation, the greater the probability for having higher-order polymer clusters present. Some experiments show that as many as 21 molecules may exist in one cluster. Consequently, the behavior of water vapor very near saturation may deviate significantly from that given by the ideal gas law. Other scientific tools are needed to treat problems of water vapor behavior near saturation, such as nucleation and condensation.

One of the recent disciplines developed to treat this type of molecular behavior is *molecular dynamics* (MD), which utilizes high-speed digital computers with large memory to perform simulations of the detailed behavior of a large number of molecules of various chemicals as a function of time. Such simulations can provide more detailed dynamical information about the behavior of matter than can the classical bulk thermodynamics, which often just provides the conditions for equilibrium. The MD technique depends on knowledge of the interaction strength between molecules, usually expressed as the interaction potential (sometimes interaction "force" is used), which specifies how molecules will interact.

One of the simplest of these potentials is the Lennard-Jones (LJ) potential, sometimes called the 6–12 potential (due to the powers of the terms in the brackets in the equation below), which is

$$V(r) = 4\varepsilon \left[\left(\frac{\sigma}{r} \right)^{12} - \left(\frac{\sigma}{r} \right)^{6} \right], \tag{3.1}$$

where $V(r)$ is the potential, ε is the depth of the potential well, σ is the distance at which the interparticle potential is zero, and r is the distance between the molecules (Fig. 3.3). Since this potential considers only the interaction between two molecules, it is called the pair potential. The gradient of the potential gives the force, and hence the dynamic state of the molecules can be calculated. The first term in the brackets represents the short-range repulsive force potential due to the requirement of Pauli's exclusion principle for overlapping electronic orbitals, which fades away quickly as r increases. The second term represents the long-range attractive van der Waals force potential, which dominates when r is sufficiently large. For very large r, there is practically no interaction and $V(r) \rightarrow 0$.

But a molecule will interact with not just one but many other molecules around it, so the real system is a many-body dynamic system and the simple pair potentials

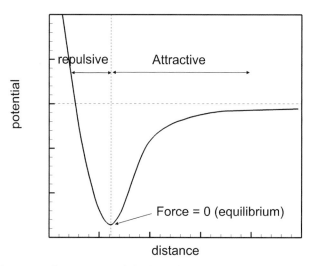

Fig. 3.3 The Lennard-Jones potential of Eq. (3.1).

such as the Lennard-Jones potential will be of limited capability in simulating realistically a more complex system. Instead of the pair potential, the many-body potential should be used. The potential functions used presently for making MD calculations typically involve the potential of interaction among many more molecules to achieve good simulation results.

While MD provides simulations that are extremely useful for the understanding of molecular structures, it is not a substitute for careful experimental measurements. This is because all simulations have certain empiricism in them. Sometimes, several different simulations using different sets of parameterizations produce similar-looking results. Real experimental data will be necessary to sort out the correct answer.

Aside from the near-saturation microphysical processes, water vapor can indeed be treated as an ideal gas without much error in most situations as far as the Earth's atmosphere is concerned. However, it must be noted that water vapor may also exist in the atmospheres of other planets or satellites in our Solar System where the pressure may be high enough that water vapor may no longer behave like an ideal gas.

Water vapor can also be supersaturated, i.e. it remains in the gaseous state even though bulk thermodynamics predicts that it should condense into liquid or sublimate into ice. Supersaturation plays a central role in the diffusion growth of water droplets and ice particles in clouds, as we will discuss in Chapter 9.

3.4 Molecular structure of ice

Ice is the solid form of water and its structure is crystalline in our atmosphere. The non-crystalline, or amorphous, form of ice is also possible but has never been observed in the atmosphere. The most prevalent ice molecular structure in the Earth's environment is *hexagonal* ice, called ice-I_h, which is the most stable form in the temperature range -80 to $0°C$. Ice-I_h is by far the most common form of ice we encounter in clouds. Those who have seen natural ice crystals, such as snowflakes, will easily notice their six-sided symmetry, and this is due to the hexagonal molecular structure, which results in the macroscopic hexagonal appearance of these particles.

3.4.1 Ice-I_h

As mentioned above, the molecular structure of ice-I_h has a hexagonal symmetry. In order to discuss such a structure, we need to define two axes, as shown in Fig. 3.4. The *c* axis is the axis normal to the basal face, and the *a* axis is the one parallel to the basal face. The six side surfaces are called the prism faces.

Molecular structures of water substance

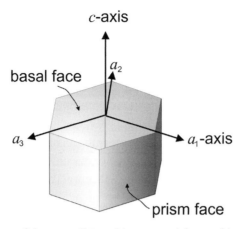

Fig. 3.4 Definitions of the crystallographic axes and faces of ice-I_h.

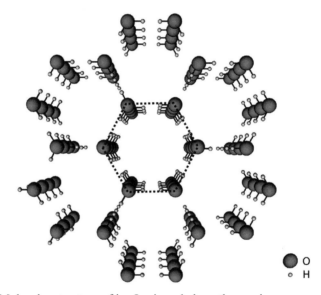

Fig. 3.5 Molecular structure of ice-I_h viewed along the c axis.

Figs. 3.5 and 3.6 show views of the hexagonal molecular structure of ice along the c axis and along the a axis, respectively. The large spheres represent the oxygen and the small spheres represent hydrogen atoms.

The hexagonal structure is most obvious when viewed along the c axis. In this view (Fig. 3.5), the water molecules line up in a close-packed hexagonal structure similar to a honeycomb. In the middle of the hexagon is a vacant shaft. But the similarity to a honeycomb stops here. When viewed along the a axis (Fig. 3.6), the oxygen atoms on a "hexagonal unit" do not line up at the same level but are

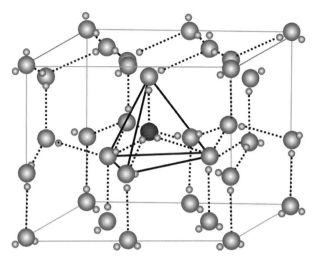

Fig. 3.6 Molecular structure of ice-I_h viewed along the *a* axis. The tetrahedron with the dark molecule in the center represents a unit cell of the ice crystal. It is seen that the center molecule has four nearest neighbors.

alternately raised and lowered. This structure is called open-puckered hexagonal rings. This arrangement of the oxygen atoms is isomorphous with the Wurtzite structure of the ZnS crystal.

The region enclosed by the thick black lines in Fig. 3.6 is called a unit cell, and forms a tetrahedron. The oxygen atom inside the unit cell (the darker sphere) is surrounded by four nearest-neighbor oxygen atoms (located at the vertices of the tetrahedron) with a mean distance of 2.76 Å.

The above structure illustrates mainly the positions of the oxygen atoms, which is easier to figure out as they are nearly the same as the H_2O molecule positions. The arrangement of the hydrogen atoms is somewhat more complicated and is subject to the Bernal–Fowler (BF) rules. The BF rules are as follows:

(1) each water molecule is oriented such that its two hydrogen atoms are directed approximately toward two of the four surrounding oxygen atoms (arranged almost in a tetrahedron);
(2) only one hydrogen atom is present on each O—O linkage; and
(3) each oxygen atom has two nearest-neighbor hydrogen atoms such that the water molecule structure is preserved.

Defects in ice-I_h

When an ice crystal strictly obeys the BF rules, it is called an *ideal ice crystal*. But most of the time this is not the case and violations of the BF rules lead to defects in

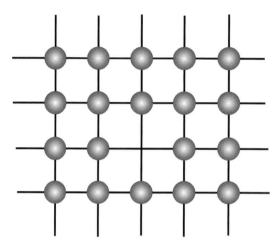

Fig. 3.7 A schematic representation of the Schottky defect of a general crystal.

ice crystals. Note that there are other defects in ice crystals in addition to those due to the violation of the BF rules.

   The following are a few common defects seen in ice crystals. Note that the illustrations associated with these defects are simplified representations for conceptual understanding only and do not correspond to real ice crystals; hence the hexagonal structure is not depicted here.

(1) *Vacancies* (also called Schottky defects). These occur when one or more water molecules are missing from the regular ice lattice (Fig. 3.7).

(2) *Interstitial defects* (also called Frenkel defects). These occur when one or more water molecules occupy irregular (interstitial) positions in the lattice (Fig. 3.8).

(3) *Chemical defects*. These occur when foreign ions are built into the lattice instead of water molecules. They often happen when ice is grown from an aqueous solution (Fig. 3.9).

(4) *Ionized states*. Normally there should be only two hydrogen atoms close to one oxygen atom in a water molecule. Sometimes another hydrogen atom is closing in near this molecule, temporarily forming an H_3O^+ ion, or one of the original two hydrogen atoms moves too far from the molecule, temporarily forming an OH^- ion. Both violate the BF rule 3. These are called ionized states (Fig. 3.10).

(5) *Orientation defects* (also called Bjerrum defects). These occur either when two hydrogen atoms occupy positions between one O—O link, so that the bond is O—H$\cdots$H—O (known as a D defect), or when there is no hydrogen present in one link and the bond becomes O$\cdots$O (known as an L defect). The orientation defects violate both BF rules 1 and 2 (Fig. 3.11).

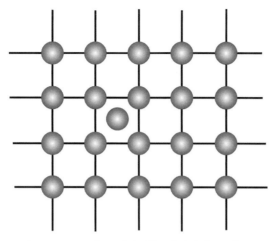

Fig. 3.8 A schematic representation of the Frenkel defect of a general crystal.

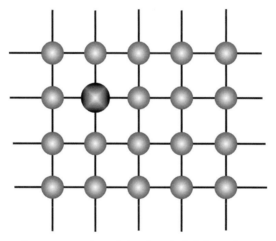

Fig.3.9 A schematic representation of the chemical defect of a general crystal. The dark and larger sphere represents a non-H_2O molecule.

(6) *Dislocations*. The defects 1–5 are called point defects. Dislocations, on the other hand, are line defects. There are two main categories of dislocations: (a) edge dislocation (Fig. 3.12) and (b) screw dislocation (Fig. 3.13).

(7) *Stacking faults*. As we mentioned above, ice crystals can be cubic instead of hexagonal (see later). When layers of cubic ice intermix with regular hexagonal layers, stacking faults result. This may happen when $T < -80°C$.

Many of these defects not specific to ice are discussed in more detail in standard solid-state physics textbooks (e.g. Kittel, 1996). Defects affect the mechanical,

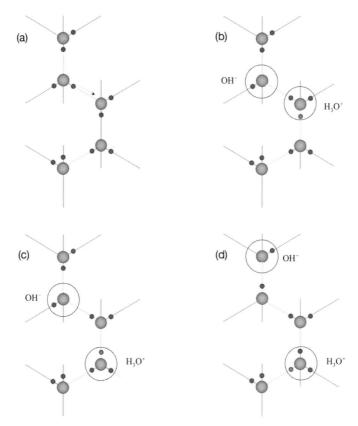

Fig. 3.10 The ionized state defect of an ice crystal. After Hobbs (1974).

electrical, and chemical properties of ice. Hobbs (1974) gave more detailed discussions on this topic.

Quasi-liquid layer on ice surface

Many studies point out that quasi-liquid layers (QLLs) (also called pseudo-liquid layers) exist on the ice crystal surface when close to the melting point. These layers may be of critical importance in cloud physics, as the surface is where the ice crystal interacts with the outside world. For example, chemical reactions between a QLL and a gas will be very different from that between the gas and the strictly latticed ice molecules. How electrification may occur in thunderclouds due to mechanisms involving ice particles may depend on the behavior of QLLs as well. Previously, it was assumed that the QLLs appear uniformly over the ice crystal surface. Recently, Sazaki *et al.* (2012) performed an experimental study on such QLLs by using advanced optical microscopy on ice and pointed out there are two types of mutually immiscible QLLs on the surface. They appear heterogeneously, move

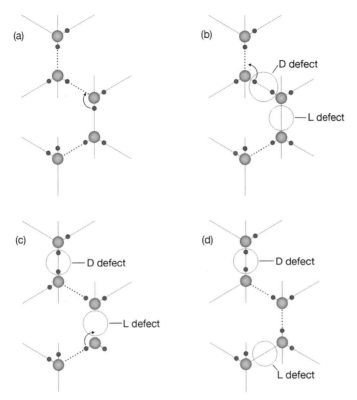

Fig. 3.11 The Bjerrum defects of an ice crystal. The D defect occurs when there are two hydrogen atoms between an O–O link. The L defect occurs when there is no hydrogen atom between an O–O link. After Hobbs (1974).

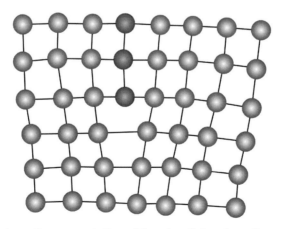

Fig. 3.12 A schematic representation of the edge dislocation of a crystal.

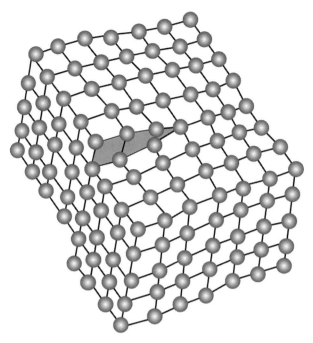

Fig. 3.13 A schematic representation of the screw dislocation of a crystal.

around, and coalesce dynamically on ice crystal surfaces. This behavior of surface melting is quite different from the conventional picture.

3.4.2 Ice-I_c

In the temperature range −100 to −80°C, a different form of ice, cubic ice or ice-I_c, may exist. Cubic ice may play a role in very high cirrus clouds whose environmental temperatures fall in this range, for example, in the upper tropical troposphere close to the tropopause and also the mesosphere, although the mesosphere is not currently thought to be important in cloud physics. Cubic ice is a metastable state crystal and will change to the stable ice-I_h. According to Murphy (2003), below about 200 K cubic ice nucleates first, then transforms to hexagonal ice over a period of minutes to days. The arrangement of the oxygen atoms in cubic ice is similar to the carbon atoms in diamond. Certain features in some halos cannot be easily explained by the refraction of sunlight through hexagonal ice but can be attributed to cubic ice.

A recent study utilizing X-ray diffraction data and Monte Carlo simulations by Malkin *et al.* (2012) showed that many previous identifications of cubic ice were probably erroneous. Instead, they suggest that the ice formed from homogeneous nucleation of supercooled water drops (which occurs at $T \leq −40°C$) is most likely an ice that consists of random stacking of ice-I_h and ice-I_c, i.e. a kind of stacking fault structure. They denote this kind of ice as ice-I_{sd}, where the subscript "sd" stands for stacking disorder.

Whether high tropical cirrus clouds consist of ice-I_h, ice-I_c or ice-I_{sd} would make some differences in estimating their impact on the climate system. For example, cubic ice has higher saturation vapor pressure and hence may produce larger ice particles, causing more fallout and greater dehydration in the tropical tropopause layer. Ice-I_{sd} has a rougher surface than other ice and may have stronger scattering ability of light that affects cloud radiative forcing.

3.4.3 Other forms of ice

There are 12 different forms of ice that have been identified so far and the two (not counting I_{sd}) discussed above are the important ones in the clouds of our atmosphere. The other ice forms mainly exist under very high-pressure conditions that are irrelevant to our atmosphere. However, there are other atmospheres in the Solar System, for example, the atmospheres of the outer planets (Jupiter, Saturn, Uranus, and Neptune), in which high pressures are possible and hence some of these ice forms may be present in their clouds.

3.5 Molecular structure of liquid water

Liquid water is probably the most familiar chemical liquid to mankind and also the most important. The human body consists of ~80% liquid water and water is crucial to almost all life forms on Earth. Yet it is harder to describe the structure of liquid water than that of water vapor or ice. Indeed, this is not just a problem about liquid water, but about any liquid in general. We know more about the physics of gases and solids than we do about liquids.

First of all, it appears that liquid water consists not only of neutral H_2O molecules but also of ions such as H_3O^+ and OH^-, and so in fact it is a mixture of all these species. Thus to understand the physical and chemical behavior of liquid water, it is in principle necessary to consider the nature of this mixture of these species.

One of the most useful tools of science to understand a phenomenon or a system is to construct an "ideal model". For gases, we have the ideal gas model – a system of perfectly random, non-interacting particles. For solids, we have the perfect crystal model – molecules sit on a perfectly ordered crystal lattice obeying all the configuration rules such as the BF rules we described for a perfect ice crystal. Both ideal models have been used profitably for us to understand the basic properties of gases and solids.

But no such "ideal liquid model" exists in the same sense as the ideal gas or perfect crystal models. The term "ideal liquid" does occur in some scientific

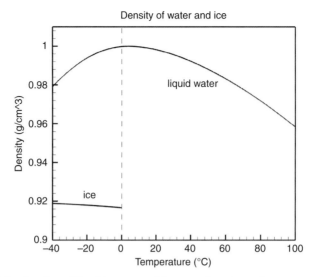

Fig. 3.14 The density of liquid water and ice as a function of temperature.

literature, but it refers either to the non-viscous (inviscid) flow condition of a liquid, which is a fluid mechanical property, or to the ideal solution related to Raoult's law (see Chapter 4). Neither refers to the basic liquid structure we are discussing here. We cannot say a liquid is 50% like a gas and 50% like a solid, although a liquid seems to have structural properties that are somewhere between those of a solid and a gas. A liquid has fairly constant volume and density similar to that of a solid, but can be deformed to fit a container of any shape and can flow easily like a gas. It also cannot withstand a shear stress, again a fluid property.

Liquid water is even more complicated than many liquids. It is the only liquid that has higher density than its solid form. If you put high pressure on ice, the ice will turn into liquid water! These bulk thermodynamic properties of water will be discussed in more detail in Chapter 4.

Fig. 3.14 shows the density of liquid water and ice as a function of temperature. At 0°C, the density of liquid water is 0.99984 g cm^{-3} whereas that of ice is 0.9167 g cm^{-3}, so ice is about 9% less dense than liquid at that temperature. As is well known, liquid water density has a maximum (0.999972 g cm^{-3}) at $T \sim 4$°C.

Such behavior of the density baffled many scientists, and many theories of the molecular structure of liquid water have been forwarded to explain it. While some theories might have seemingly explained it for a while, new experimental data keep emerging to defy them.

One of the most important techniques to study the molecular structure of matter is X-ray diffraction. The X-ray diffraction study shows that liquid water possesses short-range order but no long-range order. Fig. 3.15 shows a comparison between

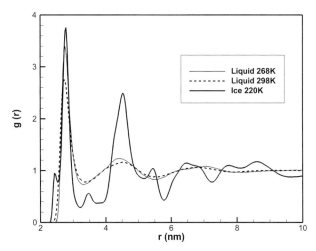

Fig. 3.15 The radial distribution function of liquid water at 268 K and 298 K, and of ice at 220 K, based on X-ray diffraction data. Data courtesy of Dr Alan Soper.

the distribution functions of the O—O link for liquid water and ice. It is clear that ice, being a crystal, exhibits more distinct peaks in $g(r)$, whereas liquid water has much less distinct peaks as r increases.

This seems to imply that water molecules may have more or less ordered structure (like ice) in a short distance but are distributed randomly over a long distance. Another interesting observation is that the peaks of supercooled water (268 K) are more distinct than that of water at room temperature (298 K).

Many theoretical models have been proposed to explain the structure of liquid water, especially the short-range order feature, and the following are a few of them.

(1) *Broken ice model.* This model suggests that liquid water consists of small broken pieces of ice clusters so that it exhibits some kind of orderliness over a short distance. But over a long distance, the orderliness disappears. The sizes (they do not have to be uniform) and/or the kinds of pieces (some can assume a quartz instead of ice-I_h structure) can be adjusted to fit the experimental data. One can use such a model to explain, for example, why liquid water has a higher density than ice. The model suggests that, instead of rigidly held (thus more open) ice lattice structure, these broken ice pieces can intrude into each other, forming a tighter, and hence higher-density, configuration.

(2) *Interstitial model.* Another way to "produce" short-range order but long-range randomness is to assume that there is a regular H bond structure but within it there are freely moving or partially bonded interstitial molecules. The configurations of these interstitial molecules can then be adjusted to fit experimental data.

(3) *Bent-bond model.* Rather than assuming a regular H bond structure in the above, one can assume that these bonds are bent at different degrees to achieve similar short-range order but long-range randomness effect.

(4) *Clathrate model.* The term clathrate initially referred to crystalline solids of gas hydrates, where water molecules surround a foreign gas molecule (e.g. N_2, O_2, CH_4 or CO_2) and form a cage like a "buckyball" (buckminsterfullerene) or soccer ball with a different number of faces. Some suggest that molecules in liquid water can arrange themselves into clathrate-like cages, with different numbers of faces of course.

(5) *Flickering cluster model.* In the crystalline ice structure, the H bonds that help to form the tetrahedral cells are considered as a quasi-steady state, i.e. they may move a little but otherwise stay in the same region. The flickering cluster model of liquid water assumes that these H bonds are not steady but keep forming and disappearing here and there. In this way, old clusters of molecules may dissolve while new clusters form in a matter of femtoseconds (10^{-15} s).

A view held by many researchers is that over a short range a liquid water molecule sometimes has four and sometimes three hydrogen bonds, so on average it has 3.5 H bonds. MD simulations using 3.5 bonds seem to give the right answers to many liquid water properties. But, as pointed out earlier, there is some empiricism involved in the MD approach, and hence the 3.5 H bond model should not be taken as the final word for the liquid water structure.

The most recent work on liquid water structure involves a new experimental technique called small-angle X-ray scattering (SAXS), which was used by Huang *et al.* (2009) for studying liquid water density fluctuation on a length scale of about 1 nm. The X-ray diffraction interaction time is on the order of an attosecond (10^{-18} s), which is 10^6 times longer than the H bond dynamics ($\sim 10^{-12}$ s); hence the SAXS data can be viewed as an instantaneous snapshot of the structure. Huang *et al.* (2009) concluded that on the time scale of X-ray scattering and spectroscopic processes, two local structural species coexist with tetrahedral-like patches of dimension of order 1 nm in dynamic equilibrium with the H bond distorted and thermally excited structure. The tetrahedral-like component is relatively insensitive to temperature whereas the H bond component continuously changes as it becomes thermally excited and expands. The two local structures indicate the coexistence of low- and high-density water. This leads to the density fluctuation in ambient water that is retained with decreasing temperature while the magnitude is enhanced. No such enhancement with decreasing temperature is observed for a normal fluid such as tetrachloromethane (carbon tetrachloride, CCl_4). MD simulations of water under ambient conditions also do not predict such enhancement.

Liquid water can be supercooled, i.e. it remains in the liquid state even though the bulk thermodynamics predicts that it should freeze into ice when the freezing point (e.g. 0°C at $p = 1$ atm) is reached. In fact, supercooling is the norm rather than the exception in atmospheric clouds and is responsible for the Bergeron–Findeisen process (see Chapter 4) and the riming formation of graupel and hail (see Chapter 11).

Problems

3.1. (a) Show that the Lennard-Jones potential in Eq. (3.1) can be written as

$$V(r) = \varepsilon \left[\left(\frac{r_{min}}{r} \right)^{12} - \left(\frac{r_{min}}{r} \right)^{6} \right],$$

where r_{min} is the distance where the potential reaches its minimum. What is the relation between σ and r_{min}?

(b) In Eq. (3.1), the quantities σ and ε are matter-specific. For water, $\varepsilon = 0.650\,1696$ kJ mol^{-1} and $\sigma = 0.316\,5555$ nm. Determine the r_{min} value for water.

(c) Derive an expression for the force between the two molecules by taking the gradient of the potential, i.e. $\vec{F} = -\nabla V$.

(d) Calculate the force at distances $r = 0.5r_{min}$ and $r = 1.5r_{min}$. Discuss the properties of these forces.

3.2. Another widely used pair potential is the Stockmayer potential:

$$V(r, \theta_1, \theta_2, \varphi) = 4\varepsilon \left[\left(\frac{\sigma}{r} \right)^{12} - \left(\frac{\sigma}{r} \right)^{6} \right] - \frac{\mu_1 \mu_2}{4\pi\varepsilon_0 r^3} (2 \cos \theta_1 \cos \theta_2 - \sin \theta_1 \sin \theta_2 \cos \varphi),$$

where μ_1 and μ_2 are the dipole moments of molecules 1 and 2, respectively, θ_1 and θ_2 are the inclination angles of the molecular dipole axis with respect to the intermolecular axis, φ is the azimuthal angle between μ_1 and μ_2, and ε_0 is the permittivity of the vacuum (not to be confused with ε). We see that this potential is the LJ potential modified by the dipole interaction. Since the water molecule is polar and has substantial dipole moment, the Stockmayer potential would predict a force field somewhat different from that predicted by the LJ potential. Can you make a general assessment of what the differences may be?

4

Bulk thermodynamic equilibrium among water vapor, liquid water, and ice

All three phases of water – vapor, liquid, and solid – are present in atmospheric clouds. As a cloud evolves, the relative proportions of water in the different phases within it will change in response to the changing cloud environment. The environment, in turn, will be changed in response to the changes in water substance. Thus the water substance in a cloud and the cloud environment are not independent of each other but are "coupled" – changes in one will lead to changes in the other. They may influence each other in many ways, but one of the most important pathways of such influences is via thermodynamics.

In this chapter, we will discuss in some detail the thermodynamic equilibrium of water of different phases: how different phases of water can turn into each other and, when this happens, what changes may result in the cloud. We will discuss the thermodynamic equilibrium not only of pure water but also of aqueous solutions, as the latter play a very important role in cloud formation.

4.1 Thermodynamic systems

Let us begin by reviewing the simple definitions of a few thermodynamic terms:

System A thermodynamic system is a specific part of the universe that is separated from the rest of the universe by *boundaries*. The "rest of the universe" will be called the *environment* of this system.

Depending on the properties of the boundary, we can have either open systems or closed systems:

Closed system A closed system is one with a boundary that does not permit mass to be exchanged between the system and its environment. Note that the definition depends solely on mass. Other quantities, e.g. energy, *can* be exchanged between the system and its environment and yet the system is still a closed system as long as the masses in to and out of the system cannot be exchanged.

Open system   An open system is, by definition, one with a boundary that allows mass exchange between the system and its environment. Exchange of mass may or may not involve exchange of energy.

Another thermodynamic system that is mentioned in the literature sometimes is the *isolated system*. This refers to a system that has no interaction with its environment at all. This of course means no exchange of masses either and is hence a special case of a closed system. The main purpose of using the concept of an isolated system is to focus on the internal structure of the system and pay attention to neither the influence of nor the influence on the outside.

State   A set of physical quantities that are used to describe the condition of a system is called a state. To describe the precise thermodynamic condition of a system, we can specify the pressure, volume, temperature, entropy, etc., of the system. Those physical quantities that we use to describe the thermodynamic condition of a system are called the *state variables*. Not every physical quantity can be called a state variable; for example, the physical quantity "heat" cannot be used as a thermodynamic state variable despite being of central importance to thermodynamics. More detailed discussions about state variables can be found in standard textbooks on thermodynamics (e.g. Reif, 1965).

Although we use the word "boundary" to represent the division between a system and its environment, we do not really require that there exists a material boundary. Sometimes a boundary can be purely conceptual or imaginary, just for the convenience of our discussion and understanding of the problem at hand, whereas in reality nothing material is present at the boundary. This point will become clearer when we study specific examples.

4.2 The first law of thermodynamics – conservation of energy

There are many so-called laws of thermodynamics. These laws are not universal laws (i.e. physical laws that are derived from certain principles and supposedly applicable to any situation, such as Newton's law of gravity) but rather they are limiting laws. They are deduced from experience – from numerous observations or experiments, but not derived from first principles. They are valid as long as there are no observations or experiments showing the opposite. We will not present a comprehensive review of thermodynamic laws here. Instead, we will focus on those few laws that are useful to us.

The first law of thermodynamics is about the conservation of energy. It basically states that the total energy of a system will remain constant as long as no energy is added or subtracted from the system. This seemingly self-obvious statement can lead to very useful conclusions when applied to some specific situations.

Another way of stating this law is that energy cannot be created nor destroyed. If you add some energy to or subtract some energy from the system, then this energy change must do something to the system and cannot disappear to nowhere. Specifically what will happen to the system depends on the thermodynamic process that the system is undergoing at the time. We will explain the thermodynamic processes by specific examples.

One point worth mentioning here is that, whereas energy is conserved if no addition or subtraction of energy occurs, the *forms* of the energy can change from one category to other categories, for example, from thermal energy to mechanical energy.

For those who are worrying about processes involving *nuclear reactions* where energies may turn into mass of particles and vice versa, let us say that the first law of thermodynamics there would include not just the conservation of energy but the conservation of energy *and* mass. With such a generalization, the law is still valid. In the context of this book, we will not worry about nuclear reactions and no such complications will occur.

4.3 Closed systems

Let us examine first a closed system. As defined above, this is a system with a boundary that does not permit mass exchange to occur between the system and its environment. Suppose this system has a total energy Q and is initially in equilibrium with its environment with a constant pressure P, as illustrated in Fig. 4.1.

So long as we do not disturb this equilibrium, the balance will be maintained. The total energy of the system will remain at Q and the pressure in the system, being in equilibrium with outside, is also P.

Now let us see what will happen if we add a little energy $đQ$ to the system. Note that we use the symbol $đQ$ instead of dQ to declare that this infinitesimal energy is not an *exact differential*, a point that will be explained later. According to the first law of thermodynamics, this little energy must do something to the system and cannot disappear to nowhere. The question is: what will it do?

Let us make this system as simple as possible so that we can understand the changes clearly. We will assume that neither chemical reaction nor phase change will occur in the system. Under those circumstances, the added energy $đQ$ may be spent by (1) warming up the system and (2) allowing the system to expand (which requires expansion work). Writing this statement in the form of an equation, we have

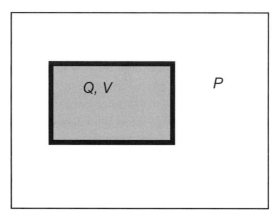

Fig. 4.1 A representation of a closed system with total energy Q and volume V in equilibrium with an environment of constant pressure P. The solid box represents the impenetrable boundary that does not allow materials to go in to and out of the system.

$$đQ = C_v dT + PdV. \tag{4.1}$$

The first quantity on the right-hand side represents the energy spent in warming up the system from a temperature T to $T + dT$. The quantity C_v is called the *heat capacity at constant volume*. It is the energy necessary to raise the temperature of the materials in the system by one degree. The second term is the expansion work done by the system against the constant pressure P of the environment, whereas dV is the increment of the volume of the system.

The two quantities on the right-hand side of (4.1) represent only what *may* happen. It is entirely possible to have a system warmed up without changing its volume (hence no expansion work is done). In this case, the system undergoes an *isochoric* (constant-volume) process. Alternatively, it is also possible that the system expands (or shrinks) without changing its temperature. In this case, the system undergoes an *isothermal* (constant-temperature) process. This is to say that the specific response of a system subjected to an external disturbance depends on the specific process that the system is undergoing.

4.4 Adiabatic process for a closed system

In the above section, we used an infinitesimal energy $đQ$ to cause the system to deviate from its equilibrium condition. The addition of $đQ$, however, is not always necessary. There is a special kind of process via which the system can change its equilibrium condition and yet the total energy Q is unchanged (so $đQ = 0$). This process is called an *adiabatic* process.

For an adiabatic process, we have from (4.1) that

$$0 = C_v dT + PdV \tag{4.2}$$

for a closed system. Eq. (4.2) says that either the two terms on the right-hand side are both zero (hence nothing really happens) or the two will have opposite signs – if one is positive, the other must be negative. This statement has important implications for atmospheric processes and we will use a specific example to illustrate.

Suppose that we have a dry air parcel (containing no water vapor) near the surface rising in the atmosphere. If the air parcel rises fast enough, there would be little time for the parcel to mix with its environment and little mass exchange occurs; hence we can approximate it as a closed system. We further assume that it does not exchange energy with its environment, so that the total energy Q of the air parcel remains constant. In effect, we have a closed system undergoing an adiabatic process and hence Eq. (4.2) applies.

We can even get some ideas about what would happen to this air parcel. As the air parcel goes up, it will experience a decrease of environmental pressure because the pressure decreases as one goes up in the atmosphere. This of course will result in expansion, as the air parcel would like to achieve mechanical balance with the outside pressure. This means that the term dV and hence the whole second term on the right-hand side of (4.2) is positive because P is always positive. Then according to our understanding of (4.2), the first term should be negative. Since the heat capacity C_v is always positive, this can only mean that dT is negative, that is, the system cools. Thus as the air parcel rises adiabatically, it expands and, at the same time, cools. Since this cooling is a result of the adiabatic ascent, it is sometimes called *adiabatic cooling*.

We can even use Eq. (4.2) to obtain a quantitative cooling rate as the air parcel ascends adiabatically. First of all, we convert the two quantities on the right-hand side to the "per unit mass" basis by dividing both sides by the total mass of the air parcel m. Thus we get

$$0 = \frac{C_v}{m} dT + P\frac{dV}{m} = c_v dT + Pd\alpha, \tag{4.3}$$

where c_v is called the *specific heat* at constant volume and α is the *specific volume* of the air parcel. Remember that our goal is to obtain a cooling rate as the parcel rises, i.e. we are looking for a quantity of the form dT/dz, where z is the vertical coordinate.

We begin by using the ideal gas law for dry air:

$$P\alpha = R_a T, \tag{4.4}$$

where R_a is the gas constant of dry air (not the *universal gas constant*). Now if we differentiate both sides of (4.4), we obtain

$$Pd\alpha + \alpha dP = R_a dT. \tag{4.5}$$

The air parcel in the above discussion contains only dry air, but our conclusions will be the same even if it contains some water vapor as long as the water vapor is of small amount (which is always the case in the Earth's atmosphere) and does not condense into liquid water or ice.

Using (4.5), we can change (4.2) into

$$c_v dT + R_a dT - \alpha dP = 0$$

or

$$(c_v + R_a)dT - \alpha dP = c_p dT - \alpha dP = 0, \tag{4.6}$$

where c_p is called the specific heat of air at constant pressure.

The pressure of the atmosphere is related to height z. Except in some specific cases (such as in regions of strong convection), the atmosphere is very nearly in hydrostatic equilibrium and the pressure–height relation is given by the hydrostatic equation:

$$dP = -\rho_a g dz, \tag{4.7}$$

where ρ_a is the air density. Upon substituting (4.7) into (4.6) and noting that $\alpha \rho_a = 1$, we have

$$\Gamma_d = -\frac{dT}{dz} = \frac{g}{c_p}. \tag{4.8}$$

This is called the *dry adiabatic lapse rate*, which is the cooling rate of the dry (or unsaturated) air parcel as it ascends. This is obviously nearly a constant (at least in the lower atmosphere, where g is nearly constant) and has the value of ~ 9.76 K km^{-1}, or roughly 10 K km^{-1}. The rate would change slightly if there is water vapor in the air parcel, but the magnitude is small enough to be negligible for the possible amount of water vapor in air.

4.5 A simple conceptual model for small cumulus cloud formation

Based on the discussions in the previous section, we can form a simple conceptual model of cloud formation for small cumulus. Consider an air parcel containing mostly dry air and a certain amount of water vapor. As the parcel rises adiabatically, it expands and cools at a rate 9.76 K km^{-1}. As the air parcel cools, the water vapor in the air parcel becomes closer to saturation because the degree of saturation becomes higher for the same amount of water vapor but colder temperature.

The degree of saturation can be measured by a quantity called the *relative humidity* (*RH*) defined as

$$RH(\%) = \frac{e}{e_{\text{sat}}} \times 100, \qquad (4.9)$$

where e_{sat} is the *saturation vapor pressure* of the water vapor. This is the vapor pressure required to make the air parcel saturated. In principle, any small excess of vapor pressure beyond saturation value will cause the excess amount of vapor to condense into liquid water or ice. If we approximate water vapor as an ideal gas, which is very nearly so in most cases in our atmosphere, then e_{sat} is only a function of temperature T. The colder the temperature, the lower is the saturation vapor pressure. This means that you do not need much water vapor to saturate the air at lower temperatures. The quantity e, on the other hand, is the actual vapor pressure present in the air parcel.

So, if the air parcel rises high enough, it will reach a level at which the temperature becomes cold enough to saturate the air parcel, and any further ascent (hence cooling) will cause excess water vapor to condense – and you will start to see cloud appearing! The level where the air parcel becomes saturated due to adiabatic ascent is called the *lifting condensation level* (LCL). The LCL can be determined graphically from adiabatic charts (such as the so-called skew-T log-P chart) for a specific sounding.

Further ascent of the air parcel above the LCL would cause more condensation (as long as there is still excess water vapor in the parcel) and hence further cloud growth. But the thermodynamic process in this stage is no longer dry adiabatic because of the release of latent heat due to the phase change of water. We will discuss this process in more detail in Chapter 12 when we treat the topic of cloud dynamics.

The above discussion is based on the assumptions that the air parcel is a closed system and the ascent is adiabatic. Neither assumption is realistic. The air parcel is not separated from its environmental air by any material boundary at all but would easily mix with the environmental air during its ascent. The mixing can be strong near the cloud boundary when the rising is rapid because the flow is turbulent. This turbulent mixing process is called the *entrainment* process – usually the drier and colder environmental air is entrained into the warmer and more humid air parcel. Thus not only material but also energy is exchanged between the parcel and its environment – thus the system is not a closed one and the process is not adiabatic. Aside from these, the presence of aerosol particles and possibly also trace gases in the air parcel may influence the condensation behavior. More discussions of this point will be given in later chapters.

4.6 Entropy

Before we turn our attention to open systems, we first introduce the concept of entropy S, which is very useful to the study of thermodynamics. We said in Sec. 4.3 that the energy Q cannot be used as a state variable because dQ is not an exact differential. If an infinitesimal quantity $dA(x, y)$ is an exact differential, then we can write

$$dA = \frac{\partial A}{\partial x} dx + \frac{\partial A}{\partial y} dy, \qquad (4.10)$$

but the nature of dQ is such that (4.10) does not apply, as we indicated before. This greatly complicated the analysis of systems in the early days of thermodynamics. Then it was found later that if we define a quantity, called the entropy S of the system, such that

$$dS = dQ/T, \qquad (4.11)$$

then magically we can write

$$dS = \frac{\partial S}{\partial x} dx + \frac{\partial S}{\partial y} dy, \qquad (4.12)$$

i.e. dS is an exact differential and hence S can be used as a state variable of the thermodynamic system.

This was originally thought of as a convenient vehicle for the sake of thermodynamic analysis. Later the Austrian physicist Boltzmann pointed out that entropy has its own physical meaning and has profound implications to the foundation of physics. He summarized his discovery by the famous Boltzmann's law:

$$S = k \ln W, \qquad (4.13)$$

where $k = 1.3804 \times 10^{-23}$ J K^{-1} is the *Boltzmann constant* and W is the number of possible configurations to arrange the particles in the system. For example, when the temperature of a perfect crystal system is at the so-called absolute zero (0 K), all molecules are supposedly frozen at fixed positions and hence there should be only one possible configuration of the particles, i.e. $W = 1$. This means, of course, $\ln W = 0$ and hence the entropy is zero. This is only true in a classical system because the requirement of quantum mechanics actually results in non-zero entropy even at 0 K, but this is beyond the scope here.

In view of (4.11), we have

$$dQ = TdS. \qquad (4.14)$$

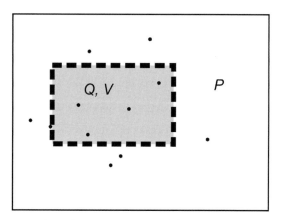

Fig. 4.2 A representation of an open system with total energy Q and volume V in equilibrium with an environment of constant pressure P. The dashed box represents the open boundary that permits materials to go in to and out of the system.

4.7 Open systems

Let us go back to Eq. (4.1), which is the conservation of energy for a closed system, and consider what changes we need to make if it is an open system. By definition, an open system allows mass to be exchanged between the system and its environment (see Fig. 4.2).

We will assume that this system contains a total of n moles of matter consisting of c components (or species) in it. Thus, when a little energy dQ is added to this open system initially in equilibrium with its environment, one more possibility – mass exchange – may occur to the system aside from the possibilities of warming up and expansion. Written in the form of an equation, this becomes

$$dQ = TdS = C_v dT + PdV - \sum_{k=1}^{c} \mu_k dn_k, \qquad (4.15)$$

that is, there is one more term on the right-hand side representing the effect of mass exchange. Here n_k is the number of moles of component k in the system. The term μ_k is called the *chemical potential* of component k. It is a measure of the tendency of that component to escape from the system. If the chemical potential μ_k of a component k in the system is greater than its counterpart in the environment, then some portion of this component will exit from the system when the equilibrium is disturbed. The chemical potential works in a similar way to the geopotential. A particle in a gravitational field will move from a position at higher geopotential to one at lower geopotential (assuming no other forces, of course). If you throw a ball off the top of a building, the ball will fall *downward* because it has a higher geopotential at the top of the building than below.

The minus sign of this last term deserves some attention. To see this more clearly, we can move this term to the left-hand side, so we would have $đQ + \mu_k dn_k$ there. Positive $đQ$ and $\mu_k dn_k$ indicates energy *gained* by the system. If there is a net exodus of particles of component k from the system, then $dn_k < 0$ and the term $\mu_k dn_k$ is negative, indicating net *loss* of energy of the system to the outside. If there is a net gain of component k of the system, then $dn_k > 0$, and the term $\mu_k dn_k$ becomes positive, i.e. net energy is *gained* by the system. Of course, it is necessary to sum over all components to obtain the net value of this term.

In a closed system, $dn_k = 0$ in (4.15) and the equation becomes the same as (4.1).

4.8 Gibbs–Duhem relation

The term $C_v dT$ that we have mentioned in various equations above is formally the change in the internal energy U of the system, that is,

$$dU = C_v dT. \tag{4.16}$$

The so-called Euler equation in classical thermodynamics can be used to calculate the internal energy. This equation is

$$U = TS - PV + \sum_{k=1}^{c} \mu_k n_k. \tag{4.17}$$

Differentiating both sides, we get

$$dU = TdS + SdT - PdV - VdP + \sum_{k=1}^{c} \mu_k dn_k + \sum_{k=1}^{c} n_k d\mu_k. \tag{4.18}$$

The above is no more than a purely mathematical operation. It merely says that a small variation in the internal energy may correspond to variations in T, S, P, V, μ, and n. But now we use the physical equation (4.15) – the conservation of energy – to eliminate some terms in (4.18). Then we obtain

$$0 = SdT - VdP + \sum_{k=1}^{c} n_k d\mu_k. \tag{4.19}$$

Dividing both sides by the total number of moles n, we obtain

$$0 = sdT - vdP + \sum_{k=1}^{c} x_k d\mu_k, \tag{4.20}$$

where

$$s = \frac{S}{n} \text{ molar entropy,}$$

$$v = \frac{V}{n} \text{ molar volume,} \qquad (4.21)$$

$$x_k = \frac{n_k}{n} \text{ mole fraction.}$$

Eq. (4.20) is the well-known *Gibbs–Duhem relation*, a very useful equation in thermodynamics.

4.9 General condition of thermodynamic equilibrium

Now we can give a more precise definition of the oft-stated thermodynamic equilibrium. There are three conditions that need to be satisfied in order for a system to be in thermodynamic equilibrium with its environment. These three are described in the following.

- Thermal equilibrium. The temperatures of all phases in the system as well as in its environment must be equal, i.e.

$$T' = T'' = \cdots = T^{(\alpha)} \qquad (4.22)$$

  for all phases involved in the equilibrium. The superscript α in (4.22) represents different phases.

For example, a water drop in the air is in thermal equilibrium if the liquid water, water vapor, and air all have the same temperature. If not, then it is not in thermal equilibrium and there will be heat flow from one part to the other until the temperatures even out completely.

- Mechanical equilibrium. The pressures of all phases in the system as well as in its environment must be equal, i.e.

$$P' = P'' = \cdots = P^{(\alpha)}. \qquad (4.23)$$

If pressures are different, then there will be motions due to mechanical imbalance until the pressures are even.

- Chemical equilibrium. The chemical potentials of all components in the different phases must be equal, i.e.

$$\mu'_k = \mu''_k = \cdots = \mu_k^{(\alpha)}. \qquad (4.24)$$

Note that this condition is slightly different from the other two in that only the chemical potentials of the same component need to be the same, whereas the equality of the chemical potentials of different components is not required. For example, consider the equilibrium of an aqueous solution of NaCl with humid air. The chemical equilibrium condition only requires that the chemical potentials of water vapor in air and the liquid water in the solution are the same, and that the chemical potentials of the NaCl in the solution and the NaCl vapor in the air are the same. It does *not* require that the chemical potentials of water and NaCl have to be the same.

4.10 Clausius–Clapeyron equation

Now we will apply the knowledge we gained from the above discussions to understand one of the most important physical phenomena in atmospheric science – the phase change of water substance.

We will first consider the equilibrium between pure liquid water and pure water vapor as illustrated in Fig. 4.3. This equilibrium system consists of *two* phases (liquid and vapor) and *one* component (H_2O). We will utilize the general conditions of the thermodynamic equilibrium to derive more concrete relations between state variables (P, V, T, etc.) during the equilibrium.

Let us begin by using the chemical equilibrium condition

$$\mu' = \mu''. \tag{4.25}$$

We have dropped the subscript for the chemical potentials because we have a one-component system ($k = 1$). Now if we change the chemical potentials, we would normally disturb the equilibrium. However, if we disturb it in such a way that

$$d\mu' = d\mu'', \tag{4.26}$$

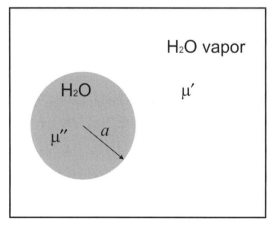

Fig. 4.3 A system consisting of a large pure water drop in equilibrium with pure water vapor. No curvature effect will be considered here.

then we can still have equilibrium, albeit not the same equilibrium state as before, but equilibrium nonetheless. Now we use the Gibbs–Duhem relation (4.20) and remember that $x_k = 1$ in the present case. Then we have for each phase

$$d\mu' = -s'dT + v'dP,$$
$$d\mu'' = -s''dT + v''dP,$$

(4.27)

and from (4.26) we get

$$-s'dT + v'dP = -s''dT + v''dP.$$

(4.28)

In deriving (4.27) and (4.28), we have used the other equilibrium conditions, namely, and $T'=T''=T$ and $P'=P''=P$. Now we can collect similar terms in (4.28) and rearrange to obtain

$$\frac{dP}{dT} = \frac{s' - s''}{v' - v''}.$$

(4.29)

Eq. (4.29) is the *Clapeyron equation*. If we multiply both the numerator and denominator on the right-hand side of (4.29) by T, and note that the definition of latent heat of evaporation per mole L_e is

$$L_e = T(s' - s''),$$

(4.30)

we get

$$\frac{dP}{dT} = \frac{L_e}{T(v' - v'')}.$$

(4.31)

Eq. (4.31) is the *Clausius–Clapeyron equation*, which prescribes the equilibrium conditions between water vapor and liquid water. Note that P in the above equation represents the vapor pressure, which is, of course, the same as the pressure in the liquid at the interface.

If we repeat the same procedure for the equilibria of the ice–vapor and ice–liquid water systems, we would obtain the same equation except that the meaning of each term is interpreted differently. In view of the common convention in the atmospheric science community that the vapor pressure is usually written as e and that the vapor pressure in the present context is about the equilibrium vapor pressure, which is the same as the saturation vapor pressure e_{sat}, we can write the Clausius–Clapeyron equation in the following forms to represent the equilibrium conditions for the vapor–water, vapor–ice, and water–ice systems, respectively:

$$\frac{de_{sat,w}}{dT} = \frac{L_e}{T(v_v - v_w)} \approx \frac{L_e}{Tv_v},\tag{4.32}$$

$$\frac{de_{sat,i}}{dT} = \frac{L_s}{T(v_v - v_i)} \approx \frac{L_s}{Tv_v},\tag{4.33}$$

$$\frac{dP_{wi}}{dT} = \frac{L_m}{T(v_w - v_i)}.\tag{4.34}$$

In the above three equations, the subscripts v, w, and i represent vapor, water, and ice, respectively; P_{wi} represents the pressure between water and ice; L_s and L_m represent the latent heat per mole of sublimation and melting, respectively. In (4.32) and (4.33), we use the fact that, under normal atmospheric conditions, the molar volume of vapor v_v is much greater than that of liquid water v_w and ice v_i at the same temperature to arrive at the last step. On the other hand, $v_w \approx v_i$ and hence we cannot do the same to (4.34).

Assuming that water vapor behaves as an ideal gas, (4.32) and (4.33) can further be written as

$$\frac{d \ln e_{sat,w}}{dT} = \frac{L_e}{RT^2},\tag{4.35}$$

$$\frac{d \ln e_{sat,i}}{dT} = \frac{L_s}{RT^2}.\tag{4.36}$$

Note that L_s and L_m are not constant but depend on temperature themselves. However, they are often treated as constant in many applications for the sake of simplicity if the temperature range is not too large.

4.11 Phase diagram for water substance

Based on Eqs. (4.32)–(4.34) we can plot charts to illustrate the equilibrium behavior of all three phases of water substance. Such charts are called *phase diagrams*. We can plot these charts in various ways, for example, *P* vs. *V* or *V* vs. *T* or even a 3D plot of *P*, *V*, and *T*, but the most useful to us here is the *P* vs. *T* plot. These three equations would define the slopes of three curves representing the equilibria, and they converge at a point, called the *triple point*, where all three phases of water are in equilibrium. The triple point is located at $T = 273.16$ K (whereas the so-called *ice point* is at $T = 273.15$ K, or 0°C, i.e. 0.01 K lower than the triple point).

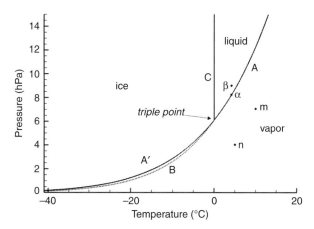

Fig. 4.4 The *P*–*T* phase diagram of water substance according to the Clausius–Clapeyron equation.

Fig. 4.4 shows this phase diagram. The three curves, A, B, and C, are drawn based on the three equations above and represent equilibrium conditions. Any pair of (*e*, *T*) values that falls right on one of the curves indicates that the vapor pressure and temperature will satisfy the equilibrium condition. For example, the point α at (8.13 hPa, 4°C) falls right on curve A, which represents the equilibrium between liquid water and water vapor, and thus indicates that $T = 4°C$ and $e = 8.13$ hPa is an equilibrium condition. This also means that 8.13 hPa is the saturation vapor pressure at 4°C.

On the other hand, the point β at (9.0 hPa, 4°C) falls not on the curve but in the region labeled "liquid". First of all, this is obviously not an equilibrium condition. Secondly, the location of the point indicates that liquid water is the favorable phase, namely, there will be the formation of more liquid water from water vapor because the current vapor pressure is supersaturated (more than necessary for saturation). The situation will only become equilibrium again when either the excess water vapor condenses into liquid water or the temperature rises (9.0 hPa is the saturation vapor pressure at a higher temperature).

How about an atmospheric condition represented by the point m at (7 hPa, 10°C) in Fig. 4.4? In this case, the point falls in the vapor region and the humidity condition is subsaturated, hence not equilibrium either. Some changes will have to be made to return the system back to equilibrium.

The number of possible routes that the above system (in fact, any non-equilibrium system) can take to return to equilibrium is infinite, but we can look at two extremes: an isobaric process and an isothermal process. The isobaric process works like this. We keep the vapor pressure constant but lower the temperature. Graphically, this means that we draw a horizontal line from the point to the left, until the line intersects curve A. At that point, the condition is again equilibrium. The temperature corresponding to this point is called the *dewpoint* temperature. The dewpoint temperature is thus the temperature at which the vapor pressure becomes saturated via the isobaric cooling process. The name is associated with the phenomenon of dew and this is explained in the following.

Our experience tells us that dew often appears in the early morning, especially near dawn. Although the actual process is more complicated, we can draw a simplified picture of how it happens. On a typical clear summer day with a certain amount of humidity in the air, the temperature is usually too high and the vapor is unsaturated. After sunset, the Earth's surface no longer receives solar radiation but continues to emit long-wave radiation (the terrestrial radiation) into space and hence cools. If there is no significant movement of the air mass, then the vapor pressure remains relatively constant and hence approximately isobaric. As the temperature of the surface becomes colder, so will the air near the ground, mainly due to conduction (although some small-scale convection can be involved as well) and the vapor comes closer to saturation. The cooling usually reaches its maximum just before dawn. The surface of rocks, grass, trees, etc. are cold enough that water vapor becomes supersaturated and excess water vapor starts to condense on these surfaces to form dewdrops. This whole process is approximately an isobaric cooling process and hence the name dewpoint.

For those who live in middle to higher latitudes, there is another phenomenon, frost, which is similar to dew but occurs at lower temperatures. A process similar to dew can explain the occurrence of frost. On a typical clear autumn day, the vapor pressure in the air is usually less than that in the summer. Although the temperature is usually lower, too, the vapor is not enough to saturate the air. Again, the same isobaric cooling process may occur if there is no significant movement of the air mass and the temperatures are cold enough to saturate the same amount of water vapor in the air. For example, the point n at (4 hPa, 5°C) in Fig. 4.4 falls in the vapor region. Upon isobaric cooling, we move the point to the left until it reaches a point on curve B, which is the curve of vapor–ice equilibrium. What will happen upon further cooling is the appearance of frost instead of dew. The temperature corresponding to this point on B is appropriately called the *frostpoint* temperature. It can only occur for vapor pressures less than that at the triple point ($e_{tr} = 6.11$ hPa).

Let us now examine curve C, which specifies the equilibrium between ice and liquid water. This curve looks almost like a vertical line, but closer examination shows that it leans slightly to the left, i.e. a negative slope. This is actually a very special phenomenon because all the other chemicals we encounter in the Earth's environment have positively sloped (lean to the right) liquid–solid equilibrium curves. For most other materials, the more pressure applied on the liquid makes the material more compact and closer to the solid state, which results in positively sloped equilibrium curves.

The negatively sloped ice–water equilibrium curve is due to the strange behavior of water – when water freezes into ice, the ice has lower density than water! It means that ice has a tendency to liquefy upon applying higher pressure. If you apply more pressure on the ice surface without changing the temperature, you will see that the resulting point is in the liquid regime, indicating that more liquid water will be produced, a somewhat counter-intuitive phenomenon.

Aside from this negative slope point, the nearly vertical curve also means that it is difficult to change one ice–liquid equilibrium state to another via pressure because any small change would involve an enormous change in pressure.

4.12 Supercooling and the Bergeron–Findeisen process

Finally, let us look at a very important phenomenon of *supercooling*. We have mentioned previously in Chapter 2 that this refers to the phenomenon that liquid water stays liquid even at temperatures below 0°C. As we have seen before, cloud droplets can stay liquid down to about −40°C. How do we determine quantitatively the equilibrium conditions between the supercooled water and water vapor? We just use the same equation for normal water–vapor equilibrium (4.32) and extend the results to $T < 0$°C. This is by extending curve A to the lower temperature range and the result is curve A′ in Fig. 4.4.

Because the curve is an extension of curve A, the vapor–liquid equilibrium behavior is the same as before. What interests us here is the equilibrium between supercooled water and ice. We see that curve A′ is located above curve B. This indicates that, at the same temperature, supercooled water requires a higher saturation vapor pressure than ice. In other words, if both supercooled water drops and ice crystals are present in the same atmospheric environment and if the environmental humidity is such that it is *saturated* for the supercooled water, then it must be *supersaturated* for the ice crystals.

It is not only possible but indeed quite common in a cloudy region in the upper troposphere that the environmental humidity is such that it is subsaturated with respect to supercooled water but supersaturated with respect to ice crystals. Often there are only supercooled water drops present in this region initially and then ice crystals are introduced into this region, for example, either by falling into it from above or by turbulent mixing. Then something dramatic happens. Since the environment is subsaturated with respect to supercooled water, the supercooled droplets will evaporate. On the other hand, the environment is supersaturated with respect to ice crystals and hence the ice crystals will grow. Thus the net consequence is that the ice crystals grow at the expense of supercooled water drops. This will continue until all supercooled water drops evaporate completely, supplying the water vapor for the growth of ice crystals. Thus, supercooled water and ice can never be in stable equilibrium.

The above phenomenon is the physical basis of the so-called Wegener–Bergeron–Findeisen process (sometimes called the Bergeron–Findeisen process, or simply Bergeron process), which plays a role in the precipitation process. The growth of ice at the expense of supercooled water was first described by Wegener and later Bergeron used that to propose that it is the process of all precipitation. Findeisen made refinements later. In this process, it was envisioned that, if ice crystals in clouds fall into a region of supercooled water in an environment subsaturated with respect to water, ice crystals will grow. Provided that the ice population is not too large, these crystals may grow to fairly large size and fall below the cloud base, melt, and turn into rain. We now know that not all precipitation occurs this way, but it may be responsible for some higher-latitude continuous rain. Certainly it can play a role, if not the major role, in the development of mixed-phase clouds.

Not only is it true that supercooled droplets and ice crystals may exist in the same environment, but it is also possible that supercooled droplets can be in contact with ice crystals directly. It has been observed that the supercooled droplets freeze quickly on the ice surface to form rimes if the ambient temperature is below freezing. This is again due to the impossibility of stable equilibrium between supercooled water and ice (though of a different kind), and the "loser" is always the supercooled drops.

4.13 Order of phase change

We have seen that the phase diagram of water substance is useful in explaining the equilibrium behavior among water substance in different phases. We even used it to make predictions of phase changes from certain starting points and seem to be successful so far. However, there are certain occasions when such a prediction will fail. One obvious example is the supersaturation phenomenon. The phase diagram predicts that, once the water vapor in an air parcel reaches saturation by cooling, condensation to form liquid will start promptly. In reality, air parcels sometimes supersaturate before condensation occurs. Fig. 4.4 clearly does not predict this behavior.

The other notable failure is the prediction of supercooling. If we start from point n at (4 hPa, 5°C) in Fig. 4.4 and cool the air parcel down isobarically, we see that the air parcel will reach curve B first. At that point (the frostpoint), as we mentioned before, the parcel is saturated with respect to ice and hence we should see sublimation occur and ice crystals start to form. Upon further cooling, however, we will reach another point on the vapor–supercooled water curve A' and we should predict that these ice crystals would then turn into supercooled water! Nobody has ever observed the phase change occur in this sequence. In fact, the opposite is true – supercooled water always forms before the ice phase.

What is the reason for this particular order of water phase change? Why does the Clausius–Clapeyron equation fail to predict the correct phase change sequence? It turns out that the Clausius–Clapeyron equation is based on the assumption that the materials in equilibrium exist in bulk quantities – for such systems, the interfaces between two phases are planes. The thermodynamics of planar interface and curved interface systems are different – the surface effects have a substantial influence on the thermodynamics of the latter.

When a new condensed phase (a small water droplet or a small ice crystal) first appears, it is always a highly curved system. In such cases, the surface effects must be included in the consideration of the thermodynamic behavior of the system. This is true for both equilibrium and non-equilibrium conditions. In Chapter 5, we will study the equilibrium between curved systems, and in Chapter 7 we will study the formation of a new condensed phase – the nucleation process. We will see that surface effects play a significant role there. Moreover, after we have studied the nucleation phenomenon, we will be able to understand why supercooled water always appears before ice!

4.14 Calculation of the saturation vapor pressures

The calculation of saturation vapor pressures is very important to atmospheric science in general and cloud physics in particular, because the prediction of much cloud behavior is sensitive to these values. In principle, we should be able to calculate the saturation vapor pressure based on Eqs. (4.35) and (4.36). But in reality the calculation is difficult because the latent heats in these equations are not constant but are functions of temperature themselves and cannot be integrated in simple forms. Discussions of the temperature dependence of latent heats are given in great detail in Pruppacher and Klett (1997, sec. 4.7). There are other complications, such as the molecular structure of ice, which may not be hexagonal at very low temperatures, and water vapor may not behave exactly as an ideal gas, etc. The common practice is to use empirical fits of experimental data such as those given in the *Smithsonian Meteorological Tables* (List, 1963). There are newer formulas by various investigators. The following is taken from Murphy and Koop (2005):

$$\ln e_{\mathrm{sat,w}} \approx 54.842763 - \frac{6763.22}{T} - 4.210\ln(T) + 0.000367T$$
$$+ \tanh[0.0415(T - 218.8)]$$
$$\times \left(53.878 - \frac{1331.22}{T} - 9.44523\ln(T) + 0.014025T\right) \tag{4.37}$$

for $123 < T < 332$ K. Note that this formula is valid for both normal and supercooled liquid water. The unit of pressure is Pa and the temperature is in K. Also from the same source:

$$e_{\text{sat,i}} = \exp\left(9.550426 - \frac{5723.265}{T} + 3.53068 \ln(T) - 0.00728332T\right) \quad (4.38)$$

for $T > 110$ K. The formulas for the latent heats that they used to arrive at (4.37) and (4.38) are:

$$L_e(T) \approx 56579 - 42.212T + \exp\{0.1149(281.6 - T)\}, \quad (4.39)$$

$$L_s(T) = 46782.5 + 35.8925T - 0.07414T^2 + 541.5 \exp\left\{\left(\frac{T}{123.75}\right)^2\right\}. \quad (4.40)$$

Problems

4.1. (a) The ideal gas law for moist air can be written as $P = \rho R_m T = \rho R_d T_v$, where P is the pressure and ρ the density of moist air, R_m and R_d are the gas constants for moist and dry air, respectively, and T_v is called the virtual temperature of the moist air. Prove that

$$T_v = \frac{T}{\left[1 - \frac{(1-\varepsilon)e}{P}\right]},$$

where $\varepsilon = M_w/M_d$ is the ratio between the molecular mass of water M_w and dry air M_d, and e is the vapor pressure.
   (b) Calculate T_v for $T = 20°C$ and $RH = 90$, 75, 50, and 25% for $P = 1000$ hPa.
   (c) What will be the values of T_v in (b) if the pressure is 900 hPa and what is the physical meaning of the results?

4.2. A two-component open system is undergoing an isobaric and isothermal process. There are 4 moles of component 1 and 3 moles of component 2. What is the ratio of the changes of chemical potentials of these two components $(d\mu_1/d\mu_2)$?

4.3. A man weighing 70 kg is skating on one foot on ice at $T = 0°C$. The blade on his skate has a dimension 20 cm × 0.5 cm. Assuming equilibrium conditions, calculate the melting point of the ice under the blade. (The density of water and ice are $\rho_w = 1$ g cm^{-3} and $\rho_i = 0.916$ g cm^{-3} at 0°C, respectively).

4.4. (a) From the Clausius–Clapeyron equation, derive an expression and then determine the value of the temperature at which the maximum difference between the saturation vapor pressures of supercooled water and ice occurs.

   (b) What would be the implication of this temperature to a system consisting of supercooled water and ice?

5

Surface thermodynamics of water substance

In the last chapter, we discussed the thermodynamic equilibrium among different phases of water substance and derived the Clausius–Clapeyron equation to describe the equilibrium quantitatively. We found that, while this equation is useful in many respects, it has problems when predicting certain phenomena related to phase change due to its omission of surface effects in the system. In actual clouds, especially in the initial stage of cloud formation, particles often have high curvatures and hence strong curvature effects, which cannot be neglected. In this chapter, we will study the surface effects and then use that knowledge to study the equilibrium behavior of cloud and precipitation particles in a more realistic way.

5.1 The interface as a phase

The interface between two bulk phases is a very special "system" that sometimes is more complicated than a bulk system. Conceptually we often simplify the interface as a geometrical surface (i.e. no thickness but has area), but in reality it must be a layer of finite thickness (albeit very, very thin). Fig. 5.1 shows a conceptual model of the interface between liquid water and water vapor for a water drop of radius a.

The vertical axis represents the density while the horizontal axis represents the distance from a fixed point in the bulk liquid. As we move from the interior of the liquid phase and get closer and closer to the interface, the density would be constant for a while and then drop very quickly as we get very close to the interface. Then the density would become the constant low value of the vapor density as we move into the bulk vapor region. The transitional zone where the density changes from liquid density to vapor density is the *interfacial layer* and its thickness is usually just a fraction of a molecular layer – very thin indeed but of finite thickness nevertheless.

In the detailed studies of surface physics and chemistry, it is necessary to treat this surface layer as having finite thickness, but in a simplified discussion, as we are doing here, it is adequate to represent this zone as a geometrical surface as long as

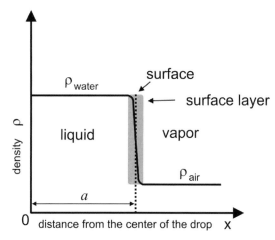

Fig. 5.1 A conceptual model of the surface layer (shaded region) of a water drop.

we take enough care to make sure that the energies and forces involved are accounted for. The usual way of defining this interface is to put it at a location such that the volumes in the transitional zone on both sides of the surface are equal. In addition, since the "constituents" on the surface are neither liquid nor vapor, we have to treat it as a separate "phase" – the *surface phase*. This geometrical surface is called the *Gibbs surface*.

Although we used the liquid–vapor interface as an example, the general concept of a surface can also be applied to solid–vapor and solid–liquid interfaces as well.

5.2 Surface tension of liquids

Those who have the experience of dropping a mercury container, such as the mercury reservoir of a barometer, would likely have noticed that the liquid mercury seems to disappear! Of course, it does not really disappear, but breaks up into many small mercury balls that roll away. Why would the small fragments of mercury roll into small balls? Here you are witnessing *surface tension* at work.

Mercury is an extreme case, but liquids generally possess surface tensions. If you shed a drop of water on a glass slide, the drop will not smear out right away but forms a quasi-hemispherical dome. That is also the action of surface tension. If you put the drop on different surfaces (for example, a steel plate or a plastic sheet), you would generally notice that the dome shapes are somewhat different. This indicates that the surface tension has something to do with the property of the substrate also and does not just solely depend on the liquid itself.

So what is surface tension? First we point out that the molecules on the surface have higher energies than those in the bulk liquid. The excess energy of the surface molecules is called the *surface free energy* E_{sfc}. The surface tension is actually the surface free energy per unit area:

$$\sigma = \frac{E_{sfc}}{A},$$ (5.1)

where A is the total area of the surface. The unit of σ is energy per unit area, which can be expressed as force per unit length; the latter is the unit of tension. Indeed, because of the higher energy of the surface molecules, the surface presents itself like a barrier and hence appears to have "tension". As a system, the surface also exhibits many common properties as other thermodynamic systems. Normally a thermodynamic system is in stable equilibrium when its energy is a minimum. For the surface system, the minimal energy state for a given environment (certain P and T) can be achieved by forming a surface with minimum surface area. The strong surface tension of liquid mercury affords itself to form a sphere when the quantity is small enough and, as we know, a sphere has the smallest surface area for a given volume. Hence, by curling up into a sphere, the mercury ball surface achieves a minimum energy state and can be in a very stable condition. Other liquids, like water, do not have surface tension large enough for them to form a sphere.

The surface tension of water is a function of temperature and impurities in it. As the temperature increases, the thermal agitation of the molecules becomes stronger and the surface tension decreases. Adding chemicals to water can also change the surface tension. There are some chemicals that serve to lower the surface tension, for example, alcohol, when they dissolve in water. There is a group of chemicals called surfactants or surface-active agents that can lower the water surface tension substantially with just a small concentration. On the other hand, inorganic salts such as sodium chloride or ammonium sulfate that are important to cloud formation are chemicals that enhance the surface tension of water.

5.3 Surface tension of solids

There is also a similar definition of the surface tension for unstrained solids, but the interpretation of its physical meaning is still being debated. This is the so-called Shuttleworth equation:

$$\gamma = F + A \frac{\partial F}{\partial A}.$$ (5.2)

According to Shuttleworth (1950), γ is the tangential stress (force per unit length) of the surface layer and F is the total Helmholtz free energy per unit area of a surface. The free energy F has been widely interpreted as the thermodynamic energy required to form unit area of new unstrained surface by cleaving, and the second term on the right-hand side of (5.2) is a term that separates the surface tension on a liquid from that on a solid. While this equation has been in wide use, a recent study by Makkonen (2012) showed that one cannot interpret F as the excess surface free energy, unlike the liquid case. It is best understood as a purely mechanical term, i.e. surface tangential stress.

We shall focus the following thermodynamic discussions on the liquid–vapor interface. Pruppacher and Klett (1997) provide more detailed discussions on the ice–vapor interface.

5.4 Mechanical equilibrium among curved interface systems

Let us now consider a practical problem for cloud physics – the equilibrium of a spherical water drop with its own vapor. We will consider only the mechanical equilibrium here, as the chemical and thermal equilibria remain the same as before.

Fig. 5.2 shows the configuration of the problem. Here a spherical water drop of radius a is located in a closed system filled with pure water vapor. The total volume V of the whole system is assumed to be constant for simplicity (it does not affect the conclusion if you would like to work with a non-constant volume situation, but the analysis will become more complex).

There are totally three phases in this system – vapor, liquid, and *surface*. We can treat each phase as a separate subsystem and utilize the conservation of energy principle (4.15) but for a closed system for each phase:

$$dU_v = TdS_v - P_v dV_v, \tag{5.3}$$

$$dU_w = TdS_w - P_w dV_w, \tag{5.4}$$

$$dU_s = TdS_s + \sigma dA, \tag{5.5}$$

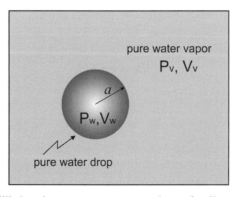

Fig. 5.2 The equilibrium between a pure water drop of radius a (and volume V_w) with pure water vapor of volume V_v.

where the subscripts v, w and s represent vapor, liquid water, and surface, respectively, and dU represents the change in the internal energy. Note that Eq. (5.5) looks different from the other two – the last term on the right-hand side has a plus sign and has no volume involved. This is, of course, due to our simplification that the surface only has area but no volume, and σdA represents the surface energy *gained* by the surface system when the drop expands by a volume increment dV_w. Because this is the gained energy, its sign is positive.

Now we sum all three equations (5.3)–(5.5) and yield

$$
\begin{aligned}
dU &= dU_v + dU_w + dU_s \\
&= T(dS_v + dS_w + dS_s) - P_v dV_v - P_w dV_w + \sigma dA \\
&= TdS - P_v dV - (P_w - P_v)dV_w + \sigma dA \\
&= TdS - (P_w - P_v)dV_w + \sigma dA.
\end{aligned} \tag{5.6}
$$

In deriving the above equation, we used the constant total volume criterion:

$$
dV = dV_v + dV_w = 0. \tag{5.7}
$$

We will now further assume that the process that the *whole system* is undergoing is adiabatic – note that each individual subsystem obviously undergoes a non-adiabatic (or diabatic) process because the changing volume of the water drop implies a phase change, which, in turn, implies the involvement of latent heats. Under this adiabatic assumption, we have $TdS = 0$ and (5.6) becomes

$$
dU = -(P_w - P_v)dV_w + \sigma dA. \tag{5.8}
$$

This equation simply says that, under the overall adiabatic assumption, any change in the internal energy of the whole system is due to the expansion (or shrinking) of the water drop and the associated change in the surface area. Under the assumption of overall thermal equilibrium, $dU = C_v dT = 0$ and (5.8) becomes

$$
(P_w - P_v) = \sigma \frac{dA}{dV}. \tag{5.9}
$$

For a spherical drop of radius a, the increments in surface area and volume are given by

$$
\begin{aligned}
dA &= 8\pi a\, da \\
dV &= 4\pi a^2 da
\end{aligned} \tag{5.10}
$$

and hence (5.9) becomes

$$
(P_w - P_v) = \frac{2\sigma}{a}. \tag{5.11}
$$

Eq. (5.11) specifies the condition of mechanical equilibrium for two phases separated by a curved interface with radius of curvature a. Unlike the mechanical equilibrium condition for a planar interface system, where the pressures in the two phases are equal, the pressures in the present case differ by an amount related to the radius of curvature and the surface tension. Eq. (5.11) indicates that the smaller the radius of curvature, the greater the pressure difference. This occurs when the droplets are very small, typically during the initial stage of droplet formation.

On the other hand, when the radius of curvature is large (i.e. large drops), the pressure difference is small. In the extreme case when $a \rightarrow \infty$ (that is, the interface becomes a plane), the pressure difference vanishes, which is precisely the mechanical equilibrium condition for a planar interface system.

Although we used water vapor and a liquid water drop as an example for discussing the mechanical equilibrium condition for a curved interface system, the conclusion (Eq. (5.11)) is valid for other materials with similar geometry. Thus for phases separated by curved interfaces, the general thermodynamic equilibrium conditions are:

- thermal equilibrium

$$T' = T'' = \cdots = T^{(\alpha)}, \tag{5.12}$$

- mechanical equilibrium

$$P'' - P' = \frac{2\sigma}{a}, \tag{5.13}$$

- chemical equilibrium

$$\mu'_k = \mu''_k = \cdots = \mu_k^{(\alpha)}. \tag{5.14}$$

In the case of no spherical surface, a would have to be interpreted as the local radius of curvature.

5.5 Contact angle and wettability

We mentioned the dome shape of a liquid drop when it sits on a solid substrate. The shapes of the dome differ for different kinds of substrate. The angle θ at which the dome intersects the substrate surface is called the *contact angle* (see Fig. 5.3). The contact angle is a function of the physicochemical properties of all materials involved – the liquid droplet, the substrate, and the gas (for example, air).

Fig. 5.3 also shows that the contact angle is the result of the vector balance of three surface tensions:

$$\sigma_{s/v} = \sigma_{s/w} + \sigma_{w/v} \cos \theta, \tag{5.15}$$

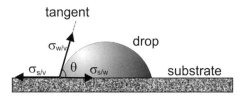

Fig. 5.3 The definition of contact angle θ.

and thus

$$\theta = \cos^{-1}\left(\frac{\sigma_{s/v} - \sigma_{s/w}}{\sigma_{w/v}}\right). \tag{5.16}$$

When $\theta = 0$, we have a substrate that is *perfectly wettable*. Conversely, when $\theta = \pi$ (i.e. when the liquid drop becomes a sphere), we have a *perfectly non-wettable substrate*. When liquid mercury rolls into a sphere on the floor, it indicates that the contact angle is π and the floor is perfectly non-wettable by mercury.

In the case of water drops, most substrates that we encounter in daily life are neither perfectly wettable nor perfectly non-wettable. That is to say, the contact angle usually lies between 0° and 180°. One of the substrate materials that produces a large contact angle for water droplets is Teflon, with $\theta \sim 100°$, and this makes Teflon a good material for certain applications. For example, so-called "non-stick" cooking utensils are usually coated with a layer of Teflon precisely because of its relatively non-wettable property; any water drops on the surface of the utensil can be easily wiped off without smearing on the surface.

The materials that result in low contact angles are called *hydrophilic* materials, whereas those resulting in high contact angles are called *hydrophobic* materials. Many hydrophilic materials are also hygroscopic, namely, they strongly absorb water vapor and water. One example of a hygroscopic material is common salt (sodium chloride, NaCl), which serves as excellent cloud condensation nuclei for cloud droplets. We will discuss this in Chapter 7 when we study nucleation phenomena.

The *wetting coefficient m* is defined as

$$m = \cos\theta. \tag{5.17}$$

For a perfectly wettable surface, $m = 1$, whereas for perfectly non-wettable surface, $m = -1$. The effect of the wettability and wetting coefficient m will become clear when we treat nucleation.

5.6 Component chemical potentials in an ideal gas mixture

Since we will discuss the equilibrium of aqueous solutions later, we need to know how to calculate the chemical potentials of different components either in a mixture or in a solution. Let us consider a simpler case – a mixture consisting of ideal gases.

From the Gibbs–Duhem relation (4.20), we can easily see that

$$\left(\frac{\partial \mu_k}{\partial P}\right)_{T,n_{j\neq k}} = v_k \quad \text{and} \quad \left(\frac{\partial \mu_k}{\partial T}\right)_{P,n_{j\neq k}} = -s_k, \tag{5.18}$$

where v_k and s_k refer to the molar volume and molar entropy of the component k. The subscript $T,n_{j\neq k}$ on the parentheses means holding the temperature T and the concentration of all other components $n_{j\neq k}$ constant while performing the partial differentiation. The meaning of the second equation is analogous. Since the k component is also an ideal gas, the first equation of (5.18) can be written as

$$\left(\frac{\partial \mu_k}{\partial P}\right)_{T,n_{j\neq k}} = \frac{RT}{P_k} \tag{5.19}$$

or

$$\partial \mu_k = RT \frac{\partial P_k}{P_k} = RT\partial \ln P_k, \tag{5.20}$$

while keeping $T,n_{j\neq k}$ constant. To obtain an algebraic relation between μ_k and other easily measurable quantities, we need to integrate (5.20):

$$\int_{\mu_{k,0}}^{\mu_k} \partial \mu_k = RT \int_1^{P_k} \partial \ln P_k. \tag{5.21}$$

Note that we integrate the pressure in the integral on the right-hand side from 1 atm to P_k, and the chemical potential on the left-hand side from $\mu_{k,0}$ to μ_k, so $\mu_{k,0}$ is the chemical potential of component k when its pressure is 1 atm. Thus

$$\mu_k = \mu_{k,0} + RT \ln P_k. \tag{5.22}$$

But since $P_k = x_k P$, where x_k is the mole fraction of component k, (5.22) becomes

$$\mu_k = \mu_{k,0} + RT \ln P + RT \ln x_k. \tag{5.23}$$

From (5.23), we can determine the chemical potential of water vapor in humid air (which is, of course, a mixture of two ideal gases – dry air and water vapor):

$$\mu_v(P, T, x_v) = \mu_{v,0}(T) + RT \ln P + RT \ln x_v. \tag{5.24}$$

The first term on the right-hand side is a function of temperature only because it is defined at $P = 1$ atm. Note that P is the total pressure of the humid air, not just the vapor pressure.

5.7 The chemical potential of water in an aqueous solution

The previous section dealt with gas mixtures. In this section, we shall calculate the chemical potential of water in an aqueous solution. By definition, an aqueous solution contains a solvent – water – and one or more solutes.

Let us first examine the case of pure water in equilibrium with pure water vapor. From (5.18) (which is applicable to any situation), we have

$$\left(\frac{\partial \mu_{w,0}}{\partial P} \right)_T = v_{w,0} \approx \text{constant} \tag{5.25}$$

because the volume (and hence molar volume) of liquid water is nearly a constant due to its near-incompressibility. Thus, we can integrate (5.25) to obtain:

$$\int_{\mu_{w,0}(0,T)}^{\mu_{k,0}(P,T)} \partial \mu_{w,0} = \int_0^P v_{w,0} \partial P \approx v_{w,0} \int_0^P \partial P, \tag{5.26}$$

and the result is

$$\mu_{w,0}(P,T) \approx \mu_{w,0}(0,T) + v_{w,0}P. \tag{5.27}$$

So we see that in the case of pure water in equilibrium with pure water vapor (and P is the vapor pressure here), the chemical potential depends linearly on P instead of ln P.

In the case of an aqueous solution in equilibrium with humid air, the chemical potential of water in the solution is required to be in equilibrium with water vapor only. The chemical potential of any component k in the solution is

$$\mu_{k,l} \approx \mu_{k,0} + RT \ln P_k, \tag{5.28}$$

where $\mu_{k,0}$ is the chemical potential of pure component k and P_k is the partial vapor pressure of this component.

5.8 Ideal and non-ideal solutions

An *ideal solution* is a solution in which the solutes and the solvents do not interact. There are in reality no such solutions because the fact that the solutes can be "dissolved" by the solvent already implies an interaction between them. However, if the concentrations of the solutes are very small, then the percentage of the solvent influenced by the solutes is fairly small and the solution behaves approximately as an ideal solution. Under this condition, the partial equilibrium vapor pressure P_k of component k over the solution is proportional to its mole fraction in the solution, i.e.

$$P_k = x_k P_{k,0}, \tag{5.29}$$

where $P_{k,0}$ is the vapor pressure of the solvent when there is no solute in it, i.e. the pure solvent. Eq. (5.29) is called *Raoult's law*. It states that the equilibrium (i.e. saturation) vapor pressure of the solvent will decrease in proportion to its mole fraction in the solution. This law is also used as a definition of an ideal solution.

In the case of an aqueous solution of salt, the saturation water vapor pressure over the solution is

$$e_{\text{sat,w}} = x_{\text{w}}\, e_{\text{sat,w},0},\tag{5.30}$$

which indicates that the saturation vapor pressure over this salt solution is lower than that over a pure liquid water surface. This implies that an aqueous solution can *resist evaporation* better than can pure liquid water because it takes less vapor pressure to maintain equilibrium (saturation). This will have important consequences for the survival of clouds in the atmosphere, as we shall see later.

The chemical potentials of water and salt in this aqueous solution are

$$\mu_{\text{w}}(P,T,x_{\text{w}}) = \mu_{\text{w},0}(P,T) + RT\,\ln x_{\text{w}}\tag{5.31}$$

and

$$\mu_{\text{s}}(P,T,x_{\text{s}}) = \mu_{\text{s}}^{+}(P,T) + RT\,\ln x_{\text{s}}.\tag{5.32}$$

The term $\mu_{\text{s}}^{+}(P,T)$ is the chemical potential of pure salt.

From the description of ideal solutions above, we can usually approximate dilute solutions as ideal solutions. Naturally, many solutions are far from ideal. In that case, we replace the mole fraction x by the *activity* a for calculating the chemical potentials. Thus for non-ideal solutions, we have

$$\mu_{\text{w}}(P,T,a_{\text{w}}) = \mu_{\text{w},0}(P,T) + RT\,\ln a_{\text{w}} \quad \text{where}\ \ a_{\text{w}} = f_{\text{w}}x_{\text{w}},\tag{5.33}$$

$$\mu_{\text{s}}(P,T,a_{\text{s}}) = \mu_{\text{s}}^{+}(P,T) + RT\,\ln a_{\text{s}} \quad \text{where}\ \ a_{\text{s}} = f_{\text{s}}x_{\text{s}}.\tag{5.34}$$

The terms f_{w} and f_{s} are called the *rational activity coefficient* of water and salt, respectively. Their values are usually obtained by experiments and are normally between 0.5 and 1.4 for most materials we encounter in the atmospheric environment.

5.9 Equilibrium between two phases separated by curved interface

5.9.1 Generalized Clausius–Clapeyron equation

Now we are ready to tackle the problem of equilibrium between two phases separated by a curved interface. Fig. 5.4 illustrates the configuration of the system we will study.

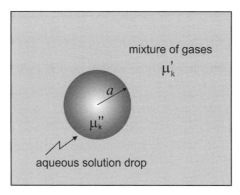

Fig. 5.4 The equilibrium between an aqueous solution drop and an ideal gas mixture.

Here we have a spherical aqueous solution drop of radius a in equilibrium with a gaseous phase, which is a mixture of different gases. We will assume that (1) the system is in thermodynamic equilibrium and (2) each phase contains several components forming a non-ideal solution/mixture.

For such a system, the chemical potential of the component k is

$$\mu_k(P, T, a_k) = \mu_{k,0}(P, T) + RT \ln a_k. \tag{5.35}$$

Dividing both sides by the temperature T, we have

$$\frac{\mu_k}{T} = \frac{\mu_{k,0}(P, T)}{T} + R \ln a_k, \tag{5.36}$$

and hence by differentiating we obtain

$$d\left(\frac{\mu_k'}{T}\right) = -\frac{h_{k,0}'}{T^2} dT + \frac{v_{k,0}'}{T} dP' + R \, d \ln a_k'. \tag{5.37}$$

The first term on the right-hand side is an exact differential and hence can be written as (see Eq. (4.10))

$$
\begin{aligned}
d\left(\frac{\mu_{k,0}(P, T)}{T}\right) &= \frac{\partial}{\partial P}\left[\frac{\mu_{k,0}(P, T)}{T}\right] dP + \frac{\partial}{\partial T}\left[\frac{\mu_{k,0}(P, T)}{T}\right] dT \\
&= \frac{v_{k,0}}{T} dP - \frac{1}{T}\left[\left(\frac{\mu_{k,0}}{T}\right) - \frac{\partial \mu_{k,0}}{\partial T}\right] dT \\
&= \frac{v_{k,0}}{T} dP - \frac{1}{T}\left[\left(\frac{\mu_{k,0}}{T}\right) - s_{k,0}\right] dT \\
&= \frac{v_{k,0}}{T} dP - \frac{h_{k,0}}{T^2} dT,
\end{aligned}
\tag{5.38}
$$

where

$$h_{k,0} = \mu_{k,0} + T s_{k,0} \tag{5.39}$$

is the molar enthalpy of component k. Therefore (5.37) can be written as

$$d\left(\frac{\mu_k}{T}\right) = -\frac{h_{k,0}}{T^2} dT + \frac{v_{k,0}}{T} dP + R \, d \ln a_k. \tag{5.40}$$

This equation is applicable to both phase 1 and phase 2. Now because the two phases are in equilibrium, we have $\mu'_k = \mu''_k$ and $T' = T'' = T$, and therefore

$$d\left(\frac{\mu'_k}{T}\right) = d\left(\frac{\mu''_k}{T}\right). \tag{5.41}$$

But we have

$$d\left(\frac{\mu'_k}{T}\right) = -\frac{h'_{k,0}}{T^2} dT + \frac{v'_{k,0}}{T} dP' + R \, d \ln a'_k \tag{5.42}$$

and

$$d\left(\frac{\mu''_k}{T}\right) = -\frac{h''_{k,0}}{T^2} dT + \frac{v''_{k,0}}{T} dP'' + R \, d \ln a''_k. \tag{5.43}$$

Therefore from (5.41) we obtain

$$-\frac{\left(h'_{k,0} - h''_{k,0}\right)}{T^2} dT + \frac{v'_{k,0}}{T} dP' - \frac{v''_{k,0}}{T} dP'' + R \, d \ln \left(\frac{a'_k}{a''_k}\right) = 0. \tag{5.44}$$

Now since this is a curved interface system, the pressures in the two phases are related by

$$P'' - P' = \frac{2\sigma}{a} \quad \text{or} \quad P'' = P' + \frac{2\sigma}{a}. \tag{5.45}$$

Hence (5.44) can be written as

$$-\frac{\left(h'_{k,0} - h''_{k,0}\right)}{T^2} dT + \frac{v'_{k,0}}{T} dP' - \frac{v''_{k,0}}{T} d\left(P' + \frac{2\sigma}{a}\right) + R \, d \ln \left(\frac{a'_k}{a''_k}\right) = 0,$$

which leads to

$$-\frac{\left(h'_{k,0} - h''_{k,0}\right)}{T^2} dT + \frac{\left(v'_{k,0} - v''_{k,0}\right)}{T} dP' - \frac{2v''_{k,0}}{T} d\left(\frac{\sigma}{a}\right) + R \, d \ln \left(\frac{a'_k}{a''_k}\right) = 0. \tag{5.46}$$

Eq. (5.46) is a form of the *generalized Clausius–Clapeyron equation*. The first two terms contains the original components of the conventional Clausius–Clapeyron equation, while the third and fourth terms represent the *curvature effect* due to the curved interface and the *solute effect* due to the impurities in the solution/mixture, respectively. In the following sections, we will explore the effects of these terms using more concrete examples.

5.10 Kelvin equation – equilibrium between a pure water drop and humid air

Let us now consider a fairly practical case in clouds: the equilibrium of a pure water drop with humid air. There are two phases (gas and liquid) and two components (dry air and water) in this system. Since we already know the conventional Clausius–Clapeyron equation, we do not need to repeat it again, just focus on the new effects. Let us assume that the equilibrium is achieved without changing the temperature and the vapor pressure so that $dP' = dT = 0$. Under such assumptions, Eq. (5.46) can be written as

$$-\frac{2v_w}{T}d\left(\frac{\sigma}{a}\right) + Rd\ln\left(\frac{a_v}{a_w}\right) = 0. \tag{5.47}$$

But since the liquid phase contains only pure water, we have $a_w = 1$. However, the gaseous phase – the humid air – consists of two components. Because the concentration of water vapor is much smaller than that of the dry air, we can treat it as an "ideal" mixture (similar to an ideal solution) and write the activity of water vapor as

$$a_v \approx x_v = \frac{e_{sat,w}}{P}, \tag{5.48}$$

where P is the total pressure of the humid air. Hence (5.47) becomes

$$-\frac{2v_w}{T}d\left(\frac{\sigma}{a}\right) + Rd\ln\left(\frac{e_{sat,w}}{P}\right) = 0. \tag{5.49}$$

Integrating (5.49), we obtain

$$-\frac{2v_w}{T}\int_\infty^a d\left(\frac{\sigma}{a}\right) + R\int_{e_{sat,\infty}}^{e_{sat,a}} d\ln\left(\frac{e_{sat,w}}{P}\right) = 0. \tag{5.50}$$

It pays to understand the range of integration. In the first integral, we integrate from $a = \infty$ (plane) to a (curved surface). The corresponding range in the second integral is from $e_{sat,w} = e_{sat,\infty}$ to $e_{sat,w} = e_{sat,a}$, from the saturation vapor pressure over a plane water surface to that over a curved water surface (such as over a spherical drop surface). The result is

$$-\frac{2v_w}{T}\left(\frac{\sigma}{a}\right) + R\ln\left(\frac{e_{sat,a}}{e_{sat,\infty}}\right) = 0,$$

which after rearranging becomes

$$e_{sat,a} = e_{sat,\infty}\exp\left(\frac{2v_w\sigma}{RTa}\right). \tag{5.51}$$

Eq. (5.51) is the famous *Kelvin equation* for the saturation vapor pressure over a curved liquid surface with radius of curvature a. It has important implications for atmospheric clouds.

For a cloud droplet of radius a, the quantities in the argument of the exponential function are all positive and the argument as a whole is positive. This means that

$$\exp\left(\frac{2v_w\sigma}{RTa}\right) > 1 \quad \text{and hence} \quad e_{sat,a} > e_{sat,\infty}, \tag{5.52}$$

i.e. the saturation vapor pressure over a liquid drop surface is always greater than that over a plane water surface! Now remember that the conventional definition of relative humidity is based on the value of $e_{sat,\infty}$: the relative humidity is 100% when $e_{env} = e_{sat,\infty}$, i.e. the reference value is the saturation vapor pressure with respect to the plane water surface. The fact that $e_{sat,a} > e_{sat,\infty}$ says that, even at $RH = 100\%$, the environment does not have enough vapor to saturate the drop. Since the environment is subsaturated with respect to the drop, the drop has no choice but to evaporate! Yet we have observed that clouds can survive stably at relative humidity considerably smaller than 100%. This reveals an important fact: most cloud droplets *cannot be pure water drops*. Otherwise they could not remain but would have to evaporate as soon as they were produced as required by the Kelvin equation. They must have some impurities in them, i.e. they are aqueous solution drops. We will examine this point in the next section.

Before we leave this subject, let us examine a few more details about the Kelvin equation. This equation is valid not just for liquid water drops but for general liquids as well. There are numerous experiments that have been carried out to check the validity of the Kelvin equation. Sambles *et al.* (1970) used an electron microscope and tested the evaporation rate of small lead and silver droplets of radii less than 0.1 μm before evaporation and found that the Kelvin equation predicts evaporation rates agreeing with experiments down to a few nanometers (10^{-9} m).

The conclusion of (5.52) is for a liquid droplet of radius a. Since a liquid droplet has a convex surface, its radius of curvature is positive, i.e. $a > 0$. How about a concave water surface such as the water surface in a glass capillary tube (see Fig. 5.5)?

In that case, $a < 0$, and we would predict from the Kelvin equation that $e_{sat,a} < e_{sat,\infty}$ and the surface would be even more resistant to evaporation than a plane water surface. Is this true? The answer turns out to be "yes", as experiments have verified. Thus, the Kelvin equation is applicable even for liquid surfaces of negative curvatures.

What is the physical reason for the vapor pressure behavior predicted by the Kelvin equation? This can be explained from a molecular point of view and it

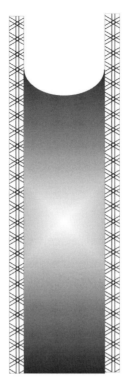

Fig. 5.5 A concave liquid water surface formed in a capillary glass tube. The radius of curvature is negative over the concave surface.

sheds light on the concept of the saturation vapor pressure. As we have indicated before, saturation implies equilibrium, which means that the number of water molecules leaving the liquid phase is the same as the number entering the liquid phase over a finite surface area of the liquid. But what influences the number of molecules leaving the liquid phase? Under definite environmental conditions, the influencing factor is the total molecular force experienced by a particular molecule at the surface. If the molecular force that holds this surface molecule in the liquid is strong, then the molecule has difficulty leaving the surface to become "airborne". In this case, the saturation vapor pressure must be low, and vice versa.

Fig. 5.6 illustrates the three situations of the liquid surface curvature: (a) a convex surface, (b) a plane surface, and (c) a concave surface. We choose an arbitrary molecule M and use it as the center to draw an imaginary sphere of radius r. We see that the amount of molecules in the liquid that lie within the sphere (we use a circle here to represent it schematically) varies in these three cases. If there are more molecules in the sphere, then M will experience greater attractive force to keep it in the liquid and the saturation vapor pressure will be low, and vice versa.

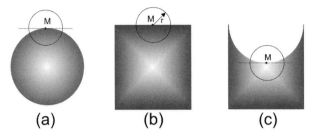

Fig. 5.6 A conceptual model of the molecular force acting on a target molecule M on (a) a spherical drop surface, (b) a plane water surface, and (c) a concave water surface.

In the plane interface case (b), we see that the volume of liquid molecules intercepted by the imaginary sphere is exactly half of the sphere, i.e. a hemisphere. In the case of convex surface (a), the liquid volume intercepted is less than a hemisphere and hence the molecule M will experience less attractive force from within the liquid – in other words, it is easier for M to leave a convex surface than a plane surface. Hence the saturation vapor pressure over the convex surface must be higher than that over a plane water surface. On the other hand, the intercepted volume for a concave surface is greater than a hemisphere, and the molecule M experiences a greater attractive force from within the liquid and it is more difficult to leave the surface. Consequently, the saturation vapor pressure is less than that over a plane interface.

5.11 Köhler equation – equilibrium between an aqueous solution drop and humid air

In the above section, we showed that pure water drops in the atmosphere (such as those in clouds) cannot survive even in an $RH = 100\%$ environment, yet we constantly see clouds forming or at least surviving even under subsaturated conditions. What makes them capable of surviving is the solute effect mentioned above. In this section, we will examine this solute effect in more detail.

Since the system is essentially the same as before except that now the liquid drop is an aqueous solution, we start from Eq. (5.47),

$$-\frac{2v_{\mathrm{w}}}{T}d\left(\frac{\sigma}{a}\right) + R\,d\,\ln\left(\frac{a_{\mathrm{v}}}{a_{\mathrm{w}}}\right) = 0, \tag{5.53}$$

and note that Eq. (5.48) is still valid in this case. We integrate (5.53) to yield

$$-\frac{2v_{\mathrm{w}}}{T}\int_{\infty}^{a}d\left(\frac{\sigma}{a}\right) + R\int_{e_{\mathrm{sat},\infty}}^{e_{\mathrm{sat},a}}d\,\ln\left(\frac{e_{\mathrm{sat,w}}}{P}\right) - R\int_{1}^{a_{\mathrm{w}}}d\,\ln a_{\mathrm{w}} = 0 \tag{5.54}$$

and next

$$-\frac{2v_w}{T}\left(\frac{\sigma}{a}\right) + R\ln\left(\frac{e_{sat,a}}{e_{sat,\infty}}\right) = R\ln a_w \qquad (5.55)$$

and finally

$$\frac{e_{sat,a}}{e_{sat,\infty}} = a_w\exp\left(\frac{2v_w\sigma}{RTa}\right) = a_w\exp\left(\frac{2M_w\sigma}{RT\rho_w a}\right), \qquad (5.56)$$

where M_w and ρ_w are the molecular mass and bulk density of water, respectively. Unlike the Kelvin equation, there is an additional factor a_w in front of the exponential function. The value of a_w can be greater or smaller than unity depending on the kind of salt in the solution. For some salts such as ammonium sulfate $(NH_4)_2SO_4$ and sodium chloride NaCl, a_w is smaller than unity and hence the right-hand side of (5.56) can be smaller than unity even if the exponential function value is greater than unity. Thus if cloud droplets contain these salts, it is possible that $e_{sat,a} < e_{sat,\infty}$, and the droplets will not evaporate even when $RH < 100\%$!

Eq. (5.56) is a form of the *Köhler equation*. In actual applications for dilute solutions, it can be approximated by

$$\ln\left(\frac{e_{sat,a}}{e_{sat,\infty}}\right) = \frac{A}{a} - \frac{B}{a^3}, \qquad (5.57)$$

where

$$A = \frac{2M_w\sigma}{RT\rho_w} \approx \frac{3.3\times10^{-5}}{T} \qquad (5.58)$$

and

$$B = \frac{3vm_sM_w}{4\pi M_s\rho_w} \approx \frac{4.3vm_s}{M_s}. \qquad (5.59)$$

Note that all the quantities in Eqs. (5.58) and (5.59) have to be in cgs units and T in K when using the approximate forms. In (5.59) v is the number of ions the solute will produce when dissolved. For example, NaCl will dissolve into two ions in water, Na^+ and Cl^-, hence $v = 2$. Eq. (5.57) can be further simplified when $e_{sat,a}/e_{sat,\infty} \approx 1$ to become

$$S_{v,w} = \frac{e_{sat,a}}{e_{sat,\infty}} = 1 + \frac{A}{a} - \frac{B}{a^3}, \qquad (5.60)$$

which is another form of the Köhler equation. Here $S_{v,w}$ is called the *saturation ratio*.

Fig. 5.7 shows the curves produced by Eq. (5.60) for drops containing various masses of $(NH_4)_2SO_4$ and NaCl at $T = 20°C$. A pure water drop case is also plotted using the Kelvin equation (curve 1). We assume that the salts are completely

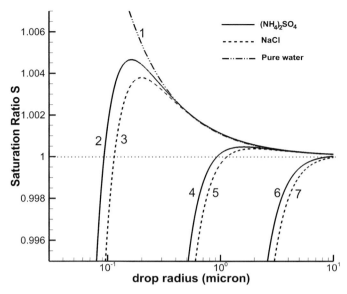

Fig. 5.7 Saturation ratio vs. drop radius for an aqueous solution drop containing various amounts of salts as determined by the Köhler equation (5.57): (1) no salt (pure water drop); (2) $(NH_4)_2SO_4$, 10^{-16} g; (3) NaCl, 10^{-16} g; (4) $(NH_4)_2SO_4$, 10^{-15} g; (5) NaCl, 10^{-15} g; (6) $(NH_4)_2SO_4$, 10^{-14} g; and (7) NaCl, 10^{-14} g.

dissolved in the drop. We first note that these curves represent equilibrium conditions – humid air in equilibrium with aqueous solution drops of various sizes and containing various salt concentrations at a fixed temperature T. The vertical axis is the saturation ratio and the horizontal axis is the drop radius. We observe that, except for curve 1 (we will come back to that), each curve has a maximum slightly above the $S_{v,w} = 1$ level, its height and position varying with the curve. The thermodynamic behavior of a system on the left-hand side of the maximum is very different from that on the right-hand side. In order to see these behaviors more clearly, we draw a schematic chart that emphasizes this maximum region (Fig. 5.8).

5.11.1 Stable equilibrium regime

We first examine point A in Fig. 5.8, which corresponds to saturation ratio 1.002 and drop radius a_1. Since this point is on the curve, the condition $(1.002, a_1)$ must be an equilibrium condition. Now, if the environment becomes more humid, say, $S_{v,w}$ becomes 1.003, what will happen to this drop? Since the humidity is now more than enough for equilibrium, the drop will grow. It will grow until its radius becomes a_2 (point C) and thence returns to equilibrium. Similarly, if the environment becomes drier so that $S_{v,x}$ becomes 0.998 (point D), the drop will evaporate to a new size a_3 and thence returns to equilibrium.

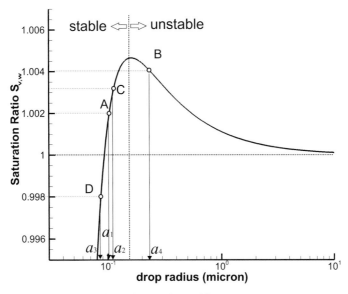

Fig. 5.8 Stable and unstable regimes of a Köhler curve.

We see that the drop is always able to return to a new equilibrium position by changing its size in response to the environmental changes. Hence the region to the left of the hump is the *stable equilibrium* regime.

5.11.2 Unstable equilibrium regime

Now examine another situation by considering an initially equilibrium condition represented by point B in Fig. 5.8 ($S_{v,w} = 1.004$, $a = a_4$), which is located to the right of the maximum. What will happen if $S_{v,w}$ becomes 1.0045? When the environment is supersaturated with respect to the drop, the drop will have to grow. But the slope of the equilibrium curve on this side of the maximum is such that, as the drop becomes bigger, it requires even less $S_{v,w}$ to maintain equilibrium. Thus the environment becomes even more supersaturated as the drop grows in size, and the drop will have to continue to grow indefinitely.

Conversely, if the environment becomes drier, say, $S_{v,w}$ becomes 1.0032, then the initial equilibrium at point B would be disturbed and the drop would have to evaporate. As it decreases in size, however, the required saturation ratio becomes even higher and the drop continues to evaporate. In fact, it will evaporate, become smaller than the size corresponding to the maximum (and hence enters the left-hand side), and stabilize at a_2 (point C). Thence it will be in stable equilibrium again.

Because the behavior of the drop is such that it keeps moving away from its equilibrium position, we call the region to the right of the maximum the *unstable equilibrium regime*.

The drop radius at the maximum is called the *critical radius*. It can be easily shown from (5.60) that the critical radius is

$$a_c = \sqrt{\frac{3B}{A}}.$$ (5.61)

5.11.3 The Kelvin curve

Finally we take a look at curve 1 in Fig. 5.7, which is the curve for a pure water drop based on the Kelvin equation. We see that the slope characteristic of the curve is similar to that in the unstable equilibrium region, except that it has no maximum. Thus when the environment becomes more humid, a pure water drop will grow indefinitely just like an aqueous solution drop. When the environment becomes drier, however, the drop will evaporate until it completely disappears! Thus a pure water drop cannot be in stable equilibrium with its environment. Moreover, we see that the curve never intersects the horizontal line $S_{v,w} = 1$, which means that a pure water drop cannot even be in equilibrium at all when the relative humidity is less than 100%! It is clear from this curve that the clouds that can remain stable in a subsaturated atmosphere cannot be made of pure water drops.

5.11.4 The effect of different solutes

Fig. 5.7 also shows the effect of different solutes. Comparing curve 3 (for a drop containing 10^{-16} g of sodium chloride) and curve 2 (for a drop containing 10^{-16} g of ammonium sulfate), we see that, for the same mass concentration, the hump for the NaCl solution drop is lower than that for the $(NH_4)_2SO_4$ solution drop. This indicates that it is easier for the former to survive than the latter in a similar environment.

It is also possible to treat the case of aqueous solution drops formed by dissolving an aerosol particle that is only partially soluble. This means that there will be an insoluble part in the drop. The general conclusion is that, for the same amount of aerosol particle mass, the equilibrium vapor pressure will be lower if the volume fraction of the soluble components is larger. In other words, the more soluble the fraction, the easier it is for the drop to stay in equilibrium with the environment.

The Köhler curves in both Figs. 5.7 and 5.8 are smooth only because we did not consider the history of the solute. If we consider the history of a salt particle starting from its deliquescence to form a solution drop, and the drop then evaporates in a less favorable environment, we would see that its behavior would not completely follow the path of the curve. Rather, a process called *hysteresis* will be involved and the actual path is more complicated. Chen (1994) has studied the details of such behavior.

5.12 Surface of ice crystals

In many respects, the behavior of the ice phase is qualitatively similar to that of the liquid phase, although quantitatively they are very different, and the theoretical treatments are not as straightforward. Part of the difficulty comes from the more complicated situation of the ice crystal surface, which we will discuss in this section.

It has been known for a long time that the surface of an ice crystal is not smooth. What appears to be a very smooth surface of a simple ice plate under a low-power optical microscope usually shows complicated surface features under a higher-magnification microscope. These rough surface features are called *steps*. The presence of steps is due to the crystalline structure of ice, which cannot be arbitrarily arranged – the water molecules must be fitted into a crystal lattice. As we have discussed in Chapter 3, dislocations occur often in ice crystals and they will eventually appear on the crystal surface and we will observe these as steps.

The steps on an ice surface vary in height and size. The smallest possible step is called an *elementary step*, which corresponds to the thickness of one molecular layer. Fig. 5.9 shows a conceptual model of an elementary step.

It shows a cross-section along the *c* axis. The basal face has a bilayer structure as explained by Hobbs (1974). The height of an elementary step is 0.37 nm, which is half of the height of a unit cell in the *c* axis direction. Such subnanometer structure has been very difficult to observe directly, although it has been the subject of theoretical discussions for a long time. It is only recently that Sazaki *et al.* (2010) have successfully observed and proved the presence of such elementary steps directly using an advanced technique called laser confocal microscopy combined with differential

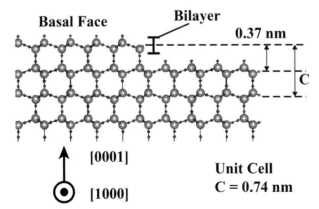

Fig. 5.9 A schematic drawing of a cross-section of an ice-I_h crystal. The large and small atoms correspond to oxygen and hydrogen, respectively. Bilayers made of water molecules are stacked in the *c* axis direction at 0.37 nm intervals. From Sazaki *et al.* (2010). Reproduced by permission of PNAS.

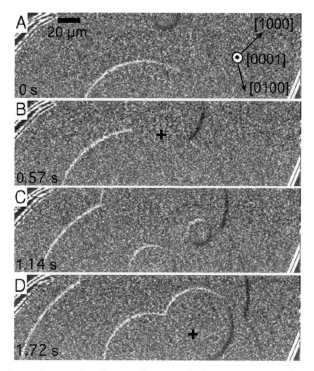

Fig. 5.10 Photomicrographs of the surface morphology on the basal face of an ice-I_h crystal. The sequence of micrographs shows the time course of a two-dimensional nucleation growth of birth-and-spread type: (A) 0 s, (B) 0.57 s, (C) 1.14 s, and (D) 1.72 s. The crosses in (B) and (D) show the regions at which the contrast between the coalesced steps disappeared. Growth conditions: $T_{sample} = -10.0°C$ and $T_{source} = -9.3°C$. From Sazaki *et al.* (2012). Reproduced by permission of PNAS.

interference contrast microscopy (LCM-DIM). Fig. 5.10 shows a sequence of photomicrographs of the surface morphology of the basal face of an ice-I_h crystal.

This sequence shows that a 2D nucleation process is proceeding on the basal face. During this process, vapor molecules land on the surface and are incorporated into the crystal lattice by diffusing toward the edge of a step. The circular ridges are the advancing elementary step "fronts". In frame (A), there are two such fronts near the middle of the frame – a smaller one in the upper and a larger one in the lower edge of the frame. At $t = 0.57$ s, the two fronts meet and coalesce and the contrast disappears completely as indicated by the "+" sign in frame (B). The merged front spreads outward and new elementary steps appear (C) on it and coalesce (D) again. Sazaki *et al.* (2010) also argued convincingly that these are indeed elementary steps and that no invisible smaller steps were present during this sequence.

Another important phenomenon on the ice crystal surface is the existence of a liquid-like layer even when the temperature is substantially below 0°C. Faraday probably was the first to suggest the existence of this layer, now commonly called a quasi-liquid layer (QLL), in 1850 when he tried to explain why two ice cubes will stick together when put in contact even though the temperature is below the melting point. This process is called *regelation*. The QLL is now thought to exist even at temperatures as low as −30°C. It is also mostly responsible for the slipperiness of the ice surface that allows skaters to skate on ice (as opposed to the previous assertions that the liquid is largely due to pressure melting due to the skater's weight or to friction due to the skate blade). Many more modern studies using various experimental methods have since confirmed the existence of the QLL (e.g. Furukawa *et al.*, 1987; Dosch *et al.*, 1995).

The presence of the QLL is important to cloud physics because this is the layer where the ice particle interacts with its environment. For example, the chemical reactions between trace gases and ice crystals in the polar stratospheric clouds are thought to happen on the QLL, whose chemical properties are different from those of an ice lattice. QLLs on graupels may also play an important role in the electrification of thunderclouds (see Chapter 14).

There are various estimates of the thickness of QLLs and they sometimes differ by two orders of magnitude. Fig. 5.11 shows the measurements by Döppenschmidt and Butt (2000) using the atomic force microscopy technique. As would be expected, the thickness of the QLL decreases with decreasing temperature. The layer thickness of ice formed from distilled water was roughly 32 nm at −1°C and 11 nm at −10°C. The temperature dependence of the thickness d could best be described by $d \propto -\log\Delta T$, where ΔT is the difference between the melting temperature and the actual temperature. They also found that the addition of salt increases the thickness of the QLL, which may have contributed to the wide discrepancies of some previous estimates. Another possibility is that the QLL may not be uniform over the surface, especially at higher temperature closer to the melting point, causing large scatter of the data points.

When the surface temperature is close to, but still below, the melting point, a phenomenon called *surface melting* occurs. Surface melting produces QLLs on the surface, which also contribute to the difficulty of observing the elementary steps directly. Sazaki *et al.* (2012) discovered that there are two QLL phases during surface melting of the basal face. The first type is the round liquid-like droplet called α-QLL, and the second is the thin liquid-like layer β-QLL, as shown in Fig. 5.12. The temperature range for the appearance of these two QLLs varied from −1.4 to −0.5°C and from −1.0 to −0.1°C, respectively, in different runs, but the β type always appears at higher temperature than the α type in a specific run. When the ice crystal is heated, α-QLLs appear first and are surrounded by elementary steps, which spread outward as indicated by the black arrows, showing that the α-QLLs function as the source of steps. The α-QLLs can coalesce if

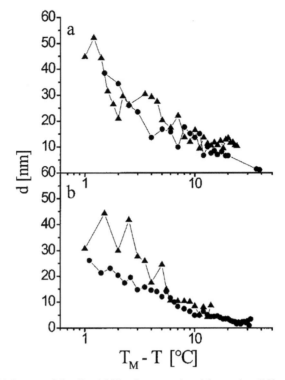

Fig. 5.11 Thickness of the liquid-like layer on ice (d) vs. the difference between the melting temperature (T_M) and the actual temperature (T). The ice samples were frozen (a) below $-30°C$ in a freezer and (b) by evaporation in a vacuum. The layer thickness is equal to the jump-in distance minus 2 nm. Triangles and circles correspond to different experiments done under the same conditions. From Döppenschmidt and Butt (2000). Reproduced by permission of the American Chemical Society.

they contact each other just like liquid droplets. Upon further heating, the thin layer β-QLL appears as indicated by the broad white arrow. The α-QLLs are generated from this thin layer as well as from the crystal surface. Eventually the thin layer covers the whole surface. The thickness of the thin layer was estimated to be several tens of nanometers. Fig. 5.13 shows a schematic summary of the surface melting process.

We have just begun to understand the surface processes on ice crystals and can expect to see more important studies in this direction in the near future.

5.13 Summary

In the last chapter and this one, we have discussed the *equilibrium* states of water substances only. We have assumed that the condensed water substance exists

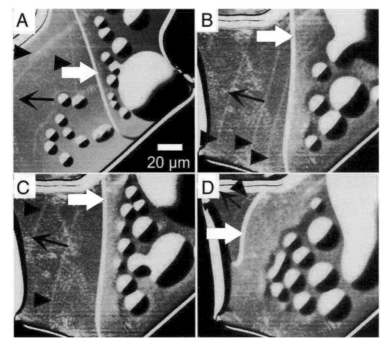

Fig. 5.12 The appearance of thin liquid-like layers (β-QLLs), indicated by white broad arrows, with increasing temperature. The temperature was −0.2°C (A) and −0.1°C (B–D). Images (B)–(D) were taken 0, 18, and 239 s, respectively, after the temperature was set at −0.1°C. Black arrowheads and black arrows indicate elementary steps and their growth directions, respectively. From Sazaki *et al.* (2012), with changes. Reproduced by permission of PNAS.

already and examined what would happen if the environment changed. In Chapter 4, we derived the Clausius–Clapeyron equation and used it to explain many phase change phenomena. We concluded that it is a useful relation but in certain cases it fails because it does not include the curvature and solute effects. In this chapter, we derived the Kelvin and the Köhler equations and showed that the existence of clouds in our atmosphere is possible only because the cloud droplets are not pure water but are aqueous solution drops. We also briefly touched upon the surface process on ice.

We have not examined how the condensed water substances are produced in the first place, which is really the process of phase change. We mentioned the phase change of water substances using the Clausius–Clapeyron equation, but that equation only predicts what the equilibrium conditions are. It does not tell us anything about the kinetic process of condensation. The same goes for the Kelvin and Köhler equations. To see how the condensed phases form, we need to study the molecular kinetic process rather than the static systems like the equilibrium cases that we have

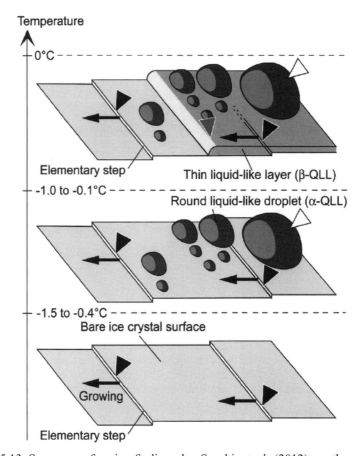

Fig. 5.13 Summary of major findings by Sazaki *et al.* (2012) on the surface melting of ice. Two types of QLLs appeared on a basal face of an ice crystal during surface melting. With increasing temperature, round liquid-like droplets (α-QLLs, white arrowheads) first appeared at −1.5 to −0.4°C. In addition, thin liquid-like layers (β-QLLs, gray arrowheads) appeared at −1.0 to −0.1°C. The appearance temperatures of α-QLLs and β-QLLs exhibited slight variations, but β-QLLs always appeared at a higher temperature than α-QLLs in the same run. These two QLL phases moved around and coalesced on the ice crystal surface. The different morphologies and dynamics of the two types of QLLs demonstrate that the two types of QLL phases had different interactions with ice crystal surfaces, suggesting the need to construct a novel picture of surface melting. Black arrowheads and arrows have the same meaning as in Fig. 5.12. From Sazaki *et al.* (2012), with changes. Reproduced by permission of PNAS.

discussed up to now. This kinetic process is called *nucleation* and is the subject of Chapter 7. Before we treat the subject of nucleation, we need to understand an important partner of the water substance in forming clouds, the aerosol, which is the subject of the next chapter.

Problems

5.1. Use the fact that

$$dG(T,P,n_k) = -SdT + VdP + \sum_{k=1}^{c}\mu_k dn_k$$

is an *exact differential* to prove that

$$\left(\frac{\partial \mu_k}{\partial T}\right)_{p,n_{j\neq k}} = -\left(\frac{\partial S}{\partial n}\right)_{T,p,n_{j\neq k}} = -s_k,$$

$$\left(\frac{\partial \mu_k}{\partial P}\right)_{T,n_{j\neq k}} = \left(\frac{\partial V}{\partial n}\right)_{T,p,n_{j\neq k}} = v_k.$$

It is important to pay attention to the subscripts here so that you understand what you are doing in terms of physical processes when you write down the equation.

5.2. Prove that

$$\left[\frac{\partial(\mu_k/T)}{\partial T}\right]_{P,n_{j\neq k}} = -\frac{h_k}{T^2},$$

where h_k is the molar enthalpy of the component k.

5.3. Show that the critical radius corresponding to the maximum of a Köhler curve is indeed given by (5.61).

5.4. (a) Derive an expression for the radius a_p of an aqueous solution drop that is in equilibrium with its environment when the saturation ratio $S_{v,w} = 1$. The radius a_p is called the potential radius.

(b) Is there any physical significance of a_p? Why is it different from the critical radius a_c?

(c) Determine a_p for a drop containing 10^{-14} g of NaCl and another drop containing 10^{-14} g of $(NH_4)_2SO_4$, and explain why there is a difference.

6

Aerosol in the atmosphere

Our atmosphere consists not only of gases but also of particulate matter. The particulate matter can be either liquid droplets or solid particles, which are collectively called aerosol particles. Strictly speaking, clouds are also a kind of aerosol because they are particulates too, but we will exclude them from this category in this book.

Many people have the impression that aerosol particles are merely a contamination in the atmosphere, but the scientific community sensed the important role of aerosol particles in the atmosphere early on. In 1875, the French scientist Paul Jean Coulier, when working with an expansion chamber, discovered that condensation of water vapor would not occur unless there were some dust particles in the air, and thus demonstrated the fallacy of the old "common sense" that water vapor would always condense as soon as it reaches saturation. The existence of *condensation nuclei* (CN) was discovered. Six years later, the Scottish scientist John Aitken independently confirmed the same thing. It is now well established that the formation of cloud droplets also needs to have these particles, and they are called *cloud condensation nuclei* (CCN) in this circumstance. Evidently, these particles should be viewed not just as contaminants but as an important component of the atmospheric physicochemical system.

Similar to the condensation of water droplets, the formation of ice in clouds also needs the presence of certain aerosol particles – called *ice nuclei* (IN) – unless the temperature is close to −40°C or colder. Both cloud condensation nuclei and ice nuclei are subsets of aerosol particles in the atmosphere. The problem of aerosol and its related cloud formation effects is currently considered as one of the most important unresolved issues in climate predictions. We will discuss more on this subject in Chapter 15.

In this chapter, we will examine the fundamental physical and chemical properties of aerosol particles as related to cloud physics. Specific properties of cloud condensation nuclei and ice nuclei are discussed in the next chapter where nucleation is the main subject. A comprehensive treatment of aerosol particles is beyond the scope of this book, but readers are referred to books such as Hinds (1982), Warneck (1988), Pruppacher and Klett (1997), and Seinfeld and Pandis (2006).

6.1 Aerosol size categories

One of the important characteristics of aerosol particles, like clouds, is their size. Dust particles near the surface are usually very large, often about 0.1–1 mm, but they do not stay very long nor can they reach high into the atmosphere. Aerosol particles, especially those present at higher levels, are generally much smaller and are able to stay much longer in the atmosphere to have important impacts. Junge (1963) divided aerosol particles into three size categories:

(1) *Aitken particles* – particles with radius $r < 0.1$ μm.
(2) *Large particles* – particles with radius $0.1 \leq r \leq 1.0$ μm.
(3) *Giant particles* – particles with radius $r > 1.0$ μm.

Obviously, large particles are singled out for a more precise size categorization because particles of this size range have important impacts on the atmosphere. For example, they are about the same size range as the visible light wavelengths (0.4–0.7 μm) and hence scatter sunlight strongly because the scattering of light is generally stronger when the scatterer size is close to the wavelength of the light. In a heavily polluted atmosphere, the large particles are so concentrated and the light scattering is so strong that they almost completely mask the scene behind them and severely reduce visibility.

Another notable property of the large particles is their impact on human health. Medical studies reveal that the hairs in our nose serve as an air filter to remove particulate matter in air when we inhale. Particles of size larger than the "large" particles, i.e. giant particles, are usually filtered out efficiently by nasal hair, while smaller particles can slip through and enter our lungs easily. Once in our lungs, however, large particles tend to become trapped inside the delicate pathways inside the lungs and cannot be exhaled out easily, whereas smaller particles, i.e. the Aitken particles, can usually slip out as easily as they got in. This behavior obviously has important implications in human health, especially respiratory diseases.

When aerosol particles serve as nuclei to nucleate cloud particles, they are called Aitken nuclei, large nuclei or giant nuclei according to their size. But we must keep in mind that not all aerosol particles serve as nuclei for cloud formation; hence these nuclei are only subsets of the particles of their respective size category.

Another aerosol particle size classification is proposed by Whitby (1978), who recognized that using a rigid cutoff size often misses the physical mechanism behind the aerosol formation. What may appear to be an insignificant inflection point on a size distribution curve may turn out to be a separation point for different modes of aerosol formation; each mode represents a size distribution that overlaps with that of other modes. Each mode may also represent a distinct mechanism regulating the aerosol size. Thus he proposed the following naming scheme where the "size" here represents the modal size of a distribution:

(1) *Nuclei mode* – particles with $d < 0.1$ μm, where d is the particle diameter; in some literature, the nuclei mode is further subdivided into nucleation mode ($d < 0.03$ μm) and Aitken mode ($0.03 \leq d \leq 0.1$ μm).
(2) *Accumulation mode* – particles with $0.1 \leq d \leq 2.5$ μm.
(3) *Coarse mode* – particles with $d > 2.5$ μm.

The term "diameter" here needs some explanation. Aerosol particles have a wide variety of shapes: some are approximately spherical but many are highly irregular (such as flakes or curled up fibers), and it is ambiguous what "diameter" really means. But since aerosol particles are floating in air, it is useful to define an "aerodynamic diameter" to represent the size of particles having different shapes and densities. The *aerodynamic diameter* is the diameter of a spherical particle having a density of 1 g cm^{-3} that has the same terminal settling velocity in the gas as the particle of interest, and this is usually what "diameter" means (see e.g. http://www.epa.gov/apti/bces/module3/diameter/diameter.htm).

   Since aerodynamic diameter is an "equivalent spherical diameter" based on particle aerodynamics, the implication of the "aerosol size distributions" based on this definition must be carefully considered when other aerosol properties are of concern. For example, the radiative property of aerosol particles is very sensitive to both particle size and shape, and hence calculations of aerosol radiative properties based solely on the aerodynamic diameter distribution will certainly lead to significant errors.

   In addition to the above terminology, particles with $d < 1.0$ μm are also called *fine particles* (FP) and those with $d < 0.1$ μm are *ultrafine particles* (UFP). Yet another scheme is to use the maximum particle diameter as a form of limit. For example, PM10 refers to particulate matter of 10 μm diameter or smaller, whereas PM2.5 refers to particulate matter of size 2.5 μm or smaller. The use of a specific naming scheme usually depends on the application, such as aerosol in global atmospheric chemistry or particulate matter in urban air pollutant regulations.

6.2 Aerosol concentration

The concentration of aerosol particles is usually expressed in terms of either number concentration (in "number per unit volume", for example, cm^{-3}) or mass concentration (for example, $\mu g\ m^{-3}$). The choice of unit usually depends on the specific application and the instrument used for the measurements. Some instruments can make number counts and some measure only the mass concentration.

Aerosol number concentration near the Earth's surface varies widely with location. In terms of number concentration, it varies from a few hundred cm^{-3} in the air over the ocean surface to well over $10^6\ cm^{-3}$ in highly polluted city air. In towns and smaller cities, the number concentrations are typically from a few 10^3 to $10^4\ cm^{-3}$. Over the mountains, the concentration is typically 10^2 to $10^3\ cm^{-3}$, and usually decreases with height in the mountains. It is our common experience that the air over oceans and higher up in mountains is "fresher" than that in a populated city, and the number concentration of aerosol particles is likely the main reason for this.

In terms of mass concentration, oceanic air is typically around $1\ \mu g\ m^{-3}$ or lower. In an urban area, the mass concentration is around a few to $\sim 10\ \mu g\ m^{-3}$. In dust storm events, the magnitude can be as high as several hundred $\mu g\ m^{-3}$.

Obviously, the span of magnitude range is higher for the number concentration (10^2–10^6) than the mass concentration (10^0–10^2). The reason is that small aerosol particles are high in number but low in total mass, whereas large particles are low in number but high in total mass. The mass concentration is largely represented by large particles whose number concentration is usually low and does not differ greatly from place to place. In contrast, a small difference in mass can mean a huge difference in the numbers of very small particles. It is of some importance to choose the right kind of concentration representation. For example, if the effect of aerosol on visibility is the main concern, then the number concentration representation would make more sense because small particles have high impact on the optical properties of the air. If it is the surface deposition of particles that is of concern, then perhaps the mass concentration is more appropriate.

6.3 Variation of aerosol concentration with height

Aerosol concentration varies not only from place to place but also with height. Although aerosol particles have sources in the atmosphere, their major source is the Earth's surface. Hence it is not surprising that the total concentration of aerosol particles in the lower atmosphere usually has a maximum near the surface and decreases with height. But the decrease is not monotonic because there are also sources and sinks of aerosol in the atmosphere. Figs. 6.1 and 6.2 show two examples of aerosol vertical distributions. It is important to note that vertical profiles of aerosol concentration are often plotted for different size categories and hence one must read the chart carefully.

Fig. 6.1 shows that the Aitken particle concentration decreases nearly exponentially with height from the surface up to about 6 km. This profile usually implies that

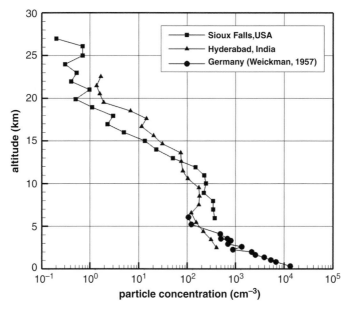

Fig. 6.1 Variation of the average number concentration of Aitken particles with height over various locations. Adapted from Junge (1963) and Pruppacher and Klett (1997), with changes.

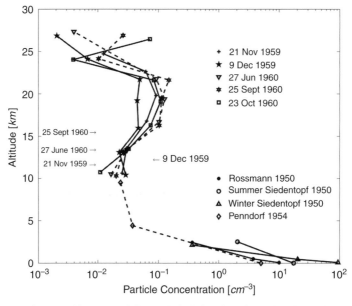

Fig. 6.2 Variation of large particles with height, showing a local maximum in the stratosphere. This layer is now called the Junge aerosol layer. Adapted from Chagnon and Junge (1961).

the surface is the source of the particles and the concentration distribution is basically controlled by the turbulent diffusional process. But this is where the exponential decrease ends. From $z \sim 6$ km up to about 10 km the concentration stays roughly constant at $\sim$50–500 cm^{-3} with 200 cm^{-3} the most frequent. Above 10 km, the Aitken particle concentration decreases again with height.

The quasi-constant phenomenon in the 6–10 km layer has been thought to indicate the existence of a "global background aerosol" of the atmosphere, which, if true, would represent an intrinsic characteristic aerosol state of the Earth's atmosphere. Recent measurements, however, have found that the aerosol concentrations change rapidly in response to long-range transport and local sources and sinks, and hence the concept of a background aerosol is now being debated.

The case of large particles looks quite different as seen in Fig. 6.2. From the surface up to the tropopause level the concentration decreases generally, but it begins to increase above the tropopause and reaches a maximum somewhere in the middle stratosphere, the precise level depending on location and season. This stratospheric maximum has been called the *Junge layer* or *stratospheric aerosol layer*. More recent measurements confirm the worldwide existence of the Junge layer, which is located in the 20–40 km level. Aerosols in the Junge layer are mostly sulfate particles. Currently, it is thought that carbonyl sulfide (COS) associated with global biogeochemical processes on the surface is the major source of sulfur in the aerosol of the Junge layer during quiescent volcanic times, whereas those volcanic eruptions reaching the stratosphere (very few) will certainly contribute also when they happen (Crutzen, 1976).

Giant particles are basically confined to the lower troposphere due to their size. One notable example of giant particles is sea salt particles over the ocean. Sea salt concentration has a maximum in the lowest few hundred meters and then decreases quasi-exponentially with height. Sea salt particles are generated by the evaporation of sea spray droplets, which we shall discuss later in this chapter.

6.4 Aerosol size distributions

The size distribution of aerosol is an important characteristic of its physicochemical state just like that for cloud particles we discussed in Chapter 2. Aerosol size distributions can be somewhat more complicated than cloud particles because of the many pathways of aerosol formation and the range of sizes that current instruments can measure (from about 0.001 μm to more than 10 μm). Multimodal distributions are not uncommon.

Fig. 6.3 shows examples of aerosol size distribution measured over the ocean surface in various parts of the Southern Hemisphere (Meszaros and Vissy, 1974). These distributions all show a peak concentration at aerosol radius close to 0.1 μm.

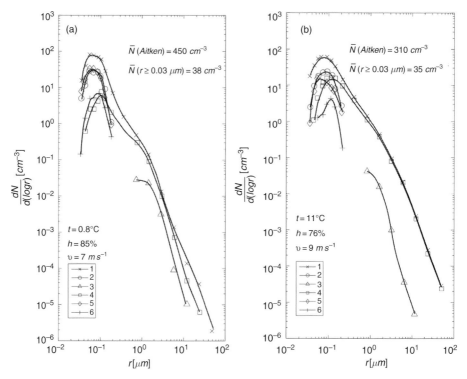

Fig. 6.3 Concentration of aerosol particles for two size ranges and size distribution of particles of different types in the range of $r \geq 0.03$ µm over two different parts of the oceans of the Southern Hemisphere: (1) all particles; (2) $(NH_4)_2SO_4$; (3) long-shaped particles; (4) NaCl; (5) non-cubic crystalline particles; and (6) H_2SO_4. The samples were taken at 12–14 h local time. Adapted from Meszaros and Vissy (1974).

In addition, there is an inflection point somewhere between 0.1 and 1 µm. These distributions can be viewed as the combination of two modes; each mode can be approximated by a gamma or log-normal distribution, as we have seen in Fig. 2.19 for the raindrop size distribution. The mode with peak close to 0.1 µm is the nuclei mode, while the "shoulder" at ~2 µm corresponds to the accumulation mode.

Some aerosol distributions show more distinct bimodal characteristic than the examples in Fig. 6.3. Some even have three distinct modes. These distributions obviously cannot be represented accurately by a single-gamma or log-normal distribution but rather require the sum of several such distributions. Jaenicke (1988) suggested the following multiple log-normal distribution:

$$-\frac{dN}{d\log r} = n^*(r) = \sum_{i=1}^{m} \frac{n_i}{\sqrt{2\pi}\log \sigma_i}\exp\left(-\frac{[\log(r/R_i)]^2}{2(\log \sigma_i)^2}\right), \qquad (6.1)$$

where m is the total number of modes in the distribution (so that, for example, for a trimodal distribution, $m = 3$), n_i is the total number of particles in mode i, r is the particle radius, R_i is the mean particle radius of the distribution, and σ_i is the standard deviation of the ith mode.

6.5 Brownian coagulation and the aging of aerosols

Most aerosol particles are much smaller than cloud droplets and can be influenced easily by the constant bombardment of air molecules and perform random motions called Brownian motion. Brownian motion is especially pronounced for particles <0.5 μm because of their small size, and they can collide and coagulate to form larger particles. This process is called Brownian coagulation. Particles larger than 0.5 μm are more massive and hence their Brownian coagulation is much less efficient. When particles of different chemical compositions coagulate, mixed particles are produced.

Now consider a cloud of aerosol particles freshly produced that may have an initial size distribution peaking at UFP size range as indicated by curve 1 in Fig. 6.4. As time goes on, particles smaller than 0.5 μm will perform rapid Brownian coagulation and become larger and larger, whereas for particles greater than 0.5 μm the coagulation is very slow and their size will not change as quickly as smaller ones. After a certain time period, the aerosol size distributions will become quasi-steady-state as indicated by curve 4, which will change very little as time goes on. This behavior is called the *aging of aerosols* and was first reported by Junge (1955).

Because of aging, the size distribution curves on the larger particle side tend to become quasi-straight lines when plotted on the double-log chart, as is the case in Fig. 6.4. This prompted Junge to suggest a power-law representation of aerosol size distribution (later called the Junge size distribution):

$$\frac{dN}{d \log r} = Ar^{\beta}, \tag{6.2}$$

where $\beta \sim -3$. Many measurements indeed substantiate this quasi-constant slope. Clearly, this only represents one side of the curve and cannot represent the complete distribution.

6.6 Physicochemical pathways of aerosol production

Aerosol particles have sources both at the Earth's surface and in the atmosphere, and that makes the tracking of aerosol sources complicated. There are generally three broad categories of physicochemical pathways of particle production.

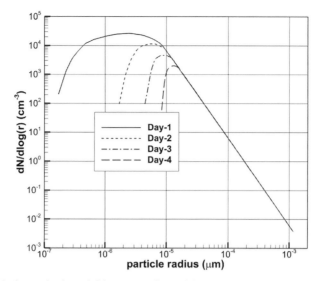

Fig. 6.4 A theoretical model interpretation of the aerosol aging phenomenon due to coagulation. Curves represent the aerosol size distribution from day 1 to day 4. Adapted from Junge (1955).

(1) *Pulverization of bulk matter* – for example, the breaking up of rocks or dirt into finer particles due to weathering, the breaking of ocean surface to form small droplets due to winds or bubble burst, or the ejection of organic droplets from combustion engine exhaust.

(2) *Reactions between different chemical species* – the reacting species can be either entirely in one phase (homogeneous reactions) or of different phases (heterogeneous reactions). When gases react to form aerosol particles, the process is called *gas-to-particle conversion* (GPC). One example of GPC is the formation of a sulfuric acid drop involving the hydroxyl radical OH as suggested by Warneck (1988):

$$SO_2 + OH \rightarrow HOSO_2 \tag{6.3}$$

$$HOSO_2 + O_2 \rightarrow SO_3 + HO_2 \tag{6.4}$$

$$SO_3 + H_2O \rightarrow H_2SO_4 \tag{6.5}$$

Up to this stage, all molecules involved are in the gaseous phase. Then in the sulfuric acid–water vapor mixture, binary nucleation can occur to form sulfuric acid droplets.

It is also possible to have *drop-to-particle conversion* (DPC), and this occurs most commonly in clouds. Many gases and particulates are soluble in water and, if absorbed by cloud droplets, become aqueous solutions during cloud drop formation. When the cloud drop evaporates, the dissolved materials may crystallize to become solid particles. If there are insoluble particles initially, they will become a part of the new particle, too. The new particle may be physically and chemically different from the original particles. This transformation of gases and particles by clouds to form new particles is often called the cloud processing of aerosol.

(3) *Condensation* – this can be either the homogeneous condensation of gas to form liquid droplets or solid particles, or the freezing of liquid into solid, or heterogeneous condensation in which gases condense on liquids or solids, or liquids freeze on solids. Homogeneous condensation processes often occur during biomass burning. Liquids contained in the tissues of vegetation may vaporize during the combustion to become vapors which then recondense into liquid or solid form when cooled later.

6.6.1 Primary and secondary aerosol

Aerosol particles that are injected into the atmosphere directly without going through other processes in the atmosphere (such as the GPC or DPC mentioned above) are called *primary aerosol particles*. Examples are windblown dusts or pollens ejected by plants. Aerosol particles that form from further processes in the atmosphere such as GPC and DPC are called *secondary aerosol particles*.

6.7 Sources of aerosol particles

Aerosol particles come from different sources, which produce particles of different chemical compositions. Once formed, these particles may be transported by atmospheric circulations and interact with gases or with each other and coalesce to form mixed particles, making the analysis of particle chemistry complicated. The following are the major sources of aerosol particles.

6.7.1 Fragmentation of land surface

Dust storms

The land surface is full of small particles that can be carried into the air by winds or by an automobile driving on an unpaved dirt road. Most of these dust particles visible to our eyes are fairly coarse, typically 0.1–1 mm in diameter, and usually cannot stay very long in the air.

Fig. 6.5 A MODIS image of Aqua on 12 March 2010 showing a dust storm carrying a large amount of dust particles from western China crossing the Yellow Sea and East China Sea. The white arrow indicates the dust plume over the East China Sea. Photo courtesy of NASA.

On the other hand, smaller dust particles, especially those in the Aitken and large particle size range, can stay for many days or longer depending on the local updraft strength and how high the particles are carried up. The vast desert regions in the world such as the Sahara in North Africa and the Gobi Desert in Northern China and Mongolia, where vegetation is either very sparse or non-existent, are especially prone to the occurrence of dust storms. Dusts in such storms can travel thousands of kilometers and inject a huge number of aerosol particles into the atmosphere. They can be seen from satellite images, such as the example shown in Fig. 6.5.

Fig. 6.6 shows an example of the size distribution of dust storm particles as measured on the surface. It is clear that, while most of the mass resides in the coarse mode or giant particles, the number concentrations of accumulation mode particles can be two or three orders of magnitude higher than that of the coarse mode. The windblown dusts from arid regions of the world are an important source of ice nuclei. Chemically, the majority of the windblown dusts are various kinds of clay particles, such as SiO_2, Al_2O_3, Fe_2O_3, and MgO.

Biomass burning

Each year, many parts of the world suffer from natural or man-made fires that destroy large areas of forests and other vegetated surfaces. These fires are

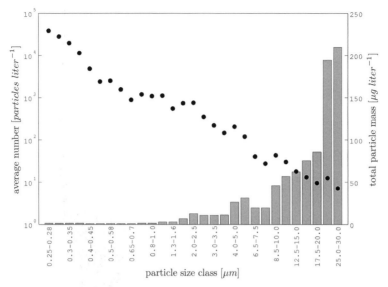

Fig. 6.6 Average number of particles (dots) and their total mass (bars) for different size classes for the whole measurement campaign. The particle masses in single size classes were calculated using the idealized estimation of a sphere particle shape and density $\rho = 2.65$ mg m^{-3}. Adapted from Hoffmann *et al.* (2008).

collectively called *biomass burning*. Forest or prairie fires are often said to be caused by lightning, although some must have been caused by careless handling of fire, such as smoking cigarettes or campfires. But some fires are set off by intentional human ignition, as in the case of biomass burning in the Amazon basin. The purpose is to clear the land for agricultural purposes. Such large-scale biomass burning can cause significant changes in local albedo, as the two photographs taken by the space shuttle crew in Fig. 6.7 show. The left image was taken during a day when there was no burning. The area had normal cloud cover but the air was transparent and the dark ocean surface can be clearly seen. The image on the right was taken during a day with large-scale biomass burning and the air became opaque due to strong reflection of sunlight by fire-produced aerosol particles. The local albedo will increase dramatically in this case.

According to Janhäll *et al.* (2010), biomass burning aerosols are dominated by accumulation mode with a count median diameter of 0.15 μm with smaller amounts of coarse and nuclei modes. The coarse mode particles consist of carbon aggregates, dust, ash, and unburned parts of the fuel, while the accumulation mode consists mostly of organic matter, with soot carbon and inorganic species making up 10% each. Of the organic matter, 40–80% is water-soluble and 20–40% consists of acid (Reid *et al.*, 2005). The exact chemistry of the particles depends on the type of fuel

Fig. 6.7 Satellite visible images of the Amazon basin area during (left) a day with no biomass burning and (right) a day with massive biomass burning. Photos courtesy of NASA.

and the burning process. But studies reveal that biomass burning particles are efficient cloud condensation nuclei and hence they may have great impact on the global climate process through their cloud formation ability (e.g. Trentmann *et al.*, 2006; Rosenfeld *et al.*, 2008).

Recent studies also reveal that biomass burning can combine with deep convective systems to create a new kind of convection system called pyrocumulonimbus (pyroCb), which can carry combustion particles into the upper troposphere and lower stratosphere. This implies that the biomass burning particles can travel a long range, further than previously thought, and thus may have a far greater impact on the global climate.

Recently, biomass burning particles have been collected in Antarctica and their origin is in central Brazil, a tropical region (Fiebig *et al.*, 2009). Thus, there is no question that tropical biomass burning particles, and whatever climatic implications they have, can reach the polar regions in a short time.

Volcanic activity

The fact that volcanic eruptions can inject large amount of particles into the atmosphere has been known from ancient times, and the story of the ash-buried city of Pompeii in Italy is familiar to many. However, the amount and the height of aerosol injection vary greatly from one eruption to another. The eruption of Mt. St Helens in Washington state, USA, on 18 May 1980 is the deadliest eruption in US history, yet its atmospheric impact is insignificant because it was a sideways eruption. The

Fig. 6.8 Eruption of Mt. Pinatubo, Philippines, on 12 June 1991. The volcanic eruption was a near-vertical one and the plumes reached the stratosphere. Image courtesy of USGS.

particle injection height was limited to the troposphere. In contrast, the eruption of Mt. Pinatubo in the Philippines in June 1991 was a straight-up eruption, and the maximum injection height was estimated at 24 km, i.e. well into the stratosphere (Fig. 6.8).

Volcanic ash contains mostly tephra bits – pulverized rocks and glass created by the eruption. These are mainly silicates and fairly large in size, so they usually settle back to the surface locally and their impact on the global climate is limited. However, there are also large amounts of much smaller particles, the most important of which are sulfate particles. Direct sampling of volcanic plumes in craters shows the existence of sulfate particles with SO_4^{2-}/SO_2 from 1 to 5% (Martin *et al.*, 2008).

Volcanic eruptions inject directly into the atmosphere not only aerosol particles but a large amount of other gases as well. The most abundant volcanic gas turns out to be water vapor, followed by CO_2 and SO_2. Among these gases, SO_2 is the precursor for the formation of sulfate aerosols. In the Mt. Pinatubo eruption, SO_2 injection is estimated at more than 2×10^7 tons, and much of this must have turned into sulfate particles and entered the stratosphere.

Human industrial activity

Human beings have been producing aerosol particles from ancient times, and most familiar to us is the burning of fuels (wood, coal, etc.) to obtain energy, which produces smoke. The afore-mentioned intentional biomass burning is also a human activity that has a long history. But the present-day human's ability to produce aerosol particles far exceeds that of our ancestors due to large-scale industrialization.

The most conspicuous particle sources due to industry are the thick smoke coming out of numerous tall stacks in the world's industrial complexes. The most common outputs from these stacks are sulfates and black carbon. SO_2 produced by burning fossil fuels is a common precursor that usually ends up as sulfate particles through DPC or GPC. Specific factories produce specific particles; for example, downwind of paper mills one finds Na_2SO_4, NH_4HSO_3, $Ca(SO_3)_2$, $NaOH$, Na_2SO_3, and H_2SO_4 (Hobbs *et al.*, 1970), while downwind of steel foundries and electric steel mills one finds metals and metal oxides.

Another familiar aerosol production due to human activity is the photochemical smog in big cities. This is mainly due to high automobile traffic volumes that produce large amounts of nitrogen oxides NO_x, which, in the presence of sunlight, produce ozone, several radicals, and some stable products. Many particulates are also produced that make the air smoggy.

6.7.2 Ocean surface process

Aerosol particles produced over the ocean surface are largely NaCl particles but there are also smaller amounts of K^+, Mg^{2+}, CO_3^{2-}, and SO_4^{2-}. Organic and biological materials are also found in marine aerosols.

The production of sea salt particles is due to the *bubble burst mechanism* (also called *sea spray mechanism*) (Fig. 6.9). Since the sea surface is constantly subject to winds, air is trapped in the surface layer to form bubbles. Sea salt aerosol particles form in two stages of the burst process. First, when a bubble rises to the top of the surface, its cap becomes a very thin film (a), which eventually ruptures to form numerous mini-droplets that fly into the air (b). Some of these film droplets eventually evaporate to leave dry sea salt particles that may be carried up higher by winds. At the same time, the water in the remnant bubble creates a jet due to the buildup of pressure at the bottom of the bubble. As the jet rises, instability grows such that the jet breaks into a few droplets that also fly into the air (c). These jet drops are usually larger than the film drops. Some of these droplets are then carried by winds to higher layers and evaporate to become dry sea salt particles (d). Some studies found that sea salt particles produced in this way range from less than 0.1 μm (the cap droplets) to more than 10 μm (jet droplets). The number of droplets and hence the number of sea salt particles produced depends on the bubble size. The bubble size presumably depends on the winds over the sea surface and also the local conduciveness conditions for wave breaking. Blanchard (1969) estimated that the

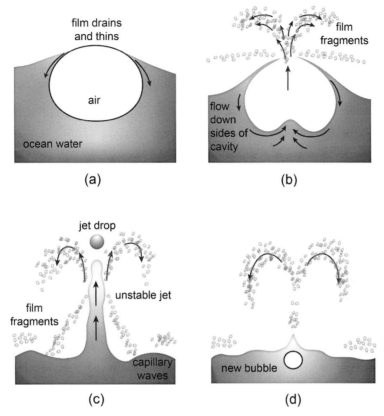

Fig. 6.9 The bubble burst mechanism of marine aerosol production. Adapted from Pruppacher and Klett (1997).

number of sea salt particles produced in this way is about 10^{28} per year, with a total mass of 300 Tg and average sea salt aerosol mass of 3×10^{-4} g.

6.7.3 Biogenic aerosols

Aerosol particles are produced by biological systems during their life cycle and plants are the main source of these biogenic aerosols. Pollens released by the flowers of various plants are spread in the air during pollination to become aerosol particles. They are typically about 10 µm or larger in size. Many plant fragments, spores of fungi, virus, and bacteria range from a few micrometers to less than 1 µm in size and thus can easily become airborne as aerosol particles. Some of these biogenic aerosol particles may serve as efficient nuclei for condensation and freezing, as we will discuss in the next chapter.

Recent measurements in the Amazon basin by Pöschl *et al.* (2010) showed that submicrometer biogenic aerosol particles are the main cloud condensation nuclei during the wet season there, in contrast to the dry season, when the biomass burning particles dominate the condensation.

6.7.4 Extraterrestrial source

The Earth's atmosphere is constantly bombarded by meteoroids, which are basically rocks of various sizes; some are larger than a few centimeters but most are probably just of dust particle size. These rocks are widely distributed in the interplanetary space in the Solar System and are attracted to the Earth by the gravitational force if their orbits happen to intercept that of the Earth. Because of their already high orbiting velocities around the Sun, they can have high relative velocities with respect to the Earth, which can be 50 km s^{-1} or greater when they enter the atmosphere. The extreme heating due to the ram pressure created by such high velocities causes these particles to burn and emit light – a phenomenon called meteors. [Note: Ram pressure is a pressure exerted on a body that is moving through a fluid medium. When the velocity of the body is high relative to the fluid, the drag is very large and shock waves develop in front of the body. Extreme compression in the shock region causes very high heating that burns the body.] Particles of sand dust to pebble size probably evaporate completely in the atmosphere, but some larger particles may break up into smaller fragments to become aerosol particles in the Earth's atmosphere. This is the extraterrestrial source of aerosol particles. The vapor of meteors can recondense to become small aerosol particles, possibly of nanometer size.

In very rare cases, meteoroids are large enough to survive the burn through the whole atmosphere and arrive at the surface with identifiable remains called meteorites. If this occurs at night, their light display against the night sky may be quite spectacular as to make the headlines in local newspapers.

Meteors usually occur in the mesosphere at an altitude somewhere between 80 and 95 km, hence the original "source" of meteoric aerosol particles must also be located at this altitude. Meteoric aerosols are known to contain elements such as Fe, Si, Mg, S, Ca, Ni, Al, Cr, Mu, Cl, K, Ti, and Co (Cameron, 1981).

There have been speculations about the possible influence of meteoric aerosol on the global climate but no definite conclusions exist. A recent high-flying aircraft study in the stratosphere in the polar vortex found that the total number of particles >0.01 μm increases with potential temperature while N_2O is decreasing, indicating that the aerosol source is in the higher stratosphere and mesosphere and that there is a downward transport of meteoric aerosol from the mesosphere (Curtius *et al.*,

2005). These nanometer-sized particles may serve as condensation nuclei to form sulfuric acid and water in the stratosphere.

6.8 Removal mechanisms of aerosol particles

The pathways via which aerosol particles are removed from the atmosphere can be divided into two broad categories – dry removal and wet removal. Dry removal mechanisms are those not involving cloud and precipitation processes, while wet removals are those related to clouds and precipitation.

6.8.1 Dry removal

Gravitational settling

Aerosol particles can stay afloat by updrafts that are strong enough to support them. If the updraft weakens or disappears, particles will go down to the surface by the pull of gravity of the Earth. The settling velocity of an aerosol particle depends on the density and size of the particle and where it is located in the atmosphere (because the air density affects the viscosity and hence the drag force on the particle). Junge *et al.* (1961) calculated the fall velocities of spheres of various sizes assuming the density is 2 g cm^{-3} and their results are shown in Fig. 6.10.

We see that particles have higher fall velocities at higher levels because of the lower air density that results in smaller drag force. Taking a particle with radius 0.01 μm as an example, its fall velocity is about 10^{-4} cm s^{-1} at $z = 10$ km but is about four times faster at $z = 20$ km. The velocity change with height is less for larger particles. For all these particles, the fall velocities are very small. We can make a quick estimate of the time required for them to fall a certain distance. For example, a particle of a few micrometers radius has a fall velocity ~1 cm s^{-1} at $z = 10$ km and it will take this particle more than 10^6 s (~12 days) to fall 1 km. A particle of radius ~0.1 μm has a fall velocity ~10^{-3} cm s^{-1} at $z \sim 10$ km and hence will take 10^9 s (~32 years) to fall 1 km! Obviously, submicrometer particles can stay in the atmosphere for a very long time.

Attachment on obstacle surfaces

Aerosol particles carried by air flow can become attached to the surface of plants, buildings or other obstacles above the ground either by their own Brownian motion, electrostatic forces, interception or inertial impaction (Fig. 6.11).

(1) *Inertial impaction* is caused by the finite inertia of the aerosol particle. When the aerosol-laden air flows past an obstacle, for example, a cylindrical rod, the air molecules will follow the streamlines, which will not intersect the rod (except

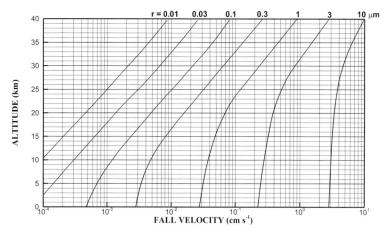

Fig. 6.10 The gravitational settling velocity of aerosol particles of various sizes as a function of altitude in the atmosphere. Adapted from Junge *et al.* (1961).

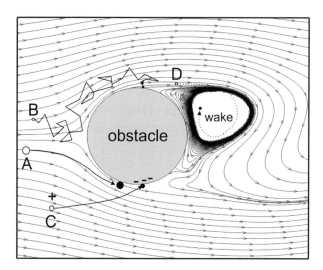

Fig. 6.11 A schematic illustrating the mechanisms involved in the dry removal of aerosol particles.

for the center streamline) but will go around it. An aerosol particle, such as particle A in Fig. 6.11, however, has finite inertia and hence its trajectory will deviate from the streamline such that it is more likely to impact on the rod and become attached. The position where the collision occurs also depends on the dimension of the particle and that effect is called interception.

Particle A collides with the obstacle in the front part. That is usually what happens to particles with larger inertia. For particles with small enough inertia, such as particle D in Fig. 6.11, it is possible that they move past the obstacle but get sucked into the wake region due to the lower pressure of the wake and become trapped there. This is called wake capture.

(2) *Brownian motion* is due to the random bombardment of the aerosol particle by air molecules that causes the particle to perform zigzag motions and become captured by the obstacle, such as particle B in Fig. 6.11. We have discussed Brownian motion previously.

(3) *Electrostatic forces* result because many particles and obstacles are electrically charged. If the signs of the charges on the particle and the obstacle are opposite, as is the case for particle C in Fig. 6.11, then the particles will be attracted to the obstacle by electrostatic forces and become captured.

Particle coagulation

We have already mentioned Brownian coagulation in which small particles collide and stick together to become larger particles. In accounting for the aerosol budget, coagulation causes particle concentration to decrease and also the coalesced particles are "removed" from their original size categories, so it should be considered as a sink of aerosol particles.

6.8.2 Wet removal

Cloud and precipitation processes are the most efficient way to remove aerosol particles from the atmosphere, and the process that removes aerosol particles is wet removal, also called precipitation scavenging. The study of precipitation scavenging was probably the most active during 1950–1970 when the testing of nuclear weapons was widely conducted in the atmosphere. Detonations of nuclear devices produce large amounts of radioactive aerosols in both the troposphere and stratosphere. The precipitation process can bring these particles down to the surface efficiently and cause radioactive pollution in the regions where fallout occurs. Later, attention was turned to particulate matter produced by industrial activities. The nuclear power plant accidents in Chernobyl, Ukraine, in 1986, and in Fukushima, Japan, in 2011, serve as a sober reminder that the threat of radioactive debris in the atmosphere is still with us.

But the present focus on wet removal is on the quantitative assessment of the aerosol impact on the global climate. Since aerosol particles have such an intimate relation with clouds – they are responsible for the formation of clouds, which are, in turn, responsible for much of the removal of aerosol particles – it is not easy to obtain a very accurate estimate. The aerosol–cloud interrelation is currently a very active research subject.

Wet removal is commonly divided into two types: in-cloud scavenging and below-cloud scavenging. Sometimes they are also called rainout and washout, respectively.

In-cloud scavenging (rainout)

In-cloud scavenging, by definition, includes all mechanisms occurring in the cloud that result in aerosol removal. The most important one is nucleation scavenging, in which aerosol particles serve as either CCN or IN for the formation of cloud drops or ice particles, and thus are removed from the atmosphere. Other in-cloud scavenging mechanisms include the attachment of aerosol particles on cloud and precipitation particles due to Brownian motion, attractive forces, and inertial impaction.

Below-cloud scavenging (washout)

Once precipitation particles (raindrops, snowflakes, graupel/hail) fall from the cloud, they can remove aerosol particles either by inertial impaction or by Brownian collision of the aerosol particles. This is called below-cloud scavenging. Discussions of precipitation scavenging will be given in Chapter 15.

Problems

6.1. It is known that the terminal velocity of a very small spherical particle of radius a can be calculated by the following equation (which will be discussed in more detail in Chapter 8):

$$u_\infty \approx \frac{2}{9} \eta_a a^2 \rho_p g,$$

where ρ_p is the particle density and η_a is the dynamic viscosity of air. The latter can be calculated by

$$\eta_a = (1.718 + 0.0049T) \times 10^{-4} \quad \text{for } T \geq 0°C,$$
$$\eta_a = (1.718 + 0.0049T - 1.2 \times 10^{-5}T^2) \times 10^{-4} \quad \text{for } T < 0°C.$$

Note that η_a is in the unit of poise (1 poise = 1 P = 1 g cm^{-1} s^{-1}) and T is in °C when using these formulas.

(a) Determine the fall velocities of particles of density 1 g cm^{-3} and radii 1, 10, and 100 μm falling in air of p = 1000 hPa and T = 20°C.
(b) Calculate the time required for these particles to fall 1 km in air.
(c) Repeat the calculations of (a) and (b) but for air with p = 200 hPa and T = −60°C.

(d) In some stormy situations, such as during a dust storm, aerosol particles are carried all the way up to the upper troposphere/lower stratosphere (UTLS). Assuming the wind speed at the UTLS level (assuming the same conditions as in (c)) is 30 m s^{-1} heading east, how far would a 0.1 µm radius particle be transported before it falls 1 km from its original level?

6.2. The Langevin equation below describes the random Brownian motion of an aerosol particle of radius a_p in air:

$$\frac{d\vec{V}}{dt} = -\beta\vec{V} + \vec{A}(t),$$

where $\beta = 6\pi a_p \eta_a / m_p$ (η_a is the dynamic viscosity of air and m_p is the particle mass). The first term on the right-hand side describes the continuous air resistance to the particle motion, and the second term describes the random force acting on the particle from air molecule bombardment. For a truly random motion, the position vector $\vec{r}$ of the particle should vanish after sufficiently long time averaging, i.e.

$$\langle \vec{r} \cdot \vec{A}(t) \rangle = 0,$$

where the angular brackets represent the long time averaging. Use this fact to prove that

$$\langle \vec{r} \rangle \propto \sqrt{t},$$

where $\langle \vec{r} \rangle$ is the mean distance traveled by the particle due to this random motion in time t.

7

Nucleation

Conventionally, the condensation of water in the atmosphere was thought to occur spontaneously once water vapor exceeded saturation (either by cooling or by addition of water vapor). Yet experimentally it has been shown that, in a very clean environment, the vapor pressure can significantly exceed the saturation value without condensation. We have mentioned this supersaturation phenomenon before. Similarly, the conventional wisdom of freezing is that liquid water will freeze into ice once the temperature drops to below 0°C, yet experiments and observations show that supercooling occurs very frequently, and that certainly is the case for atmospheric clouds.

The fundamental process of phase change is *nucleation*. The condensation of liquid water from expansive and highly random water vapor does not occur in a large quantity in its initial stage but first appears as small clusters of molecules that have properties close to bulk liquid. Similarly, the freezing of liquid water to form ice also occurs as small clusters of molecules that have quasi-ice properties. Once these clusters stabilize and continue to grow, we will then see bulk water and ice phases forming. The formation of these quasi-condensed phase clusters is the nucleation process and is the main subject of this chapter.

There are two broad categories of nucleation processes:

(1) *Homogeneous nucleation* – new phases form from a mother phase consisting of pure substance without any foreign substances involved. Water droplets nucleating from pure water vapor (homogeneous condensation), ice crystals from pure water vapor (homogeneous deposition), and ice crystals from pure liquid water (homogeneous freezing) are all examples of homogeneous nucleation processes.

(2) *Heterogeneous nucleation* – new phases form from a mother phase consisting of pure substance with foreign particles involved. The formation of a liquid droplet on a salt particle and an ice crystal on a dust particle are both examples of heterogeneous nucleation processes.

As we shall see, it is nearly impossible to form cloud droplets in our atmosphere via homogeneous nucleation; rather, heterogeneous nucleation is mostly responsible. Cloud droplets form initially on some suitable particles, called *cloud condensation nuclei* (CCN), and grow to larger sizes when the environmental conditions are favorable. The origin of ice crystals in clouds is less clear. There is a possibility that ice crystals form directly on suitable particles, called ice nuclei (IN), without going through the liquid phase. It is also possible that liquid droplets form first via heterogeneous nucleation to become aqueous solution drops and then freeze into ice without involving *additional* foreign substances. Observational study of ice initiation in our atmosphere is currently an active research area.

In the following, we will treat the subject of the homogeneous nucleation of water droplets in a more quantitative manner. The classical reference on this subject is Dufour and Defay (1963). The subjects of heterogeneous nucleation of water drops and the nucleation of ice will be treated more qualitatively, as they are both more involved and as yet not as clearly developed.

7.1 Homogeneous nucleation of water drops

Water vapor is usually regarded as an ideal gas in most meteorological studies. This assumption is justified when the humidity in air is relatively low. When the humidity is low, water vapor molecules are far apart from each other and hence do not interact significantly, which is the condition for an ideal gas.

When humidity is close to saturation, however, water vapor molecules are close to each other and interact strongly, and the whole vapor starts to exhibit non-ideal gas properties. Under the near-saturation condition, vapor molecules tend to form clusters of various numbers. These are called molecular clusters. A single molecule is sometimes called a *monomer*. A cluster consisting of two molecules is called a *dimer*, a cluster of three molecules a *trimer*, etc. In general, a cluster of i molecules is called an i-mer. Such a state is, in fact, a polymer state. It has been measured that, under certain near-saturation conditions, the value of i can be greater than 180.

The homogeneous nucleation process can be understood qualitatively in the following manner. When humidity is very low, most vapor molecules are monomers and interact little with each other. As the humidity increases, the number of vapor molecular clusters with higher i index also increases; many of them may collide and coagulate to form a quasi-droplet but may disintegrate again due to thermal agitation; so they are essentially fluctuating in to and out of a real water droplet (i.e. liquid) state. When the humidity reaches a certain threshold, these quasi-liquid drops stabilize and become real water drops. This whole process proceeds without involving other non-water materials.

It was a surprise that experiments showed that in the environment of the Earth's surface the quasi-liquid drops cannot be stabilized even at relative humidity significantly exceeding saturation. Instead, a relative humidity of several hundred percent is necessary for the nucleation to occur when no foreign particles are involved in the process.

The following simple treatment, called the *classical theory of homogeneous nucleation*, can explain why this is so. This theory does not utilize statistical mechanics but uses quasi-bulk thermodynamics to approach this problem. Even though it is somewhat simple-minded, it nevertheless gives a reasonable description and even some semi-quantitative predictions of nucleation behavior.

We begin by studying the population of molecular clusters mentioned above. We shall call the molecular clusters that are not yet activated to become a liquid droplet an *embryo*. While an embryo is not yet a real liquid drop, we will nevertheless assume that it has physical properties similar to those of a liquid droplet, including the curvature effect that we have discussed before. Once an embryo is activated, it will be called a *germ*. A germ is a *bona fide* water droplet and its growth or dissipation will be governed by the bulk thermodynamics of water.

7.2 The population of embryos

Now let us examine the probability of germ formation. Molecular clusters (embryos) can be activated to become water droplets (germs) by environmental fluctuations that can cause additional water molecules to join the embryo. Apparently, those embryos with cluster size closer to a germ would have greater probability to become activated. On the other hand, it takes energy to form such embryos. We shall call the energy necessary to form an *i*-mer embryo (an embryo consisting of *i* molecules) the *free energy of i-mer formation*.

One of the basic assumptions of the classical nucleation theory is that the population of *i*-mer embryos follows the Boltzmann distribution:

$$N_i = N_{\text{sat,w}} \exp\left(\frac{-\Delta F_i}{kT}\right), \tag{7.1}$$

where N_i is the number concentration (i.e. the population) of *i*-mer embryos, $N_{\text{sat,w}}$ is the total number of molecules in this system, ΔF_i is the free energy of formation of an *i*-mer, and k is the Boltzmann constant. Eq. (7.1) says that, at a given temperature, N_i decreases with increasing ΔF_i, i.e. the higher the required free energy to form an *i*-mer, the smaller the *i*-mer concentration. If $\Delta F_i = 0$, then (7.1) says that $N_i = N_{\text{sat,w}}$, which is equivalent to saying that there is no need for energy to form a single molecule. This is, of course, because we assume that these molecules are already in existence at the beginning.

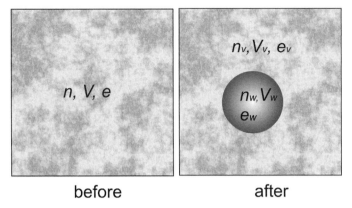

before   after

Fig. 7.1 The thermodynamic configuration of the system before and after the formation of a spherical *i*-mer.

So ΔF_i controls the population of *i*-mers, but what controls ΔF_i? Intuitively, we expect that this free energy should depend on the degree of saturation – the closer to saturation, the less energy would be required to form clusters, and vice versa. In the following section, we will derive an equation to express the free energy of *i*-mer formation quantitatively.

7.3 Free energy of *i*-mer formation

How do we determine the free energy of *i*-mer formation? One way is to determine first the total energies before and after the formation of the *i*-mer. The difference between the two energies is the free energy necessary to form the *i*-mer.

So we now consider the configuration of our problem as shown in Fig. 7.1. The system as a whole is assumed to be a closed system. This means that the total amount of materials in the system remains unchanged. We further assume that the total volume and temperature remain constant as well. Before the *i*-mer formation, the vapor pressure is e_1, total number of moles of water molecules is n, and total volume is V. After the *i*-mer formation, some vapor molecules become part of the quasi-liquid *i*-mer embryo. Therefore, the vapor pressure should decrease to a new value, e_2, which corresponds to a new vapor concentration n_v. On the other hand, the number of water molecules in the *i*-mer is n_w and the pressure inside the embryo is P_w.

Under such an isochoric and isothermal process, the appropriate free energy to represent the *i*-mer formation process is the Helmholtz free energy:

$$F = U - TS = \sum_{k=1}^{c} \mu_k n_k - PV. \qquad (7.2)$$

Hence, we can write down the free energy before and after the *i*-mer formation as follows:

- before

$$F_1 = n\mu_{v,1}(e_1) - e_1 V = (n_v + n_w)\mu_{v,1}(e_1) - e_1 V, \qquad (7.3)$$

- after

$$F_2 = n_v \mu_{v,2}(e_2) + n_w \mu_w(P_w) - e_2(V - V_w) - P_w V_w + \sigma\Omega. \qquad (7.4)$$

In the above equations, we have utilized the criteria of constant volume (so that $V_v = V - V_w$) and closed system (constant total mass or number of moles, so that $n = n_v + n_w$). Note also that the last term on the right-hand side of (7.4) represents the surface free energy, which was absent before the creation of the *i*-mer.

Under the usual condition, the number of vapor molecules is much larger than that of the *i*-mer, hence the influence of the *i*-mer formation on the vapor pressure and vapor chemical potential is fairly small. Thus we can write

$$\mu_{v,1} \approx \mu_{v,2} = \mu_v, \quad e_1 \approx e_2 = e. \qquad (7.5)$$

The energy of *i*-mer formation is the difference between F_1 and F_2:

$$\Delta F_i = F_2 - F_1 = n_w(\mu_{w,P_w} - \mu_{v,e}) - V_w(P_w - e) + \sigma\Omega, \qquad (7.6)$$

where μ_{w,P_w} and $\mu_{v,e}$ represent the chemical potentials of water molecules in the *i*-mer (at pressure P_w) and in the vapor (at pressure e), respectively. This can be further simplified by recalling the Gibbs–Duhem relation (5.20):

$$0 = sdT - vdP + \sum_{k=1}^{c} x_k d\mu_k, \qquad (7.7)$$

which, for the embryo in the present situation (isothermal, hence $dT = 0$, and component number $k = 1$), is simply

$$v_w dP = d\mu_w. \qquad (7.8)$$

We integrate (7.8) and obtain

$$v_w \int_e^{P_w} dP_w = \int_{\mu_{w,e}}^{\mu_{w,P_w}} d\mu_w. \qquad (7.9)$$

Hence,

$$v_w(P_w - e) = \mu_{w,P_w} - \mu_{w,e}. \qquad (7.10)$$

We note that

$$V_w = n_w v_w. \qquad (7.11)$$

Now substituting (7.10) into (7.6), we get

$$\Delta F_i = n_w \left[\mu_{w,e} + v_w(P_w - e) - \mu_{v,e} \right] - V_w(P_w - e) + \sigma\Omega$$
$$= n_w \left[\mu_{w,e} - \mu_{v,e} \right] + \sigma\Omega = \Delta F_{i,\text{vol}} + \Delta F_{i,\text{sfc}}, \quad (7.12)$$

i.e. the free energy can be divided into two terms: one for the increase in volume of the embryo and the other for the creation of surface. The surface energy part can be written down more specifically. If we assume that the embryo is spherical, then the surface free energy is

$$\sigma\Omega = 4\pi a_i^2 \sigma, \quad (7.13)$$

where a_i is the radius of the *i*-mer.

Now let us examine the volume free energy $\Delta F_{i,\text{vol}} = n_w [\mu_{w,e} - \mu_{v,e}]$. When the embryo is in equilibrium with the vapor, $\mu_{w,e} \neq \mu_{v,e}$, rather it is $\mu_{w,P_w} = \mu_{v,e}$. This is of course due to the curvature effect of the embryo. From (7.10), we know that

$$\mu_{w,e} = \mu_{w,P_w} - v_w(P_w - e) = \mu_{w,P_w} - v_w \frac{2\sigma}{a_i}, \quad (7.14)$$

so that the volume free energy is

$$\Delta F_{i,\text{vol}} = n_w \left(\mu_{w,P_w} - v_w \frac{2\sigma}{a_i} - \mu_{v,e} \right) = -n_w v_w \frac{2\sigma}{a_i}. \quad (7.15)$$

But the Kelvin equation (5.51) says that

$$e_{\text{sat},a} = e_{\text{sat},\infty} \exp\left(\frac{2v_w\sigma}{RTa} \right)$$

and hence

$$\frac{2v_w\sigma}{a} = RT \ln\left(\frac{e_{\text{sat},a}}{e_{\text{sat},\infty}} \right) = RT \ln S_{v,w}, \quad (7.16)$$

and so we can write (7.15) as

$$\Delta F_{i,\text{vol}} = -n_w RT \ln S_{v,w}. \quad (7.17)$$

Considering that the relation between the universal gas constant R and the Boltzmann constant k is $R = N_{Av}k$, where N_{Av} is the Avogadro number, we can rewrite (7.17) as

$$\Delta F_{i,\text{vol}} = -n_w N_{Av} kT \ln S_{v,w} = -\frac{V_w}{v_{\text{mole}}} kT \ln S_{v,w} = -\frac{4\pi a_i^3}{3v_{\text{mole}}} kT \ln S_{v,w}, \quad (7.18)$$

where v_{mole} is the *molecular volume* (not molar volume) – the volume of each individual water molecule.

Putting (7.18) and (7.13) into (7.12), we obtain the free energy of *i*-mer formation as

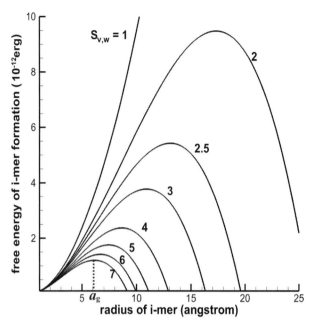

Fig. 7.2 The energy of *i*-mer formation at various saturation ratios as a function of
the *i*-mer radius at $T = 0°C$.

$$\Delta F_i = -\frac{4\pi a_i^3}{3 v_{\text{mole}}} kT \ln S_{v,w} + 4\pi a_i^2 \sigma. \qquad (7.19)$$

Using (7.19), we can determine the values of ΔF_i as a function of a_i at given
temperature T and saturation ratio $S_{v,w}$.

Fig. 7.2 is the chart showing ΔF_i as a function of a_i at $T = 0°C$ and various
saturation ratios. We see that the free energy of *i*-mer formation is not a monotonic
function of the embryo radius a_i. Rather, for each $S_{v,w}$ (except the case where
$S_{v,w} = 1$, which we will discuss later), it has a maximum. Taking the curve $S_{v,w} = 2$
as an example, we observe that ΔF_i increases first as the embryo size is increasing
(meaning that the free energy of *i*-mer formation is greater for larger embryos)
initially. It reaches a maximum and then decreases as the size increases. This indicates
that, once the embryo reaches a size corresponding to the energy maximum, any
further growth would require less energy to achieve. Thus, the maximum represents
a kind of energy barrier: if an embryo can overcome this barrier, it will become
activated – in other words, it becomes a *germ*, a *bona fide* water droplet, and the
growth after that should obey the common liquid-phase physics. The embryo size
corresponding to this maximum is the germ size a_g.

From (7.19), we can easily derive a_g and the free energy of germ formation (i.e.
the energy at the maximum) for a given temperature and $S_{v,w}$. They are

$$a_g = \frac{2\pi M_w \sigma}{RT\rho_w \ln S_{v,w}} \qquad (7.20)$$

and

$$\Delta F_g = \frac{\sigma\Omega}{3} = \frac{16\pi M_w^2 \sigma^3}{3(RT\rho_w \ln S_{v,w})^2}, \qquad (7.21)$$

respectively.

We observe that, as the saturation ratio increases, both a_g and ΔF_g decrease. This can be interpreted as follows. As $S_{v,w}$ increases, the energy barrier to form a germ decreases and the population of i-mers that can be in equilibrium with the environment increases. Even a relatively small embryo can be stabilized to a germ without fluctuating back to the cluster state. In other words, it is easier to have nucleation when $S_{v,w}$ is high.

Now let us examine the special curve for $S_{v,w} = 1$. This is essentially equivalent to the situation that $RH = 100\%$. This special curve has no maximum; rather, the value of ΔF_i goes higher with increasing embryo size. This indicates that there can never be a stable size and the free energy of germ formation is infinity! This says that homogeneous nucleation of water droplets is impossible at $RH = 100\%$. Note that this is different from what is implied by the Kelvin equation, as we discussed in Chapter 5. There we saw that a pre-existing drop will evaporate even in a $RH = 100\%$ environment. Here we see that the drop will not even be able to form in such an environment.

Therefore, in order to have homogeneous nucleation at all, it is necessary to have $RH > 100\%$. In fact, the required humidity is much greater, because the probability of forming a drop in an environment only slightly greater than saturation is very small. It is customary to define a *critical nucleation rate* of 1 droplet cm^{-3} s^{-1} and determine the necessary $S_{v,w}$. We will not derive the nucleation rate but simply write it down (Fletcher, 1962):

$$J \approx B N_{sat,w} \exp\left(-\frac{\Delta F_i}{kT}\right), \qquad (7.22)$$

where B is the net rate at which an embryo of critical size gains one molecule through collisions with surrounding vapor molecules. Using this and the related equations of the free energy as derived above, we can make an estimate of the saturation ratio required to produce a critical nucleation rate. Since precise values of B are not available but can be estimated roughly, the result certain comes with some errors. But surprisingly it is not too sensitive to B. For $J = 1$ droplet cm^{-3} s^{-1}, the value of $S_{v,w}$ turns out to be about 4.4. Even if we adopt different values of J as the critical nucleation rate, for example, between 10^3 and 10^{-3} droplets cm^{-3} s^{-1}, the corresponding values of $S_{v,w}$ are 4.0 and 4.8. Therefore, we can probably say that, in order to have homogeneous nucleation in our atmosphere, it is necessary to have a relative humidity several times higher than 100%, a condition clearly non-existent in the natural atmosphere.

A similar treatment can also be carried out for the homogeneous nucleation of ice, for the case of either homogeneous deposition or homogeneous freezing, and the conclusion is similar. Ice nucleation is more sensitive to temperature, and the conclusion is that it is difficult to form ice particles via a homogeneous nucleation process, unless the ambient temperature is lower than about −40°C. At that temperature, the probability calculations show that spontaneous homogeneous freezing of a pure water drop into ice is very likely to occur. This condition applies to the formation of high cirrus clouds or the anvil of deep convective storms. For middle and low clouds, most ice particles occur at much warmer temperatures and those are formed via heterogeneous ice nucleation.

Thus we conclude that the formation of middle and low clouds in the Earth's atmosphere is chiefly due to heterogeneous nucleation. Only at higher altitude would we see the occurrence of homogeneous nucleation, which, if it happens, is most likely homogeneous freezing of liquid water.

7.4 Molecular dynamics simulation of homogeneous freezing of pure water

The treatment in the preceding section utilized the classical bulk thermodynamics methods. Even though there are still points that are up for debate – for example, the assumption that the inactivated embryo is spherical and has similar properties as a liquid drop – the classical model seems to give plausible results. There are also treatments using statistical mechanical techniques, but the results are more or less similar.

Bulk thermodynamics, however, cannot provide instantaneous details of how water molecules interact to form a germ. This is made possible by molecular dynamics simulations. In the following we will use an MD simulation of the homogeneous freezing of pure water to illustrate the nucleation by molecular interaction. This is based on the work by Matsumoto *et al.* (2002), which was the first successful attempt to simulate the homogeneous freezing for an unbounded water system. They used an empirical water model called TIP4P, which is a rigid planar Lennard-Jones four-site interaction potential model. They used various numbers of molecules in their simulations and the results shown here used 512 molecules.

The water temperature is assumed to be 230 K, which is below the spontaneous freezing temperature, so homogeneous freezing is possible. In Fig. 7.3(a) a snapshot of the water molecules is shown at $t = 208$ ns. This represents the common random state of the liquid (albeit supercooled). The brighter bonds (two examples are

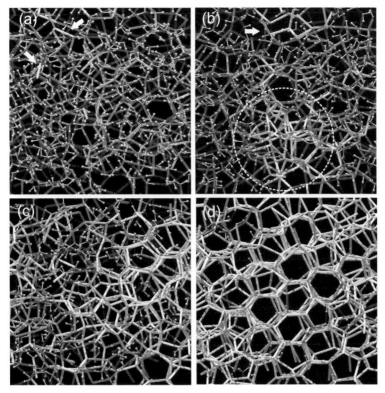

Fig. 7.3 Four snapshots of a molecular dynamics simulation of the homogeneous nucleation of ice-I_h from pure liquid water at $T = 230$ K. See the text for the description of each panel. Photos courtesy of Dr Masakazu Matsumoto.

indicated top left by small arrows) represent long-lasting hydrogen bonds. These bonds randomly appear and disappear. To form a stable ice-like structure, it would be necessary to have a higher population of such bonds. This occurs at $t = 256$ ns (Fig. 7.3b) when suddenly a cluster with long-lasting hydrogen bonds appears in the region indicated by the dashed circle at bottom center. This cluster starts to exhibit ice-like structure, which quickly spreads to other regions at $t = 290$ ns (Fig. 7.3c). Eventually the molecules in the whole domain arrange themselves in an ice crystal structure at $t = 320$ ns (Fig. 7.3d). In the short time between 256 and 320 ns, the potential energy of the system decreases very rapidly and the number of long-lasting hydrogen bonds increases at a similar pace.

The above simulation was performed to understand the general homogeneous freezing in the interior of a bulk liquid. Since homogeneous freezing may occur in high clouds, which would start with very small cloud droplets, the surface effect could be very important. A relevant question about the homogeneous freezing of such small droplets is this: Does the nucleation begin in the bulk liquid or near the

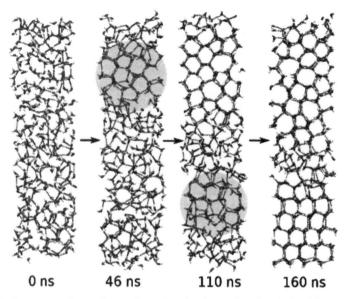

0 ns   46 ns   110 ns   160 ns

Fig. 7.4 Sequence of snapshots of a molecular dynamics simulation illustrating the preference for the homogeneous nucleation of ice in the subsurface layer of pure liquid water. From Vrbka and Jungwirth (2006). Reproduced by permission of the American Chemical Society.

surface? If the nucleation begins near the surface, then the freezing data would strongly depend on the drop size and surface contamination. To investigate this question, Vrbka and Jungwirth (2006) performed an MD simulation using thin slabs of liquid water configuration so that the surface layer is well represented. They chose the freezing temperature of 250 K, as they found that was when the nucleation rate is fastest. Their simulations show that the nucleation starts in the subsurface, and the result of one of the trajectories is shown in Fig. 7.4.

Four snapshots are shown here. At $t = 0$, water molecules are still in the random liquid state. At $t = 46$ ns, nucleation starts spontaneously in the layer just below the surface (shaded region) and propagates downward. At $t = 110$ ns, another nucleus forms spontaneously at a deeper layer. Both nuclei continue to grow at $t = 160$ ns and a random layer between the two nuclei can still be seen. At $t = 200$ ns (not shown), the slab is completely frozen. However, the quasi-liquid layer is always present at the surface. Interestingly, the ice formed by their homogeneous freezing simulations is preferentially the metastable cubic ice (ice-I_c) although pockets of ice-I_h also occur.

Why does the freezing start in the subsurface layer? There are two possible reasons. First, the freezing is accompanied by an increase in volume and the surface region can accommodate the volume increase better than the aqueous bulk. Second, the presence of a large electric field at the air–water interface due to preferential orientation of surface water molecules may also help to start the nucleation in the subsurface layer.

7.5 Heterogeneous nucleation

By definition, heterogeneous nucleation must involve materials other than water. These materials normally present themselves as small aerosol particles. The particles that are responsible for facilitating the formation of cloud droplets are called *cloud condensation nuclei* (CCN), whereas those responsible for the formation of ice particles are called *ice nuclei* (IN), as we have already mentioned in Chapter 6.

Whether an aerosol particle is more suitable to serve as a CCN or IN depends on its physical and chemical properties. Aside from the intrinsic properties of the particles, the environmental conditions are also very important in deciding whether the particle will play the role as CCN or IN. For example, some particles may serve as CCN in a certain warmer temperature range and as IN at colder temperatures.

7.6 Cloud condensation nuclei

What kinds of particles are suitable to serve as CCN? From our discussion of the Köhler equation in Chapter 5, we know that an aqueous solution drop can survive even in an unsaturated environment due to the solute effect whereas a pure water drop will evaporate even at $RH = 100\%$. Thus if a particle is made of material that can be dissolved in water so as to form an aqueous solution drop, then apparently it will be a good CCN. One example of such a material is common sea salt NaCl. NaCl not only dissolves easily in water but absorbs water strongly. It deliquesces easily in a humid summer day because of absorbing water vapor in air. A tiny salt particle suspending in air can absorb water vapor to form a small aqueous solution droplet in humid conditions.

How many particles in air can serve as CCN? The answer depends on the environmental humidity condition. To understand this situation, let us briefly look at how CCN concentration is measured. The instrument used to measure CCN concentrations is called a CCN counter. The main idea is to create a supersaturated section in the counter where droplets will form on CCN. This can be achieved by keeping a wetted surface in the counter to serve as the water vapor source and allow the temperature to vary along the path of air. An air sample entering the counter will experience a relative humidity gradient along the path because the temperature changes. When the air is subsaturated, the air sample would look transparent (i.e. dark when not illuminated). When the air becomes supersaturated, however, some aerosol particles become activated to form droplets and appear to be bright when illuminated. The number of

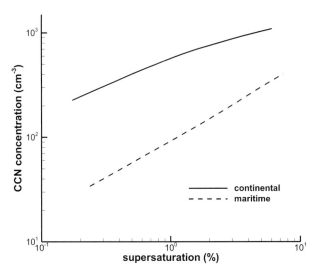

Fig. 7.5 Typical CCN concentration as a function of supersaturation in continental and maritime air masses. Adapted from Twomey and Wojciechowski (1969).

droplets represents the number of CCN (assuming one droplet contains one CCN). As the supersaturation increases, even less effective particles will be activated to form droplets, and hence the counter will register more CCN. Thus it is understandable that the number concentration of CCN generally increases with supersaturation. Fig. 7.5 shows an example plot of the CCN concentration as a function of supersaturation.

While the CCN concentration varies with humidity, the typical CCN concentrations are roughly: $10-10^2$ cm^3 over ocean and 10^2-10^3 over land, based on many measurements. Thus, the CCN concentration over land is about one order of magnitude greater than over ocean. In Chapter 2 we saw that the drop concentrations are generally higher in continental clouds and lower in maritime clouds, and the main reason is because there are more CCN over land, as we have just discussed. Needless to say, in highly polluted urban areas, the CCN concentrations can be 2–3 orders of magnitude greater than that of the general overland concentration.

7.6.1 Chemical properties of cloud condensation nuclei

Chemically, those particles with ability to attract and accumulate water molecules onto their surfaces (called *hydrophilic*) are better CCN than others. These particles are called *hygroscopic* particles. Note that hygroscopic particles are not necessarily water-soluble, but water-soluble particles are necessarily hygroscopic.

The most common CCN in our atmosphere are $(NH_4)_2SO_4$ and NaCl, and both are water-soluble. Here we are talking about the main composition of these particles,

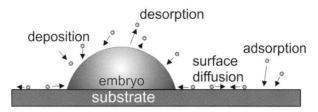

Fig. 7.6 A schematic illustrating various molecular processes during the heterogeneous nucleation of a liquid water drop.

but we have to keep in mind that most aerosol particles in the atmosphere are mixed particles, as already noted in Chapter 6. Indeed, there are also insoluble particles that serve as CCN and some of them are probably derived from the soil.

7.7 Nucleation of water drops on a plane substrate

7.7.1 Plane substrate

The detailed treatment of theories about heterogeneous nucleation on the surface of a real cloud condensation nucleus is beyond the scope of this book. Fletcher (1962) and Pruppacher and Klett (1997) have provided more in-depth treatments based on the earlier work of Volmer (1939). Here we will first examine a simplified case: nucleation of water drops on a pure substrate with a plane surface. Fig. 7.6 is a schematic showing the various mechanisms involved in the nucleation process.

Here we see that there are four mechanisms associated with the nucleation:

(1) *Desorption* – the mechanism by which water molecules leave an embryo and become vapor molecules. This is analogous to evaporation, where water molecules leave the surface of a water drop.
(2) *Deposition* – water molecules land on and become part of the embryo.
(3) *Adsorption* – water molecules land on the substrate.
(4) *Surface diffusion* – water molecules land on the surface of the substrate and diffuse randomly. Some of the diffusing molecules will combine with the embryo and become part of it.

The first two mechanisms are present in the homogeneous nucleation also, but the last two mechanisms – adsorption and surface diffusion on the substrate – operate only in heterogeneous nucleation. This is the main difference between homogeneous and heterogeneous nucleation. The adsorption and surface diffusion allow the vapor molecules to land on a pre-existing surface and diffuse toward the nucleating site on this surface, unlike the case of homogeneous nucleation where any "landing" implies creating a new surface and the necessary energy for forming it.

The free energy of germ formation for nucleation of a liquid droplet on a plain substrate is

$$\Delta F_{\mathrm{g,s}} = \frac{4\pi a^2 \sigma}{3}\left[\frac{(2+m)(1-m)^2}{4}\right] = \frac{16\pi M_{\mathrm{w}}^2 \sigma^3}{3\left(RT\rho_{\mathrm{w}}\ln S_{\mathrm{v,w}}\right)^2} f(m), \qquad (7.23)$$

where

$$f(m) = \left[\frac{(2+m)(1-m)^2}{4}\right], \quad m = \cos\theta, \qquad (7.24)$$

is the factor influenced by the wettability of the surface. Comparing (7.23) and (7.21) makes it clear that the difference in the energy of germ formation between homogeneous and heterogeneous nucleation is in the factor $f(m)$.

When the surface is perfectly wettable (the contact angle $\theta = 0$), we have $m = 1$ and hence $f(m) = 0$. Consequently, the energy of germ formation is zero. This simply states that the "surface" is already here and is perfectly wettable so that no energy barrier is present for incorporating vapor molecules into the surface.

On the other hand, when the surface is perfectly non-wettable (the contact angle $\theta = 180°$), $m = -1$ and $f(m) = 1$. Consequently, under this situation, the energy of germ formation (7.23) is identical to that for homogeneous nucleation (7.21). Thus particles with perfectly non-wetting surfaces are useless as far as heterogeneous nucleation of water drops is concerned because they do not offer any energy advantages for facilitating the process.

The nucleation rate can be determined as follows. The number of germs is

$$n(a_{\mathrm{g}}) = n_{\mathrm{sat,w}}\exp(-\Delta F_{\mathrm{g,s}}/kT). \qquad (7.25)$$

Following Fletcher (1962), if we approximate the surface area of a germ as πa_{g}^2 and the rate of impact (per unit area) of vapor molecules as $e_{\mathrm{v}}/\sqrt{2\pi\dot{m}_{\mathrm{w}}kT}$, where e_{v} is the vapor pressure and $\dot{m}_{\mathrm{w}}$ is the mass of a water molecule, then the nucleation rate per unit area of the substrate is

$$J \approx \left(e_{\mathrm{v}}/\sqrt{2\pi\dot{m}_{\mathrm{w}}kT}\right)\pi a_{\mathrm{g}}^2 n_{\mathrm{sat,w}}\exp\left(-\Delta F_{\mathrm{g,s}}/kT\right). \qquad (7.26)$$

7.7.2 Curved substrate

The above results pertain only to substrates with a plane surface. For small nuclei, the surface may have significant curvature and that should influence the contact angle and hence the nucleation rate. For a spherical CCN with radius r_{N}, the free energy of germ formation is given by (Fletcher, 1962):

$$\Delta F_{g,s} = \frac{16\pi M_w^2 \sigma^3}{3(RT\rho_w \ln S_{v,w})^2} f(m,x),$$ (7.27)

where

$$x = \frac{r_N}{a_g}$$ (7.28)

and

$$f(m,x) = \frac{1}{2}\left\{1 + \left[\frac{1-mx}{\phi}\right]^3 + x^3\left[2 - 3\left(\frac{m-x}{\phi}\right) + \left(\frac{m-x}{\phi}\right)^3\right] + 3mx^2\left(\frac{m-x}{\phi} - 1\right)\right\},$$ (7.29)

with

$$\phi = \sqrt{1 + x^2 - 2mx}.$$ (7.30)

The nucleation rate is given by

$$J \approx \left(e_v/\sqrt{2\pi \dot{m}_w kT}\right) 4\pi^2 r_N^2 a_g^2 n_{\text{sat,w}} \exp\left(-\Delta F_{g,s}/kT\right),$$ (7.31)

where $\Delta F_{g,s}$ is given by (7.27).

7.7.3 Size dependence of cloud condensation nuclei

Many studies have shown that the effectiveness of a particle to serve as a CCN depends not only on its chemical composition but also on its size. It is found that large particles are usually better CCN than Aitken particles if their chemical compositions are the same. Thus, in principle, we need to know both the chemical composition and size distribution of aerosol particles in order to assess their overall nucleation efficiency. An interesting question (and also an important one for modeling the aerosol effect in climate processes) is this: Is the chemical composition or the size more significant in deciding whether or not an aerosol particle is a good CCN?

A recent study demonstrates that the size of aerosol particles is more important than chemical composition (Dusek *et al.*, 2006). In this study, four samples of aerosol particles were taken from air that originated from three different continental air masses and one maritime air mass. The size distributions and the CCN efficiency for each size bin (= CCN/C_N, where C_N is the total particle concentration) are measured. Note that the size-resolved CCN efficiency represents the effect of the chemical composition. Detailed sensitivity studies were performed on these aerosol samples in order to understand the relative importance of the shape of the size distribution and CCN efficiency. The researchers formed a base case by calculating

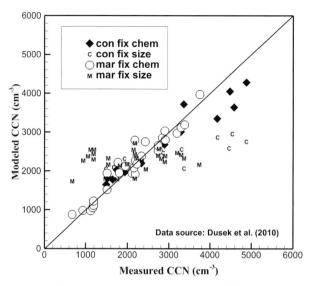

Fig. 7.7 Modeled CCN concentration vs. measured CCN concentration in a study
to determine the relative importance of the size and chemical nature of CCN. Data
taken from Dusek *et al.* (2006).

a time series of 6 h averaged total CCN concentration for the entire measurement
period and the size-resolved CCN efficiency. Next, they formed two modified CCN
time series: (i) using fixed CCN efficiencies (campaign average) but variable size
distributions, and (ii) using a fixed size distribution (campaign average) but variable
CCN efficiencies reflecting variable chemical composition. Then correlation coefficients were calculated for the base series and the two modified series.

  Fig. 7.7 shows one set of their results assuming a fixed supersaturation of 0.4%.
Here all three continental samples are lumped together. For perfect correlation
between the two series ($R^2 = 1$), all data points would lie on the diagonal straight
line. The data point distribution here shows that, for the case of fixed chemical
composition (diamonds and circles), though there is some scatter, the correlation is
excellent ($R^2 = 0.93$), indicating that size distribution alone can explain almost all the
temporal variation in total CCN concentration. On the other hand, the variation of
CCN efficiency (i.e. chemical composition, those data points labeled as "C" and "M")
can hardly explain the temporal variation of the total CCN concentration ($R^2 = 0.27$).
Obviously, the effect of aerosol size far outweighs that of the chemical composition.

7.8 Electrical effect on the nucleation of liquid water

Given that water molecules are highly polar in nature, it is to be expected that an
external electric field should have a significant impact on the heterogeneous

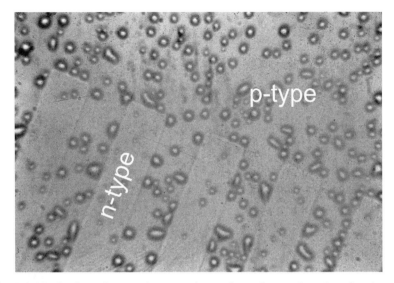

Fig. 7.8 Nucleation of water drops on the surface of a p–n junction showing the preference of water nucleation on the p-type surface. From Wang *et al*. (1980).

nucleation of liquid water. An experimental study carried out by Wang *et al*. (1980) indeed confirmed this conjecture. They used a clean surface of a p–n junction formed by doping boron (the p-type) and phosphorus (the n-type) on the surface of a silicon wafer. They placed the p–n junction on a cold stage cooled by a thermoelectric device and passed water vapor over the junction. The experiments were performed in a subfreezing environment. At a certain temperature when the vapor is supersaturated, water droplets formed on the surface of the junction. They observed that larger and more numerous droplets appear on the p-type surface than on the n-type surface, and Fig. 7.8 shows such an example. It appears that the p-type material provides a more efficient surface for nucleation to occur. The exact mechanism is still unclear, but there is little doubt that it has to do with the surface electric field.

When the p–n junction was connected to a battery so as to increase the electric field strength on the boundary of the junction, the nucleation proceeds even faster, which corroborates the conjecture that the electric field has substantial impact on water nucleation.

7.9 Ice nuclei

Just like the formation of cloud drops, ice particles in clouds do not just form simply by the freezing of liquid water or sublimation (deposition) of water vapor unless the temperature is close to or lower than −40°C. They, too, need the help of some special aerosol particles, called ice nuclei (IN), to form. Because ice can form either from freezing or sublimation, the IN can, in principle, be divided into four different categories according to their mode of action:

(1) *Deposition nuclei* – the IN that serve to form ice directly from the deposition of water vapor without going through the liquid phase.
(2) *Condensation freezing nuclei* – the IN initially serves as a CCN to form a drop at $T < 0°C$ and the drop freezes during the condensation.
(3) *Immersion nuclei* – the IN that serve to form ice by freezing liquid water. The particle is thought to be immersed in the liquid already (but does not serve as a CCN) before the nucleation occurs.
(4) *Contact nuclei* – the IN that cause liquid to freeze upon contact. In this case, the particle is thought to be outside of the liquid initially.

In reality, however, it is difficult to determine which mode is really working from field observations of IN unless under very special situations. Field observations of IN are normally done by examining particles residing in snow crystals collected either in clouds or on the ground. We may see particles residing in the center of a snow crystal, but we cannot readily be sure which mode activated the nucleation. This is because soon after the nucleation, ice will grow around the nucleus and it will be difficult to distinguish one mode from the others.

7.10 The chemical composition of natural ice nuclei

The most common IN turns out to be just the clay minerals, usually containing silicate and with variable amounts of water trapped in the mineral structure, but very small sizes: from 0.1 μm to a few micrometers in diameter. Clay minerals such as

- illite $(K,H_3O)(Al,Mg,Fe)_2(Si,Al)_4O_{10}[(OH)_2,(H_2O)]$,
- kaolinite $Al_2Si_2O_5(OH)_4$, and
- vermiculite $(Mg,Fe,Al)_3(Al,Si)_4O_{10}(OH)_2·4H_2O$

have been identified. Laboratory nucleation experiments have also proved that they can act as IN.

The source of these minerals is the surface soil, and in arid regions of the world they are only loosely bound to the surface. Hence it is commonly thought that they are carried up by strong upward motion (such as dust storms) in arid regions to the middle and upper troposphere and transported thousands of miles by winds. Some of them then serve as IN to initiate ice particles in clouds and precipitation, and thus are carried back to the surface thereafter. Such large-scale and long-range transport of surface dust particles has been observed frequently by satellites. Fig. 6.5 shows such a case that occurred in March 2010. Indeed, many studies have found that some IN collected over Japan and northwest USA originated from the Gobi Desert in

Northern China and Mongolia, whereas IN over Western Europe often originate from the Sahara Desert of Africa.

Other sources of IN include volcanic ashes and urban man-made pollutants. However, the effectiveness of these materials as ice nuclei is not entirely clear. Whereas the laboratory studies seem to confirm that some volcanic materials can be reasonably good IN, field studies show that the IN concentrations do not necessarily increase following certain volcanic eruptions. Similarly, while there are some materials in urban pollutants (especially metal oxides from sources like steel mills or aluminum works) that have been confirmed by laboratory experiments to be effective IN, there are field studies showing that the upwind side of an industrial area often contains more IN than the downwind side. One suggestion to explain this is that there are materials in urban pollution that act to deactivate existing IN.

Other kinds of particles can also serve as IN. One of the more recently studied IN categories is the biogenic IN. Szyrmer and Zawadzki (1997) gave a summary of the biogenic and anthropogenic IN. The recent interest in biogenic IN is mainly out of concern for the impact of biomass burning in places such as the Amazon basin. Biomass burning is a more primitive method of preparing land for farming but it generates a large amount of biogenic particles that may have environmental impact. The effectiveness of some biogenic particles to act as IN was proved in laboratory experiments a long time ago, but whether or not they serve any important role in ice formation in clouds is unclear. However, there are studies showing that they are more abundant and ubiquitous than previously thought. A recent study performed by Prenni *et al.* (2009) suggests that the contribution of local biological particles to ice nucleation is increased at higher atmospheric temperatures, whereas the contribution of dust particles is increased at lower temperatures.

A very special kind of biogenic particles, though usually not considered as an important IN source in clouds, nevertheless plays an interesting role in artificial snow-making. This is the nuclei associated with a rod-shaped bacterium called *Pseudomonas syringae*, which is a plant pathogen that causes a wide variety of plant diseases (Fig. 7.9).

One of the most notable impacts it causes on plants is frost damage on the surface. Normally plant surfaces (leaves or bark) do not freeze readily even if the ambient temperature drops substantially below freezing. However, when the surfaces, such as leaves, are infested with *P. syringae*, they may freeze at a temperature as warm as −1.8°C, and frost will appear on the surface. The frost is due to the nucleation ability of the Ina (ice nucleation-active) proteins contained in the bacterium (Lee *et al.*, 1993). This is by far the best ice nuclei known to us and such a good capability did not go unnoticed by the commercial sector. It turns out that the bacterium can be caused to "shed" the ice nuclei so that a large quantity of them can be collected and used for artificial snow-making by ski resorts.

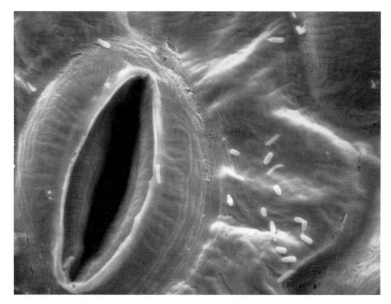

Fig. 7.9 Some *Pseudomonas syringae* bacteria infesting a bean leaf. The bacteria are a few micrometres in size. Photo courtesy of Prof. Gwyn Beattie.

7.11 Ice nuclei concentrations

The basic idea behind the measurement of the IN concentration is to take in a sample of aerosol-containing air and allow it to enter a saturated or supersaturated environment (usually achieved by lowering the temperature), and count the ice crystals formed in it. Since the relative humidity increases with decreasing temperature, it is easy to understand that the IN concentration so measured usually increases as the temperature becomes colder. Fig. 7.10 shows an example.

In contrast to the relatively abundant CCN, the IN concentration in the atmosphere is much smaller, usually around a few tens per liter. This is at least 10^3 less than the typical concentration of CCN, indicating the relative scarcity of IN. This also implies that, unlike the much more abundant CCN, it is possible to influence ice formation in the atmosphere by artificially injecting IN into clouds to influence the ice concentration and eventual cloud development. We will discuss this topic further at the end of Chapter 15.

Another important dissimilarity between the CCN and IN concentrations is that the former varies greatly over different locations while the latter seems to hold at a relatively constant level everywhere. Even in very remote places, such as Antarctica, the order of magnitude of IN concentrations does not differ very much from that at low-latitude locations (Bigg and Hopwood, 1963).

However, a phenomenon call an IN storm occurs sometimes, which causes the IN concentration to change suddenly. Fig. 7.11 shows such an example.

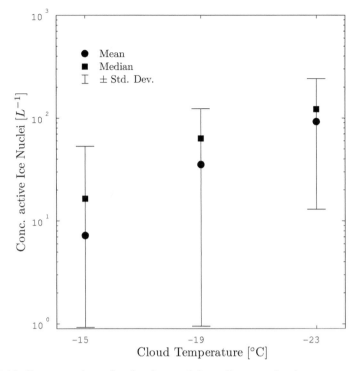

Fig. 7.10 Concentration of active ice nuclei per liter vs. cloud temperature. The values for the mean and median are shown, and the standard deviation is indicated by the bars. Adapted from Castro *et al.* (2004).

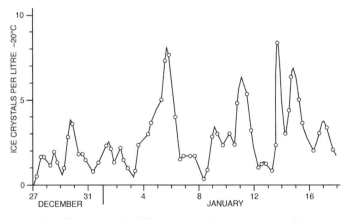

Fig. 7.11 Typical fluctuation in IN concentration as measured in Antarctica at $T = -20°C$. Adapted from Bigg and Hopwood (1963).

In contrast to the land, the world's oceans do not appear to be an important source of IN.

7.12 Criteria for effective ice nuclei

Materials that make good IN have properties quite different from those that make good CCN. The following is a list of some known properties that are necessary for a particle to be an effective IN. Detailed studies of why these are necessary are so far lacking, and the explanations given below are suggestive but not definitive.

- *Water insolubility* – Water-soluble materials can make good CCN but not IN. The known good ice nuclei are all water-insoluble. This may be due to the difficulty of establishing an ordered crystalline structure on a substrate that is water-soluble. It would be easier to construct an ordered structure on a surface whose properties remain fixed during the nucleation.

- *Size* – Both field observations and laboratory experiments show that "large" particles ($r_p > 0.1$ μm) are better IN than Aitken particles. A recent study by Lüönd *et al.* (2010) using kaolinite particles as the immersion mode IN found that the median freezing temperature increases from $-35°C$ for 200 nm kaolinite particles to $-33°C$ for 800 nm particles, again confirming this efficacy.

- *Hydrophobicity* – In contrast to CCN, where hygroscopicity and hydrophilicity are the important virtues, hydrophobic materials are more suitable for IN. Virtually all "good" ice nuclei found so far are also hydrophobic.

- *Chemical bond similarity* – Since the H bond plays such an important part in ice molecular structure, it is thought that substrates that also contain H bonds or bonds of similar nature would also help to promote ice nucleation. There are materials of this kind that are indeed effective IN and activate ice nucleation, such as metaldehyde $(CH_3CHO)_4$, which has an ice nucleation threshold of $T \sim -0.4°C$ (Fukuta, 1963), and cholesterol $C_{27}H_{46}O \cdot H_2O$, which has a threshold of -1 to $-2°C$ (Fukuta and Mason, 1963).

- *Crystallographic similarity* – Because ice is a crystal with a definite structure, it seems logical to expect that ice can form with relative ease on a substrate whose crystallographic structure is similar to that of ice, i.e. wurtzite. One such material is silver iodide (AgI), which is a yellow crystalline solid and also of wurtzite structure. Experimentally, it was found that AgI can nucleate supercooled water drops at $\sim -5°C$, and it has been used extensively in cold cloud seeding operations (see Chapter 15 for more details).

- *Active sites* – Owing to the success of the above idea about the crystallographic similarity of AgI to ice as an indication of the efficacy of IN, one might tend to think that the purer the AgI crystal one can make, the better IN it will be. Surprisingly, experiments showed that this is not the case. It turns out that AgI crystals that contain suitable amount of impurities on the surface can be more effective IN than the pure crystal. This may be due to the following mechanism: When the crystal surface is pure, every site is just as effective as every other site, and consequently water vapor tends to be spread fairly uniformly over the surface, preventing any site from developing an ice germ quickly. On the other hand, if there are impurities on the surface, these impurity sites (which are often high-energy sites and are called "active sites") become the preferential locations for ice germs to form, and subsequent vapor flux tends to be concentrated in these locations to make the germs grow faster.

7.13 Ice multiplication

It is usually accepted that one CCN nucleates one water droplet, and it seems to be reasonable to expect that IN would do the same. This means that the concentration of ice crystals should be about the same as that of IN in a cloudy air sample. Surprisingly, many earlier measurements found that the concentration of ice crystals is much more than that of the IN by a factor from a few to as much as 10^4–10^5. This phenomenon is variously called *ice multiplication* or *ice enhancement* or *secondary ice production*.

The multiplication process can occur very rapidly. Hobbs and Rangno (1990) reported that ice concentrations can increase from less than 0.01 l^{-1} to more than 100 l^{-1} within 10 min in clouds with cloud-top temperatures warmer than $-10°C$.

This phenomenon implies that either one IN can nucleate many ice crystals or there are other mechanisms responsible for extra crystal production. Most theories are inclined toward the latter. A few ice multiplication mechanisms have been suggested by investigators, as now outlined.

7.13.1 Fragmentation of ice crystals

Anyone who has ever handled ice crystals understands how fragile the crystals are, especially the delicate dendrites. Hence it is intuitively appealing that ice crystals may break up into many fragments due to turbulence in clouds or when an ice crystal collides with other particles or when the crystal is melting. Ice crystals may break up during the sampling process as well. There are indeed cloud crystal samples containing crystal fragments, so this mechanism is possible. However, questions linger on whether this mechanism is efficient enough to produce the observed magnitude of ice multiplication. Presently there is no accurate estimate of the enhancement ratio due to ice fragmentation.

More recent and more careful measurements, however, indicate that some of these fragments are likely an artifact of the sampling device, especially the inlet design of the sampler, which causes the ice crystals, especially the large ones, to shatter and break into many small pieces when sampled (McFarquhar *et al.*, 2007; Jensen *et al.*, 2009). The new measurements suggest that the natural small ice crystal concentration is typically one to two orders of magnitude lower than that inferred from such a device.

7.13.2 Shattering of freezing drops

During the freezing of a water drop, the process may proceed very rapidly, such that there is not enough time for the drop to achieve local thermal and mechanical equilibrium. Consequently, pressure may build up inside the drop, causing the freezing drop to shatter into numerous frozen drop fragments and ice splinters. This process has been experimentally observed by various investigators, but the numbers of ice splinters produced per shatter are usually just a few.

7.13.3 Hallett–Mossop mechanism

This process may occur when a supercooled droplet collides with an ice particle (e.g. a graupel), the droplet becomes rime on ice but also ejects copious ice splinters. Hallett and Mossop (1974) found that the process operates effectively at about −3 to −8°C for a drop size greater than 24 μm and an impact velocity of the drop of 1.4–3 m s^{-1}. The maximum production occurs at $T \sim -5°C$ and impact velocity ~ 2.5 m s^{-1}. If the temperature of the ice surface is warmer than −3°C, the drop would just spread over the surface without splinters, whereas at a temperature colder than −8°C the drop would freeze rapidly on the surface without splinter also. The enhancement factor of this mechanism has ranged from a few to about 10 if it happens.

The ice multiplication issue has not been completely resolved yet. Some natural enhancement is still possible, but judging from the recent studies it appears that the large enhancement factors observed previously are likely due to a sampling artifact and this process is probably less significant than previously thought. It is increasingly felt by the research community that aerosol size distribution and composition should play the main role in determining the ice concentration.

Problems

7.1. The energy of *i*-mer formation in the homogeneous nucleation process of a water drop is

$$\Delta F_i = 4\pi a_i^2 \sigma - \frac{4\pi a_i^3}{3v_w} kT \ln S_v.$$

Prove that

(a) the size of the germ (activated embryo) is given by

$$a_g = \frac{2\pi \sigma v_w}{kT \ln S_v},$$

(b) the free energy of germ formation is $\Delta F_g = \frac{1}{3}\sigma\Omega$, where Ω is the surface area of the germ.

7.2. Compare Eqs. (7.26) and (7.31). What conclusion can you obtain about the advantages or disadvantages of a plane surface substrate and curved substrate for the nucleation?

8

Hydrodynamics of cloud and precipitation particles

In the previous chapters, we have discussed the basic thermodynamics of cloud and precipitation particles and how they are initiated by nucleation, but we have not discussed their motion. None of these particles are truly stationary in the cloud – they are heavier than air so that they are always falling relative to the air. If a cloud appears to be floating in the air without apparent motion, it is only because the updraft in the cloud supports the fall of the cloud particles and gives us the illusion that the cloud is not moving vertically.

The motion of cloud and precipitation particles has great impact on the behavior of the cloud system. It influences how fast vapor can be accumulated on these particles, how fast these particles evaporate when they fall, how they collide and coalesce with each other and aerosol particles, and how they break up when acted on by sufficient force. All these processes ultimately determine how fast a cloud will develop or dissipate and how much precipitation will be produced by the cloud.

The motion of a cloud or precipitation particle in a viscous medium – the air – causes a flow field around the particle. The study of the flow fields around such particles belongs to the discipline of hydrodynamics, which is the main subject of this chapter.

8.1 Basic equations governing the flow past an obstacle

The study of the hydrodynamics of cloud and precipitation particles is centered on the flow past an obstacle. In this section, we will discuss the basic equations relevant to this problem.

First of all, the air is treated here as a continuous fluid and hence the *continuity equation* applies:

$$\frac{d\rho_a}{dt} = -\rho_a \nabla \cdot \vec{u}, \tag{8.1}$$

where ρ_a is the air density and $\vec{u}$ is the air velocity. The continuity equation is basically a statement of the conservation of mass of air. We assume here that no "new" air is produced and no air is taken away from the flow. Under such a "no source or sink" condition, the change in the air density (following the air particles, since we are using the total derivative here) must be due to the divergence of the flow.

In motions on the small time scales (typically a few seconds to minutes) that we are dealing with here, the change of air density is relatively unimportant and we usually assume that ρ_a is constant. Under this assumption, the motion of the air behaves as if the air is *incompressible*. The continuity equation becomes

$$\nabla \cdot \vec{u} = 0, \tag{8.2}$$

i.e. the flow must be non-divergent.

Next, the equation of motion of the air is

$$\frac{d\vec{u}}{dt} = \frac{\partial \vec{u}}{\partial t} + (\vec{u} \cdot \nabla)\vec{u} = -\frac{\nabla p}{\rho_a} + \vec{g} + v\nabla^2 \vec{u}. \tag{8.3}$$

This is the Navier–Stokes equation, which is based on Newton's second law of motion for a fluid. It is customary to express the law of motion in terms of "force per unit mass" in fluid dynamics. So, the first term is the acceleration of the fluid per unit mass, which should be equal to the sum of all external forces per unit mass acting on the fluid given on the right-hand side.

Let us examine the physical meaning of each term of (8.3). The term $d\vec{u}/dt$ is the acceleration of the fluid particle; $\partial \vec{u}/\partial t$ is the local change of fluid velocity, i.e. the change of fluid velocity with time at a fixed point; $(\vec{u} \cdot \nabla)\vec{u}$ is the inertial term representing the inertial force per unit mass due to the fluid motion; $-\nabla p/\rho_a$ is the pressure gradient force per unit mass; $\vec{g}$ is the gravitational force per unit mass; and $v\nabla^2 \vec{u}$ is the viscous force per unit mass. In the following discussions, however, we will ignore the "per unit mass" phrase when discussing these forces.

A further simplification is usually made to (8.3). The gravitational force $\vec{g}$ is considered as a constant here for the small-scale motion we are studying and it can be formulated as the gradient of the geopotential. Consequently, it can be combined with the pressure gradient term to simplify the equation, i.e.

$$-\nabla p/\rho_a + \vec{g} = -\nabla p'/\rho_a, \tag{8.4}$$

where p' is called the *dynamic pressure*. Hereafter in this chapter, the term "pressure" will always mean dynamic pressure and the prime symbol will be dropped. With this simplification, (8.3) can be rewritten as

$$\frac{\partial \vec{u}}{\partial t} + (\vec{u} \cdot \nabla)\vec{u} = -\frac{\nabla p}{\rho_a} + v\nabla^2 \vec{u}. \tag{8.5}$$

Eqs. (8.2) and (8.5) form the theoretical foundation of the flow field discussions to follow in this chapter.

8.2 Flow characteristics and Reynolds number

Before we embark on the solutions of the flow fields, it is useful to review some fundamental characteristics of flow past obstacles. The flow fields past obstacles of many different shapes and sizes, in fluids of different viscosities, and at different flow speeds, have been observed by using the flow visualization technique. The geometrical shape (including also the surface roughness) plays an important role in deciding the general pattern of the flow field, so that, for example, the flow pattern past a sphere looks quite different from that past a flat plate. Even if we confine our attention to the flow past obstacles of the same shape, we would see that the flow field varies with the flow speed, the size of the obstacle, and the fluid viscosity.

But for flow past an obstacle of the same shape, it is found that the flow patterns would look the same in many cases even though the obstacle sizes or flow speeds are quite different. It turns out that the flow pattern depends on a special combination of the obstacle size, flow speed, and fluid viscosity, called the *Reynolds number*, defined as

$$N_{Re} = \frac{Ud}{v},\tag{8.6}$$

where U is the general flow speed, d is the characteristic dimension of the obstacle, and v is the kinematic viscosity of the fluid. (In this book we use N_{Re} for Reynolds number, but Re is often used in other work.) For flow past obstacles of the same shape, e.g. spheres, the flow patterns would look the same as long as their flow Reynolds numbers are the same. Note that the Reynolds number is a *dimensionless number*, as the units of the numerator and the denominator on the right-hand side of (8.6) are both (length^2/time).

In addition, it is in general true that when N_{Re} is large the flow pattern is more irregular and even turbulent, whereas the flow pattern is more laminar when N_{Re} is small. Fig. 8.1 shows the experimental flow patterns past a sphere at different N_{Re} that illustrates this behavior. The same sphere was used for the experiments and N_{Re} was changed by changing the flow speed. We see that when N_{Re} is small, the flow field looks very smooth (laminar) and the field is nearly symmetric in the up- and downstream regions. There is no sign of eddies in the flow. As N_{Re} increases, the flow gradually becomes more asymmetrical. While the upstream region changes relatively little, the downstream region has obvious changes; at $N_{Re} = 25.5$, a pair of standing eddies start to appear. These eddies are symmetric with respect to the centerline, are attached to the rear of the sphere, and their patterns are steady (i.e. will not change with time). As N_{Re} increases further, the size of the eddies increases

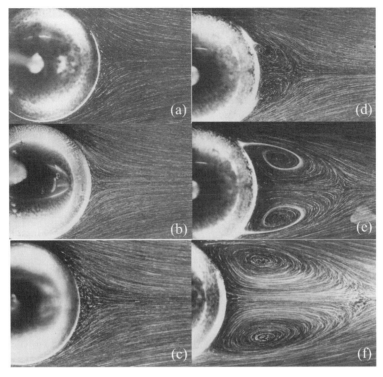

Fig. 8.1 Experimental flow pattern in the wake region immediately behind a sphere at various Reynolds numbers: (a) $N_{Re} = 9.15$, (b) 17.9, (c) 26.8, (d) 37.7, (e) 73.6, and (f) 118. The flow is from left to right. From Taneda (1956). Reproduced by permission of the Physical Society of Japan.

to about the same size as the sphere itself and eventually greater than the sphere size when $N_{Re} > 300$.

The figure only shows the steady-state flow fields. At higher Reynolds number, the flow becomes unsteady. At a fixed point, the flow velocity may change with time. When N_{Re} increases beyond 400, a phenomenon called *eddy shedding* occurs – eddies are no longer attached to the sphere but are shed from the sphere on alternate sides of the centerline. The shedding is initially periodic but gradually becomes turbulent when N_{Re} increases further.

Physically, N_{Re} is a measure of the relative importance of the inertial force vs. the viscous force in (8.5). When the inertial force is much bigger than the viscous force, the Reynolds number is high and the flow tends to be more turbulent. On the other hand, when the viscous force is much bigger than the inertial force, the Reynolds number is low and the flow tends to be more laminar. We can show that N_{Re} represents the ratio of inertial to viscous force easily using the simple case of flow past a sphere. We will perform a scaling of the two forces by noting that the scale for the fluid speed is U and the scale for the characteristic dimension is the diameter of the sphere d. Since both ∇ and ∇^2 in (8.5) are spatial operators, they can be scaled as $1/d$ and $1/d^2$, respectively. Then we see that

$$N_{\mathrm{Re}} = \frac{|(\vec{u} \bullet \nabla)\vec{u}|}{|v\nabla^2\vec{u}|} = \frac{\left(U\frac{1}{d}\right)U}{v\left(\frac{1}{d^2}\right)U} = \frac{Ud}{v}. \tag{8.7}$$

The kinematic viscosity is related to another quantity η, called the *dynamic viscosity*, by the relation

$$v = \eta/\rho_{\mathrm{a}}. \tag{8.8}$$

Hence the Reynolds number can be written as

$$N_{\mathrm{Re}} = \frac{2\rho_{\mathrm{a}}Ua}{\eta}. \tag{8.9}$$

8.3 Hydrodynamic behavior of falling cloud drops

We can make rough estimates of the Reynolds numbers associated with the fall motions of cloud and raindrops. The exact values of N_{Re} depend on the pressure levels. A typical cloud droplet radius is 10 μm and its terminal fall velocity is typically a few centimeters per second; the Reynolds number of its fall motion in air is on the order of ~ 0.1. At such a low N_{Re}, the flow field is very laminar, steady, and without eddies. A typical drizzle drop has a radius of 250 μm, and the Reynolds number of its fall motion in air is on the order of 100. The flow field is steady but with standing eddies attached to the drop. A typical raindrop has a radius of about 1 mm (1000 μm) and the associated N_{Re} is on the order of 1000. The flow field is highly unsteady with a turbulent wake.

In the following, we will discuss some theoretical studies of the flow fields associated with falling cloud drops. We will confine our discussions of mathematical derivations to steady flow cases as it is complicated to treat unsteady flow using analytical methods. Consequently, the analytical solutions derived below are applicable mainly to cloud drops.

Since we will consider only the steady flow cases, the local change term $\partial \vec{u}/\partial t$ in (8.5) is assumed to be zero and the equations to be considered here are

$$(\vec{u} \bullet \nabla)\vec{u} = -\frac{\nabla p}{\rho_{\mathrm{a}}} + v\nabla^2\vec{u} \tag{8.10}$$

and

$$\nabla \bullet \vec{u} = 0. \tag{8.11}$$

Since the drops we will consider here are relatively small, they are quite spherical as verified by experiments. In addition, to simplify the study further, we will first approximate the theoretical problem by flow past a rigid sphere. This approximation is justified as long as we are concerned with the external flow field since the circulation inside these small drops has a fairly small impact on the external field. If we are concerned with the internal circulations, of course, the rigid-sphere assumption cannot be adopted. We will consider the theoretical problem of flow past a liquid sphere later.

The theoretical treatment of the flow field for a sphere falling in air is equivalent to that of air passing a stationary sphere at the same relative velocity, but the latter is much easier to formulate when we use spherical coordinates. For flow past a rigid sphere of radius a, the relevant boundary conditions are

$$
\begin{aligned}
\vec{u} &= 0 && \text{at } r = a, \\
\vec{u} &= U_\infty \hat{e}_z && \text{at } r \to \infty,
\end{aligned}
\tag{8.12}
$$

where U_∞ is the fall speed of the sphere and $\hat{e}_z$ is the unit vector along the fall direction. The first condition (the *inner condition*) simply says that the flow velocity must vanish on a rigid surface (also called the *non-slip condition*). The second condition (the *outer condition*) says that the flow velocity sufficiently far away is a constant and will not be influenced by the presence of the sphere.

Eqs. (8.10)–(8.12) form the core equations for the steady flow past a rigid sphere. But instead of solving these equations directly, it is customary to express these equations in non-dimensional form. The benefit of doing so is that it renders the solutions more general instead of applicable only to a sphere of radius a falling at a speed U_∞ in air of density ρ_a. The non-dimensionalization is accomplished by the following substitutions:

$$
\vec{r}' = \frac{\vec{r}}{a}, \ \vec{u}' = \frac{\vec{u}}{U_\infty}, \ t' = \frac{t}{a/U_\infty}, \ p' = \frac{p}{\rho_a U_\infty^2}, \ \nabla' = \frac{\nabla}{a},
\tag{8.13}
$$

where all the primed quantities are dimensionless.

After this operation, (8.10)–(8.12) are transformed to non-dimensional forms. These non-dimensionalized equations are (after dropping the primes, with the understanding that all quantities and operators are now dimensionless):

$$
(\vec{u} \cdot \nabla)\vec{u} = -\nabla p + \frac{2}{N_{\text{Re}}} \nabla^2 \vec{u},
\tag{8.14}
$$

$$
\nabla \cdot \vec{u} = 0
\tag{8.15}
$$

and

$$\vec{u} = 0 \qquad \text{at } r = 1,$$
$$\vec{u} = 1 \cdot \hat{e}_z \qquad \text{at } r \to \infty. \tag{8.16}$$

The new boundary conditions (8.16) now apply to the general case of steady flow of fluid of any density at any velocity past a sphere of whatever size, not just limited to the one with radius a. Eq. (8.14) makes it clear that the flow field is a function of the Reynolds number.

There are no known general analytical solutions of (8.14)–(8.16), and in general numerical methods are necessary to solve these equations without further simplifications. The main difficulty appears to be the presence of a nonlinear term – the inertial force term – in (8.14). However, there exist a few further approximations that render these equations solvable by analytical methods. In the following we will present some of these simplified solutions that are relevant to cloud physics problems and discuss their properties. We will discuss numerical solutions in later sections.

8.3.1 Streamfunction formulation of flow fields

All the analytical solutions to be summarized below are *steady-state* and *axisymmetric* with respect to the flow axis in the z direction. The latter point says that the flow is of two-dimensional nature. Fig. 8.2 shows the schematic configuration of the flow past a sphere in spherical coordinates.

A special formulation of the flow field utilizing the streamfunction ψ takes advantage of these two properties and is most suitable to deal with this type of flow problem. Unlike velocity $\vec{u}$, which is a vector, the streamfunction is a scalar

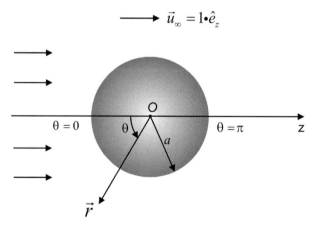

Fig. 8.2 Configuration of the theoretical problem for flow past a sphere.

function for the two-dimensional problems we are dealing with here. The contours of the streamfunction are called *streamlines*. The flow velocity vectors are tangents to the streamlines. In the case of steady flow fields, the streamlines coincide with the trajectories of the fluid particles.

In spherical coordinate systems, the streamfunction ψ is defined as

$$u_r = -\frac{1}{r^2\sin\theta}\left(\frac{\partial\psi}{\partial\theta}\right),$$
$$u_\theta = \frac{1}{r\sin\theta}\left(\frac{\partial\psi}{\partial r}\right),$$

(8.17)

where u_r and u_θ are the radial and tangential component of the velocity $\vec{u}$, respectively. When solving Eq. (8.14) we are actually dealing with two equations – one for u_r and one for u_θ. The chief benefit of using the streamfunction representation of the flow equation is that the Navier–Stokes equation (8.14) can be reformulated using ψ as the dependent variable instead of $\vec{u}$. Since ψ is a scalar function, we will be dealing with only one equation instead of two. Because of this benefit, all the analytical solutions to be discussed below are based on this formulation.

8.3.2 Stokes flow

As mentioned above, the main difficulty of solving (8.14) analytically is due to the nonlinear inertial force term. The simplest approximation to remove this difficulty is to ignore the inertial force term altogether:

$$(\vec{u}\cdot\nabla)\vec{u} = 0.$$

(8.18)

The resulting flow field due to this approximation is called the *Stokes flow*. In reality, of course, the inertial force term is zero only when $\vec{u}$ is zero. But the inertial force term is very small compared to the pressure gradient and viscous force terms when N_{Re} is small (for example, when $\vec{u}$ or the particle is very small) and hence this approximation is justified. This means that the Stokes flow is a good approximation for low-Reynolds-number flow.

The Navier–Stokes equation thus can be simplified to become

$$\nabla p - \frac{2}{N_{\mathrm{Re}}}\nabla^2\vec{u} = 0.$$

(8.19)

Using the streamfunction formulation, (8.19) can be written as

$$E^4\psi = 0,$$

(8.20)

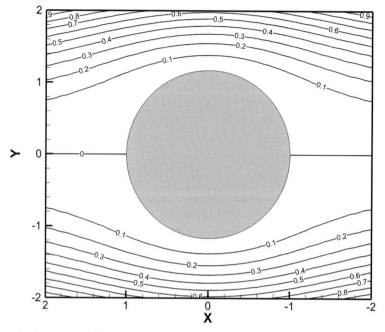

Fig. 8.3 Streamlines for Stokes flow past a rigid sphere for $U_\infty = 1$ and $a = 1$.

where

$$E^4 = E^2(E^2), \quad E^2 = \frac{\partial^2}{\partial r^2} + \frac{\sin\theta}{r^2}\frac{\partial}{\partial\theta}\left(\frac{1}{\sin\theta}\frac{\partial}{\partial\theta}\right), \tag{8.21}$$

is a special operator.

While it is convenient to non-dimensionalize the equations for the convenience of solution techniques, it is easier to understand the solutions themselves with the dimensions retained. Thus, in the following discussions, we will plug the units back in.

Eq. (8.20) can be solved analytically and the solution is

$$\psi_{St} = \frac{U_\infty a^2 \sin^2\theta}{4}\left(\frac{2r^2}{a^2} - \frac{3r}{a} + \frac{a}{r}\right). \tag{8.22}$$

The flow pattern (as revealed by the contour map of the streamfunction, i.e. the streamline pattern) is shown in Fig. 8.3.

We see that the Stokes flow field is rotationally symmetric with respect to the flow axis. It also possesses front-and-aft symmetry so that the upstream and downstream flow fields are mirror images of each other. This latter symmetry can be easily seen from (8.22) due to the nature of $\sin^2\theta$ in the solution. There are also no eddies anywhere in this flow field.

The velocity components can be obtained using (8.17):

$$u_r = U_\infty \cos\theta \left(1 - \frac{3a}{2r} + \frac{a^3}{2r^2} \right),$$

$$u_\theta = -U_\infty \sin\theta \left(1 - \frac{3a}{4r} - \frac{a^3}{4r^2} \right).$$

$$(8.23)$$

It is easily verified that (8.23) satisfies the incompressibility assumption $\nabla \cdot \vec{u} = 0$. The pressure distribution around the sphere is given by

$$p = p_\infty + \frac{3\eta a U_\infty \cos\theta}{2r^2},$$

$$(8.24)$$

where p_∞ is the pressure far away from the sphere. On the sphere surface, the pressure maximum occurs at $\theta = 0$ ($\cos\theta = 1$). This point is called the *front stagnation point* and the region around this point is a high-pressure region. The presence of this high-pressure region is not limited to the Stokes flow but is a general feature of flow past spheres and plates. It prevents incoming particles (for example, other cloud droplets or aerosol particles) from penetrating and colliding with the drop in this region. More discussions on this point will be made when we discuss collision growth.

One of the important quantities that we need to determine is the *drag force* of the air acting on the sphere. This drag force can be calculated using the following integral:

$$F_D = -2\pi a^2 \int_0^\pi \left[p \cos\theta + \eta \left(\frac{\partial u_\theta}{\partial r} \right) \sin\theta \right]_{r=a} \sin\theta \, d\theta,$$

$$(8.25)$$

and the resulting drag force due to the Stokes flow field is

$$F_{D,St} = 6\pi\eta a U_\infty.$$

$$(8.26)$$

This Stokes drag is a good approximation for the drag that the air exerts on a falling cloud droplet of less than a few tens of micrometers in diameter.

Drag force on a very small sphere

While the Stokes flow is in general very useful when dealing with small spheres falling at low velocities, such as cloud drops and other small spherical particles falling in air, care must be taken when we deal with *really* small spheres, i.e. those spheres of size comparable to the *mean free path* λ of air molecules (about 10^{-5} cm or 0.1 μm at the surface level). To such small spheres, the air is not a continuum but rather an ensemble of individual molecules, and once in a while the spheres could be located in a space without any air molecules, i.e. spheres are "slipping" in space. Consequently, the drag experienced by these spheres will be smaller than that by larger spheres and a correction is necessary. This correction factor is called the *Stokes–Cunningham slip correction factor* f_{SC} such that

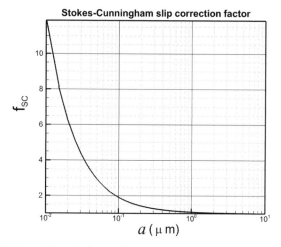

Fig. 8.4 The Stokes–Cunningham slip correction factor as a function of sphere radius a according to Eq. (8.28).

$$F_{D,St} = \frac{6\pi a\eta U_\infty}{f_{SC}}. \tag{8.27}$$

There are various formulations of f_{SC} and here we use the expression given by Davies (1945):

$$f_{SC} = 1 + \left[1.257 + 0.400\exp\left(-\frac{1.1}{N_{Kn}}\right)\right]N_{Kn}, \tag{8.28}$$

where N_{Kn} is the *Knudsen number* defined as

$$N_{Kn} = \frac{\lambda}{a}. \tag{8.29}$$

Fig. 8.4 shows the variation of f_{SC} with a at sea level assuming $\lambda = 0.068$ μm. It is clear that the correction is substantial when $a \leq 1$ μm.

8.3.3 Oseen flow and Carrier's modification

The Stokes approximation completely ignores the inertial force, which obviously cannot be valid when the Reynolds number becomes substantially greater than unity. Experimental measurements show that Stokes drag significantly underestimates the real drag for higher N_{Re}. One attempt to remedy this deficiency is called the Oseen flow.

Instead of totally ignoring the inertial force, Oseen flow assumes that the inertial force can be approximated by

$$(\vec{u} \cdot \nabla)\vec{u} \approx (\vec{U}_\infty \cdot \nabla)\vec{u}. \tag{8.30}$$

Now $\vec{U}_\infty$ is a constant vector, not a variable any more, hence the nonlinear inertial force is linearized by Oseen's approximation. This linearization of the Navier–Stokes equation renders the analytical solution possible and the resulting streamfunction is

$$\psi_{Os} = \frac{U_\infty a^2 \sin^2\theta}{2} \left(\frac{r}{a} - 1\right)^2 \left[\left(1 + \frac{3}{16}N_{Re}\right)\left(1 + \frac{a}{2r}\right) + \frac{3}{16}N_{Re}\left(1 + \frac{a}{r}\right)^2 \cos\theta\right]$$

(8.31)

and the drag force computed using (8.25) is

$$F_{D,Os} = 6\pi\eta aU_\infty\left(1 + \frac{3}{16}N_{Re}\right).$$

(8.32)

The flow field of (8.31) no longer possesses fore-and-aft symmetry, as the existence of the factor $\cos\theta$ in the square brackets indicates, but it is still rotationally symmetric with respect to the flow axis. In addition, one can solve the equation $\psi_{Os} = 0$ and obtain a solution, indicating the presence of standing eddies whose region is enclosed by $\psi_{Os} = 0$ streamlines. We recall that this standing eddy feature is absent in Stokes flow even at high Reynolds numbers. Thus, on this point alone, the Oseen flow is superior to the Stokes flow when high N_{Re} cases are concerned. Fig. 8.5 shows the Oseen flow field at $N_{Re} = 20$.

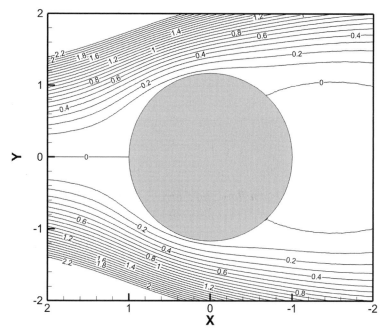

Fig. 8.5 Streamlines for Oseen flow past a rigid sphere for $U_\infty = 1$, $a = 1$ and $N_{Re} = 20$.

Experimental measurements, however, show that Oseen drag (8.32) significantly overestimates the drag force on a sphere. This is mainly caused by the use of $\vec{U}_\infty$ in place of $\vec{u}$ in the representation of the inertial force term everywhere, which is indeed an overestimate. The standing eddies shown in Fig. 8.5 are also much larger than experimentally observed.

Carrier (1953) proposed to revise the Oseen drag (8.32) as

$$F_{\text{D,Ca}} = 6\pi\eta a U_\infty \left(1 + \frac{3c}{16} N_{\text{Re}}\right), \tag{8.33}$$

to empirically correct the drag force with the empirical factor c determined by experiments. Carrier determined the value of c by performing experimental measurements for flow past a flat plate and found that $c \sim 0.43$. The same value seems to fit the flow past a sphere as well for $N_{\text{Re}} \leq 40$. Note that Carrier's approach only corrects the drag force but does not provide correctional information about the flow field itself.

8.3.4 Potential flow

At still higher Reynolds number, the inertial force will become much larger than the viscous force and hence the latter may be ignored for some cases. Note that at high Reynolds numbers the flow tends to become turbulent, which is intrinsically unsteady, hence the steady-state assumption fails and (8.14) is no longer valid. However, in the context of falling cloud and precipitation particles, the unsteady behavior mainly occurs in the rear of the sphere, whereas the upstream portion of the flow field remains approximately steady.

Therefore, if only the upstream flow field is of concern, then a simplification can be made such that the viscous force is ignored, i.e.

$$\nu\nabla^2\vec{u} = 0. \tag{8.34}$$

The resulting flow is called the *potential flow*. The reason for this terminology is the following. From vector identity, we know that

$$\nu\nabla^2\vec{u} = \nu[\nabla(\nabla \bullet \vec{u}) - \nabla \times (\nabla \times \vec{u})] = 0. \tag{8.35}$$

Since we are dealing with incompressible fluid, $\nabla \bullet \vec{u} = 0$, hence we have

$$\nabla \times \vec{u} = 0. \tag{8.36}$$

The curl of $\vec{u}$ is called the *vorticity*. Eq. (8.36) states that the vorticity is zero everywhere in the flow field, so that the flow is *irrotational*.

Now any vector $\vec{u}$ that satisfies (8.36) can be expressed as the gradient of a scalar potential function ϕ,

$$\vec{u} = -\nabla\phi, \tag{8.37}$$

because of the vector identity

$$\nabla \times \nabla\phi = 0. \tag{8.38}$$

This is why this flow field is called potential flow.

From the incompressibility condition, we have

$$\nabla \bullet \vec{u} = -\nabla \bullet \nabla\phi = -\nabla^2\phi = 0 \tag{8.39}$$

or simply

$$\nabla^2\phi = 0, \tag{8.40}$$

which is the Laplace equation, and the solution methods of the Laplace equation are well known (see e.g. Lorrain and Corson, 1970).

Under this inviscid assumption, the Navier–Stokes equation (8.14) becomes

$$(\vec{u} \bullet \nabla)\vec{u} = -\nabla p. \tag{8.41}$$

But to solve (8.41) completely to obtain a specific solution, we cannot use the boundary conditions (8.12) or (8.16). The reason is that the viscous effect is absent in the potential flow scenario and the non-slip condition $\vec{u} = 0$ on the surface of the sphere cannot be satisfied by this flow. We replace that inner boundary condition by requiring that the normal component of the velocity vanishes on the surface of the sphere:

$$\vec{u} \bullet \hat{e}_n = 0 \quad \text{at } r = a, \tag{8.42}$$

where $\hat{e}_n$ is the unit vector normal to the sphere surface. The outer boundary condition remains unchanged. With this modification, we obtain the solution for the potential function,

$$\phi = U_\infty \cos\theta \left(r + \frac{a^3}{2r^2} \right), \tag{8.43}$$

and by (8.37) the velocity of the flow,

$$\vec{u} = -U_\infty \cos\theta \left(1 - \frac{a^3}{r^3} \right)\hat{e}_r + U_\infty \sin\theta \left(1 + \frac{a^3}{2r^3} \right)\hat{e}_\theta, \tag{8.44}$$

where $\hat{e}_r$ and $\hat{e}_\theta$ are the unit vectors in the r and θ directions, respectively. The streamfunction of the potential flow can be obtained by plugging (8.44) into (8.17) and solving for ψ. The result is

$$\psi_{Po} = \frac{U_\infty r^2 \sin^2\theta}{2} \left(1 - \frac{a^3}{r^3} \right). \tag{8.45}$$

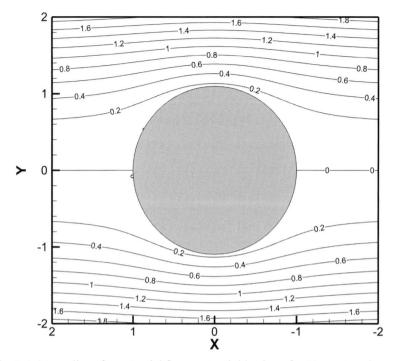

Fig. 8.6 Streamlines for potential flow past a rigid sphere for $U_\infty = 1$ and $a = 1$.

Obviously the potential flow field is rotationally symmetric with respect to the flow axis and possesses fore-and-aft symmetry also, just like the Stokes flow. And, like the Stokes flow, it does not have standing eddies either. (Note, however, that Stokes flow is not irrotational, unlike the potential flow.) Fig. 8.6 shows the potential flow field.

Under the potential flow assumption, the inertial force term can be written as

$$(\vec{u} \cdot \nabla)\vec{u} = \nabla\left(\frac{u^2}{2}\right) - \vec{u} \times (\nabla \times \vec{u}) = \nabla\left(\frac{u^2}{2}\right) \tag{8.46}$$

because of the irrotational nature. Thus (8.41) can be written as

$$\nabla\left(\frac{u^2}{2}\right) = -\frac{\nabla p}{\rho_a}, \tag{8.47}$$

which can be rearranged to become

$$\nabla\left(\frac{u^2}{2} + \frac{p}{\rho_a}\right) = 0 \tag{8.48}$$

or

$$\frac{u^2}{2} + \frac{p}{\rho_a} = \text{constant} \tag{8.49}$$

upon integration. Eq. (8.49) is called *Bernoulli's law*, which is often cited to explain some properties of incompressible fluids. It states that the pressure and velocity extrema are opposite to each other: where pressure is a maximum, the velocity is a minimum, and vice versa. While this is derived for the inviscid fluid we are dealing with here, it is also approximately true for other flows. For example, in the Stokes flow, the pressure maximum occurs in the front stagnation area where the velocities are indeed minimal.

Another notable property of the potential flow is that the net drag force is zero. Thus while the potential flow field is useful sometimes in representing the upstream flow of a falling spherical drop, it cannot be applied to obtain the drag acting on the drop.

8.3.5 Hadamard–Rybczynski flow past liquid spheres

The above three analytical solutions are applicable only to flow past a rigid sphere where the boundary conditions (8.16) or (8.42) hold true. However, cloud drops are liquid spheres and the boundary conditions mentioned above are no longer true, and so we need to examine the flow past liquid spheres in order to correctly understand the real nature of the flow fields past liquid drops. An analytical solution of flow past a liquid sphere analogous to Stokes flow is called the *Hadamard–Rybczynski flow*. Thus this solution is applicable to small falling cloud droplets with $N_{Re} \leq 1$. The following is an outline of this solution.

The main difference between the flow past a rigid sphere and a liquid sphere is that the liquid surface can move in response to the external air flow whereas a rigid sphere cannot have surface motion. Thus the difference is in the boundary conditions. Because of this difference, though, we need to consider the flow in two regions: inside and outside of the drop.

For a small liquid drop falling in air, the Stokes approximation should be valid for both flows inside (internal flow) and outside (external flow) of the drop. Thus we can use (8.20) for both regions,

$$E^4 \psi_{HR,in} = 0 \tag{8.50}$$

and

$$E^4 \psi_{HR,ex} = 0, \tag{8.51}$$

where $\psi_{HR,in}$ and $\psi_{HR,ex}$ represent the streamfunction of the internal and external flow, respectively. So we note that there are two simultaneous equations of motion to be solved.

But the boundary conditions become much more complicated. There are now five conditions, as follows.

(1) The external boundary condition remains the same. Writing this condition in terms of the streamfunction instead of velocity, we have

$$\psi_{HR,ex} \rightarrow -\frac{U_\infty r^2 \sin^2\theta}{2}. \tag{8.52}$$

(2) The velocity vector should not penetrate the liquid–air boundary. This requires that the normal velocity component inside and outside of the drop be zero,

$$u_{r,in} = u_{r,ex} = 0 \quad \text{at } r = a, \tag{8.53}$$

which can be translated into the boundary condition of the streamfunction as

$$\psi_{HR,in} = \psi_{HR,ex} = 0 \quad \text{at } r = a. \tag{8.54}$$

(3) The tangential velocity components should be continuous across the drop surface:

$$\frac{\partial \psi_{HR,in}}{\partial r} = \frac{\partial \psi_{HR,ex}}{\partial r} \quad \text{at } r = a. \tag{8.55}$$

(4) The tangential stress should also be continuous across the drop surface:

$$\frac{\partial}{\partial r}\left(\frac{1}{r^2}\frac{\partial \psi_{HR,in}}{\partial r}\right) = \kappa \frac{\partial}{\partial r}\left(\frac{1}{r^2}\frac{\partial \psi_{HR,ex}}{\partial r}\right) \quad \text{at } r = a. \tag{8.56}$$

where $\kappa = \eta_a/\eta_l$ is the ratio between the air and liquid viscosities.

(5) The vertical stress should be continuous as well:

$$p_a - 2\eta_a \frac{\partial}{\partial r}\left(\frac{1}{r^2 \sin\theta}\frac{\partial \psi_{HR,in}}{\partial \theta}\right) + \frac{2\sigma}{a} = p_l - 2\eta_l \frac{\partial}{\partial r}\left(\frac{1}{r^2 \sin\theta}\frac{\partial \psi_{HR,ex}}{\partial \theta}\right) \quad \text{at } r = a, \tag{8.57}$$

where the $2\sigma/a$ term is the excess pressure inside a spherical drop as compared to the pressure outside that we have discussed in Chapter 5.

Now with two second-order differential equations with five boundary conditions, one may wonder whether or not this becomes an overspecified problem because only four boundary conditions are needed to completely specify the solutions. It turns out from hindsight that (8.57) is not necessary for the solutions but the solutions nevertheless satisfy (8.57) automatically. Hence there is no overspecification problem.

Solving (8.50) to (8.56) we obtain the following solutions:

$$\psi_{HR,ex} = \frac{U_\infty r^2 \sin^2\theta}{2} \left[1 - \frac{a(3+2\kappa)}{2r(1+\kappa)} + \frac{a^3}{2r^3(1+\kappa)} \right], \tag{8.58}$$

$$\psi_{HR,in} = \frac{U_\infty r^2 \sin^2\theta}{4} \left(\frac{\kappa}{1+\kappa} \right) \left[\frac{r^2}{a^2} - 1 \right], \tag{8.59}$$

where $\kappa = (\eta_a/\eta_w)$. We see that the external flow streamfunction (8.58) looks similar to the Stokes streamfunction (8.23) but modified by the terms involving κ. For rigid spheres, $\kappa \to \infty$ and (8.58) converges to (8.23) and (8.59) becomes zero.

For cloud droplets falling in air, $\kappa \sim 10^{-3}$, and the terms containing κ in (8.58) are

$$\frac{a(2+3\kappa)}{2r(1+\kappa)} \sim \frac{a}{r}, \quad \frac{a^3}{2r^3(1+\kappa)} \sim \frac{a^3}{2r^3}, \tag{8.60}$$

and therefore the external flow field is not too much different from the case of a rigid sphere. However, (8.59) is not zero and so there exists an internal circulation in the drop. The circulation is exactly the Hill's spherical vortex described by Hill (1894). It is a circulation with fore-and-aft symmetry just like the Stokes flow except that it is confined within the sphere (see Fig. 8.7).

Many experiments show that there is no internal circulation when the flow speed is very small, while (8.59) predicts the existence of it regardless of the flow speed. This could be due to shortcomings in either the theory or the experimental techniques. There seems to be no definite conclusions on this matter at this time. One tricky point is that the presence of surfactant on the drop may have a large influence on the surface stress and cause errors in the flow field measurements.

The drag force on a sphere due to the Hadamard–Rybczynski flow is

$$F_{D,HR} = 6\pi\eta_a a U_\infty \left(\frac{1 + \frac{2}{3}\kappa}{1+\kappa} \right), \tag{8.61}$$

which reduces to the Stokes drag when κ is zero for the case of a rigid sphere because $\eta_{rigid} \to \infty$. One consequence of (8.61) is that $U_{\infty,liquid} > U_{\infty,rigid}$.

8.3.6 Numerical solutions of flow past spherical drops

Analytical solutions are useful in providing us with some basic understanding of the nature of flow fields, but they often have crude assumptions that render them not very useful in real applications. Out of the four cases mentioned above, the Stokes and Hadamard–Rybczynski solutions are probably the most useful to us in the context of treating small falling cloud droplets, small frozen drops, and spherical aerosol particles, and the analytical flow fields are indeed close to those observed.

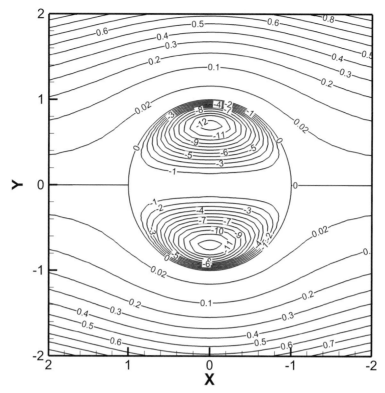

Fig. 8.7 Streamlines for Hadamard–Rybczynski flow past a liquid sphere given by Eqs. (8.58) and (8.59) for $U_\infty = 1$, $a = 1$ and $\kappa = 10^{-3}$.

To obtain flow fields for larger drops, we have to resort to numerical solutions. Here we can have the luxury of including both inertial and viscous forces without serious simplifications and solve the complete set of equations (8.14)–(8.16). It is beyond the scope of this book to discuss the numerical techniques in solving the Navier–Stokes equation, and we will only discuss numerical results that are relevant to cloud physics problems.

Fig. 8.8 shows the numerical results of Le Clair *et al.* (1972) for flow past a spherical drop for $N_{Re} = 30$. Fig. 8.8(a) shows the streamline pattern of the external flow, which is obviously asymmetric in the fore-and-aft direction. A small standing eddy can be seen downstream whose size agrees well with experimental results. Fig. 8.8(b) shows the internal flow streamline pattern, which is also asymmetric unlike that shown in Fig. 8.7 of the Hadamard–Rybczynski flow. This is naturally due to the inclusion of the inertial term. Fig. 8.9 shows the case for $N_{Re} = 300$. It is seen in Fig. 8.9(a) that the size of the standing eddy is somewhat larger than the diameter of the sphere at this N_{Re}, agreeing with experimental results. The internal flow, shown in Fig. 8.9(b), shows not only the

(a)

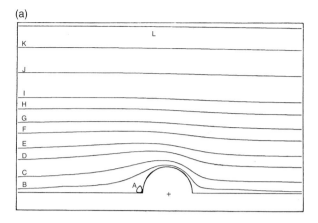

(b)

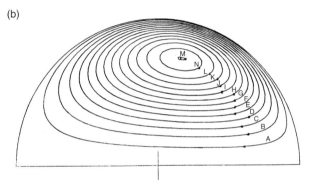

Fig. 8.8 Numerical streamlines for flow past a liquid sphere at $N_{Re} = 30$. (a) External flow. (b) Internal flow. From Le Clair *et al.* (1972). Reproduced by permission of the American Meteorological Society.

asymmetric feature but also a small eddy in the downstream (lower left) corner which is a counter-eddy caused by the standing eddy in the external flow.

The internal circulation of falling water drops has been subject to experimental studies. Fig. 8.10 shows three examples of the internal circulation pattern of falling water drops.

Fig. 8.11 shows the experimentally measured internal circulation velocity as a function of drop diameter. This velocity is just a fraction of the terminal fall velocity of the drop, as would be expected.

8.4 Flow past large drops

The above numerical solutions pertain only to the steady-state Navier–Stokes equation and hence they are only applicable to relatively small drops. As mentioned previously, experimental observations show that for a moving sphere with $N_{Re} \geq 400$ the flow

(a)

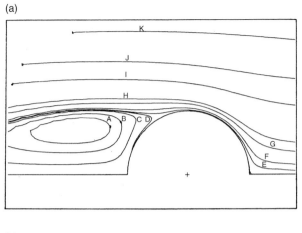

(b)

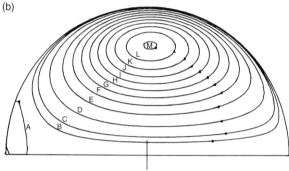

Fig. 8.9 Numerical streamlines for flow past a liquid sphere at $N_{Re} = 300$. (a) External flow. (b) Internal flow. From Le Clair *et al.* (1972). Reproduced by permission of the American Meteorological Society.

starts to become unsteady and eddy shedding occurs, and the solutions for the steady-state Navier–Stokes equation are no longer valid. Thus far there seem to be no numerical solutions of the time-dependent Navier–Stokes equation for unsteady flow past a liquid drop with $N_{Re} \geq 400$. At present, the information about the behavior of falling large water drops comes mostly from experiments.

8.4.1 Drag coefficients for falling water drops

The drag force acting on a falling precipitation particle depends on both its Reynolds number and its shape. In order to discuss the drag force in a more general way, it is convenient to define a *drag coefficient* as

$$C_D = \frac{F_D}{\left(\frac{1}{2}\rho_a U_\infty^2\right) A_c}, \tag{8.62}$$

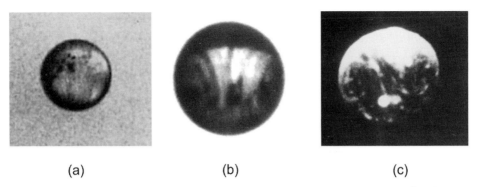

(a)   (b)   (c)

Fig. 8.10 Flow patterns of the internal circulation of a drop: (a) 630 μm radius (Le Clair *et al.*, 1972); (b) 800 μm and (c) 6 mm in diameter (Szakáll *et al.*, 2009). Reproduced by permission of the American Meteorological Society.

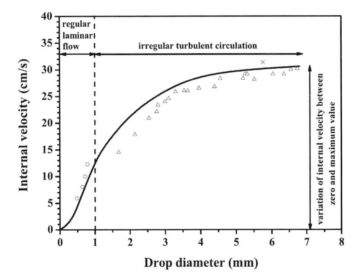

Fig. 8.11 Internal velocity of drops as a function of radius. Experimental data: average values (circles); maximum values determined with the Beaulieu film camera (triangles); and maximum value determined with the Redlake Locam II film camera (cross). Theoretical values (solid line) calculated according to Le Clair *et al.* (1972). From Szakáll *et al.* (2009). Reproduced by permission of the American Meteorological Society.

where A_c is the cross-sectional area of the body normal to the fall direction. The value of C_D can be calculated theoretically (when it is possible) or measured experimentally. For Stokes flow past a sphere, it is easy to show that

$$C_{D,St} = \frac{24}{N_{Re}} \tag{8.63}$$

or

$$\frac{C_{D,St}\,N_{Re}}{24} = 1. \tag{8.64}$$

For non-Stokes flows past spheres (or quasi-spheres), the drag coefficient would not satisfy (8.64), of course, but the drag force can be expressed as

$$F_D = 6\pi\eta_a a U_\infty \left(\frac{C_D N_{Re}}{24}\right). \tag{8.65}$$

For non-spherical particles, (8.62) has to be used to obtain the drag force. Drag coefficients are usually obtained by experimental measurements.

Eq. (8.65) clearly indicates that the drag force is a function of both the drag coefficient and the Reynolds number. When the drop size increases, the drag force is also increasing generally, but the drag coefficient usually decreases until the drop reaches a certain size.

Fig. 8.12 compares the drag coefficients as a function of Reynolds number among various theoretical results and experimental measurements for flow past spheres. The results of Beard and Pruppacher (1969) and Gunn and Kinzer (1949) are experimental data, and the other two curves are Stokes and Oseen drag coefficients. Stokes and Oseen drags are acceptable when $N_{Re} \leq 1$, but deviate more and more as N_{Re} increases. Oseen drag increasingly overestimates whereas Stokes drag increasingly underestimates.

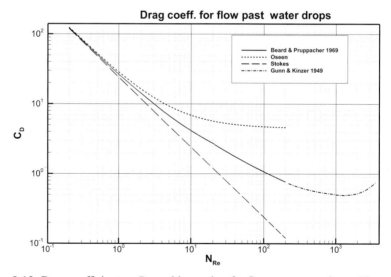

Fig. 8.12 Drag coefficient vs. Reynolds number for flow past water drops. The solid curve is based on the experiments of Beard and Pruppacher (1969). The dashed-dotted curve is based on Gunn and Kinzer (1949). The dashed curve is the Stokes drag, whereas the dotted curve is the Oseen drag for flow past rigid spheres.

Beard and Pruppacher (1969) provided the following empirical fits for their results:

$$C_{\mathrm{D}} = (24/N_{\mathrm{Re}})(1 + 0.102N_{\mathrm{Re}}{}^{0.955}) \quad \text{for } 0.2 \leq N_{\mathrm{Re}} \leq 2, \quad (8.66)$$

$$C_{\mathrm{D}} = (24/N_{\mathrm{Re}})(1 + 0.115N_{\mathrm{Re}}{}^{0.802}) \quad \text{for } 2 \leq N_{\mathrm{Re}} \leq 21, \quad (8.67)$$

$$C_{\mathrm{D}} = (24/N_{\mathrm{Re}})(1 + 0.189N_{\mathrm{Re}}{}^{0.632}) \quad \text{for } 21 \leq N_{\mathrm{Re}} \leq 200. \quad (8.68)$$

For $N_{\mathrm{Re}} > 220$ we can fit the Gunn and Kinzer (1949) data by the following polynomial:

$$\begin{aligned}
C_{\mathrm{D}} = {} & 8.594 \times 10^{-1} - 9.028 \times 10^{-4} N_{\mathrm{Re}} + 8.546 \times 10^{-7} N_{\mathrm{Re}}{}^{2} \\
& - 3.971 \times 10^{-10} N_{\mathrm{Re}}{}^{3} + 9.431 \times 10^{-14} N_{\mathrm{Re}}{}^{4} - 8.666 \times 10^{-18} N_{\mathrm{Re}}{}^{5} \\
& \text{for } 220 \leq N_{\mathrm{Re}} \leq 3549.
\end{aligned}$$

$$(8.69)$$

The increasing trend of the drag coefficient in Gunn and Kinzer's curve for Reynolds number greater than ~ 1300 is due to the increasing flattening of the drop that causes the greater drag force.

Close inspection of the original experimental drag data reveals that the drag force as a function of N_{Re} can be broadly divided into three regimes: (1) $N_{\mathrm{Re}} \leq 20$; (2) $20 < N_{\mathrm{Re}} \leq 400$; and (3) $N_{\mathrm{Re}} > 400$. The slope of the drag curve (not shown) remains relatively steady (though not constant) within each regime, but undergoes a slight but abrupt change at the boundary between two regimes. The reason for this change is likely due to the behavior of the eddy. At $N_{\mathrm{Re}} \sim 20$, the onset of standing eddies occurs, and at $N_{\mathrm{Re}} \sim 400$, the shedding of eddies starts. Apparently, the standing eddies have a significant impact on the drag force.

Experimental measurements show that the drag forces on rigid spheres and liquid drops are almost indistinguishable up to $N_{\mathrm{Re}} \sim 700$. Beyond that, the drag force on a liquid drop becomes increasingly larger than that on a rigid sphere at the same Reynolds number. This is mainly due to the distortion of the liquid drop shape, which begins to deviate from a spherical shape at higher N_{Re}. A liquid drop at this range of N_{Re} has a flattened bottom and hence increased cross-sectional area normal to its fall direction, and that enhances the drag force.

At this point, we only have some measurements of the drag forces on large falling drops. We do not yet have detailed knowledge of their flow fields, but we do know that the wakes must be quite turbulent.

8.4.2 Fall behavior of raindrops

The fall behavior of small water drops (cloud droplets and drizzle drops) is relatively simple – we expect the drops to fall steadily as the steady flow fields imply. But as soon as the unsteadiness sets in – starting with $N_{Re} > 400$ – the fall behavior becomes more complicated. For medium-sized drops (from ~ 600 μm to 1 mm in radius), the periodic eddy shedding will likely make these drops fall in a non-straight manner, although this behavior has not been studied systematically. For even larger drops ($a > 1$ mm), the distortion of the drop shape and the oscillation, canting, and breakup become important.

Drop shape as a function of size

The distortion of the shape of raindrops from spherical has impacts not only on their fall behavior but also on their detection by dual-polarization radars, which can distinguish the difference between spherical and non-spherical particles (see the discussion in Chapter 2). As we have seen before, larger drops tend to become more flattened. If we define an axis ratio $AR = b/a$, where a and b are the lengths of semi-axis in the horizontal and vertical directions, then we would expect that AR decreases with drop size. Fig. 8.13 shows the measured AR as a function of the equivalent drop diameter D_0 as measured in laboratory experiments.

We see that, as the drop size increases, the drop becomes more flattened and AR indeed decreases. The curve has been fitted by Chuang and Beard (1990) as

$$AR = 1.101668 - 0.09806\,D_0 - 2.52686\,D_0{}^2 + 3.75061\,D_0{}^3 - 1.68692\,D_0{}^4$$

$$(8.70)$$

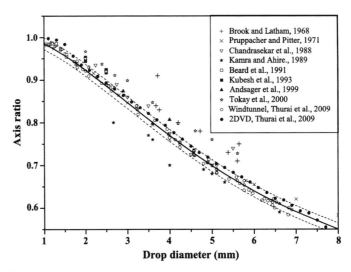

Fig. 8.13 Axis ratio as a function of equivalent drop diameter derived from different experiments. From Szakáll *et al.* (2010). Reproduced by permission of Elsevier B.V.

for $0.1 \leq D_0 \leq 0.9$ cm (D_0 has to be in centimeters). There are also values derived from radar observations that confirm the general trend. However, the data in Fig. 8.12 pertain to the equilibrium shapes of the drops because they are obtained in laboratories with relatively calm environments. In contrast, the radar observations pertain to real storm conditions, which are very turbulent. Under the turbulent conditions, we expect that the *AR* would be somewhat different because of oscillations of the drops. This is discussed in the next subsection.

Oscillation of large falling raindrops

Unlike a small droplet, which remains spherical, water drops larger than ~ 500 μm in radius have been observed to oscillate. The oscillation amplitude is initially small for smaller drops but increases quickly as the drop size increases. Depending on the modes of oscillation, the axis ratio *AR* can be greater or smaller than the equilibrium values as given by (8.70). For example, if the oscillations are in the transverse mode, then the horizontal extent of the drop appears to increase and consequently *AR* decreases when averaged over a period. On the other hand, a vertically oriented oscillation would make the drop elongate in the vertical direction and would make the time-averaged *AR* greater than the equilibrium value.

Drop oscillations can have many modes also, but oscillation in the fundamental mode is by far the predominant mode. Fig. 8.14 shows the frequency of observed

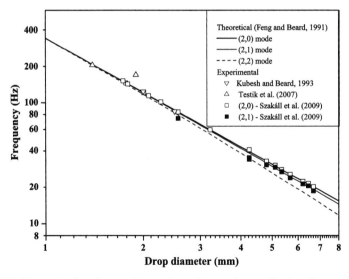

Fig. 8.14 Theoretical and experimental results on the oscillation frequency as a function of drop size. From Szakáll *et al.* (2010). Reproduced by permission of Elsevier B.V.

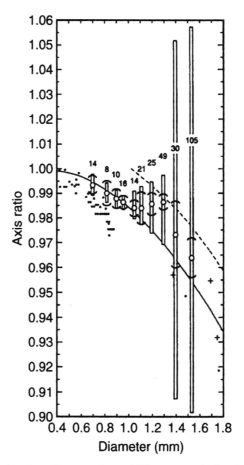

Fig. 8.15 Axis ratio of small water drops falling at terminal velocity as measured by Beard *et al.* (1989a,b, 1991) and Beard and Kubesh (1991). The solid curve is the axis ratio from the numerical model of Beard and Chuang (1987), and the dashed curve shows the postulated shift in axis ratio based on polarization radar and disdrometer measurement in rain (Goddard and Cherry, 1984). From Beard *et al.* (2010). Reproduced by permission of Elsevier B.V.

two-lobed oscillation of drops as a function of diameter. It is seen that, the larger the drop, the smaller the frequency of this mode of oscillation.

Field observed *AR* values by dual-polarizing radar and disdrometer (dashed curve in Fig. 8.15) are generally higher than the equilibrium curve, indicating that the drops in actual storms are generally more spherical. This discrepancy suggests that drops must be oscillating vigorously during their fall in the storm although the mode of oscillation cannot be determined from this figure. Tokay and Beard (1996) considered the three possible causes of drop oscillation: (1) drop–drop collision, (2) turbulence and shear in air, and (3) eddy shedding due to the fall motion of the drop. They concluded that the response of the drop to the eddy shedding is the most powerful cause for the oscillations. It is probably not a coincidence that the observed oscillation starts at a drop diameter of 1 mm, which is also close to the onset of eddy shedding. The changing *AR* due to the drop oscillation can have a great impact on the rainfall estimation using the radar differential reflectivity (Z_{DR}) technique. For example, a 30% decrease in Z_{DR} from 1.14 to 0.80 dB corresponds to a decrease of rainfall rate from 10 to 2 mm h^{-1}, a 5-fold change.

Drop canting

In the presence of environmental wind shear, the flattened bottom of a falling raindrop would not be parallel to the horizontal plane as in quiet air but would tilt to one side. This tilting behavior is called *canting*. Canting is thought to have a significant impact on microwave transmission. Field observations show that the canting angles are distributed about 0.48°, with a standard deviation of 1.77°, and over 82% of the measured angles are within ±2.25°.

Drop breakup

Large raindrops can break up due to either their own instability or their collision with other particles. Drop breakup will influence the drop size spectra and collision growth; hence there is some importance associated with this behavior.

Drops suspended in the quiet air in wind tunnel experiments can be as large as 4.5 mm in equivalent radius before breaking up. Drops in more turbulent air in rain storms possibly break up at even smaller size, but there is no direct observation.

Raindrops can break up either due to collision between the drop and other particles (another drop or ice particle) or due to the growth of hydrodynamic instability in an isolated drop. We will only describe some observations about the latter in this section. Collision breakup will be discussed in Chapter 10. There are two modes of the individual drop breakup, as follows.

- *Dumbbell breakup* – the drop develops a thin neck in the middle (hence the name dumbbell), possibly due to the oscillatory shear deformation, and subsequently breaks into two or more parts. This mode produces relatively large drop fragments.
- *Bag breakup* – the drop develops into the shape of a bag, with the opening of the bag facing downward (Fig. 8.16). This can occur due to the very strong hydrodynamic pressure acting on the bottom surface and causes the drop to be blown up like a bag. The "opening" of the bag is a ring of thicker water, whereas the "bag" part is a bubble-like structure formed by thin water. Upon breakup, the ring breaks into several larger fragments and the bag breaks into many small droplets.

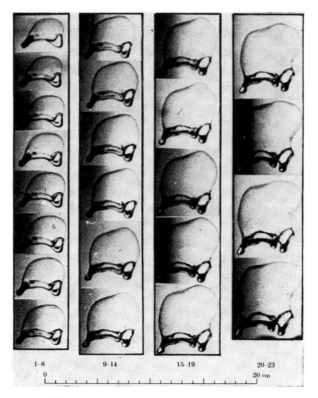

Fig. 8.16 A large falling water drop becomes unstable and forms an expanding bag supported on a toroidal ring of liquid. The bag eventually bursts, producing large numbers of small droplets, and the toroid breaks up into several large drops. The photographs are taken at intervals of 1 ms. From Mason (1971). Reproduced by permission of Oxford University Press and the Royal Meteorological Society (UK).

8.5 Hydrodynamic behavior of falling ice particles

8.5.1 Fall patterns of ice particles

Except for small frozen drops, most ice particles are not spherical and their fall behaviors are much more complicated than that of small spherical drops. Even small frozen drops behave quite differently than do spherical liquid drops. The difference is immediately clear in wind tunnel experiments. When a small liquid droplet is suspended in the wind tunnel, it usually appears to be very still. If we then cool the wind tunnel to a subfreezing temperature, the drop will become supercooled but without any change in the flow behavior. The supercooled drop can be turned into a frozen drop by introducing a stream of contact nuclei into the tunnel that makes the drop frozen. Upon freezing, the flow behavior of the drop changes immediately. The frozen drop would spin rapidly in the air and we can see the flashing sparkles if we shine light into the drop. This is because of the crystalline structure of ice in the frozen drop, which is not a true sphere but has a non-smooth crystalline surface.

The fall behavior of ice crystals and snowflakes has been studied in field observations. When ice crystals are small, they fall in a steady manner. Very small ice particles ($N_{Re} \leq 1$) probably fall with random orientations. In the intermediate range ($1 < N_{Re} \lesssim 100$), hexagonal ice plates fall steadily with their basal planes oriented horizontally. For planar ice crystals with narrow branches, the upper N_{Re} limit of steady motion is ~200. Ice columns in the intermediate range ($1 < N_{Re} \lesssim 40$) fall steadily with their long axes horizontal.

As crystals grow larger, their fall become unsteady and eddy shedding starts to occur in the downstream region. Eventually they may fall in fairly complicated manners. Most of them develop secondary motions that may occur simultaneously. For example, falling planar ice crystals can perform rotational, oscillatory, and zigzag translational motions that are familiar to those who have seen snowfall. Columnar ice crystals can develop fritting, oscillatory, and rotational motions. Snow aggregates likewise perform many complicated secondary motions.

In wind tunnel experiments, conical graupel tend to fall with apex pointing upward, and presumably this is how they fall in clouds. They may also develop secondary motions such as spiral fall, axial rotation, and pendulum swing.

Most hailstones are quasi-spheroidal and their fall behavior is influenced by the surface conditions. There are many experimental measurements of the drag force on hail, and the results show that it varies greatly, depending not only on the size but also on whether the hail surface is smooth or rough or soaked.

8.5.2 *Flow fields around falling ice particles*

Both experimental measurements and theoretical calculations of flow fields around falling ice crystals are difficult to perform due to the complex shapes of the crystals. There are only a few laboratory experiments and numerical computations available so far. We will briefly discuss numerical results of flow fields around falling ice columns, hexagonal plates, broad-branch crystals, and graupel, and compare them with experiments when available.

Flow fields around vertically falling columnar ice crystals

Fig. 8.17 shows the numerical flow fields for flow past columnar ice crystals for $N_{Re} = 40$ and 70 as expressed by the streaks of tracer particles. The tracer particle plot is a snapshot of the flow field at a certain instant. This is not to be confused with the trajectories of particles, but in the steady flow case, it is the same as streamlines and trajectories. In this figure the diameter/length ratio of the column with $N_{Re} = 70$ is smaller than the one with $N_{Re} = 40$. Experimentally, the flow field at $N_{Re} = 40$ is

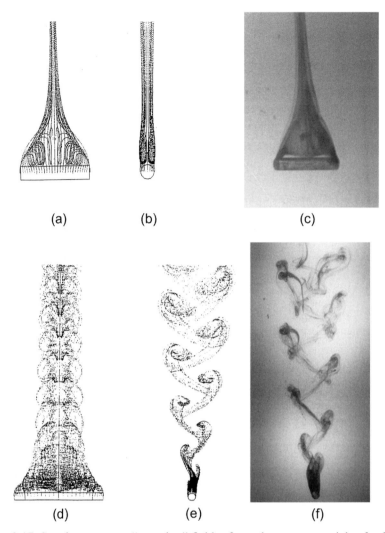

Fig. 8.17 Streak pattern, or "snapshot" field, of massless tracer particles for flow past an ice column. (a) Broad-side view, $N_{Re} = 40$. (b) End view, $N_{Re} = 40$. (c) Experimental photograph of a falling short cylinder at $N_{Re} = 40$. (d) Broad-side view, $N_{Re} = 70$. (e) End view, $N_{Re} = 70$. (f) Experimental photograph of a falling short cylinder at $N_{Re} = 70$. Photos courtesy of Dr K. O. L. F. Jayaweera. Adapted from Wang and Ji (1997).

steady and that at $N_{Re} = 70$ is unsteady. Numerical calculations reproduce these fields well.

Based on these numerical results, Wang and Ji (1997) determined the drag coefficients of flow past cylinders of finite length. The results are shown in Fig. 8.18. The diameter to length ratios are indicated in the figure. Here the drag coefficient is defined as

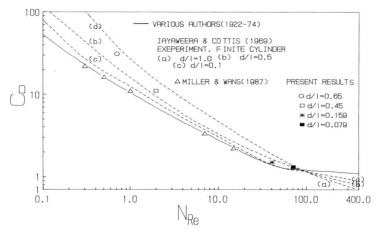

Fig. 8.18 Drag coefficients for flow past cylinders of various d/l ratios. The solid line and the triangles are for infinitely long cylinders. It is seen that as N_{Re} increases, the cylinder becomes longer and the drag coefficient becomes closer to that of an infinitely long cylinder. From Wang and Ji (1997). Reproduced by permission of the American Meteorological Society.

$$C_D = \frac{D}{\rho U_\infty^2 \alpha}, \tag{8.71}$$

where D is the drag and α is half of the cylinder's cross-sectional area normal to the flow. Clearly, the drag on a finite cylinder differs from that on an infinitely long cylinder. The difference is the greater the smaller the Reynolds number. This is because the aspect ratios of the cylinders with larger Reynolds numbers are closer to those of infinitely long cylinders and hence their drag coefficients are closer, too.

The drag coefficients calculated here can be fitted by the following expression:

$$\log_{10} C_D = 2.44389 - 4.21639A - 0.20098A^2 + 2.32216A^3, \tag{8.72}$$

where

$$A = \frac{\log_{10} N_{Re} + 1.0}{3.60206}. \tag{8.73}$$

This formula is valid within the range $0.2 < N_{Re} < 100$.

Flow fields around vertically falling hexagonal ice plates and broad-branch crystals

While there appears to be no systematic laboratory experimental studies for the flow fields around falling planar ice crystals, those for simple circular disks are available and can be used as approximations to hexagonal ice plates. Here we use the results of Willmarth *et al.* (1964) to illustrate the main features. Very small plates ($N_{Re} \leq 1$)

tend to fall with their initial launching orientation because the viscous effect dominates the flow. But in the intermediate range $(1 < N_{Re} < 100)$, disks fall with their base plane oriented horizontally. Any other orientations will generate a torque that will reorient the disk back to the horizontal position. Because of plate geometry, standing eddies occur in flow past planar ice crystals at Reynolds numbers smaller than that due to ice columns. Pitter *et al.* (1973) performed numerical calculations of flow past an ice plate using a thin oblate spheroid as an approximation and showed that standing eddies appear at $N_{Re} = 1.5$, in agreement with laboratory experiments. Eddy shedding occurs downstream of a falling disk at $N_{Re} \geq 100$.

Figs. 8.19 and 8.20 show examples of numerical flow fields past hexagonal ice plates for both steady ($N_{Re} = 20$) and unsteady ($N_{Re} = 140$) cases. The shape of an exact hexagon with uniform thickness was used for these calculations. We see symmetric standing eddies in the wake region for the steady case and asymmetric shedding eddies in the unsteady case. The drag coefficients computed from these flow fields are close to that for flow past thin oblate spheroids of comparable dimensions, indicating that the latter is a good approximation of hexagonal plates. The computed drag coefficients for flow past hexagonal plates can be fitted into the following empirical formula:

$$C_D = \left(\frac{64}{\pi N_{Re}}\right)\left(1 + 0.078 N_{Re}^{0.945}\right), \tag{8.74}$$

valid for the range of N_{Re} between 0.2 and 150.

Broad-branch crystals are also a common form of ice and snow crystals. No quantitative measurements about the flow fields around a falling broad-branch

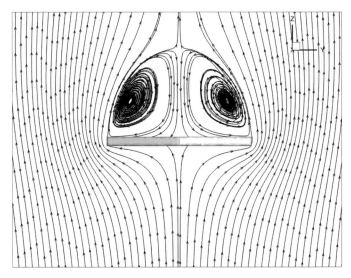

Fig. 8.19 Streamlines for steady flow past a hexagonal plate at $N_{Re} = 20$, based on the calculations of Hashino *et al.* (2010).

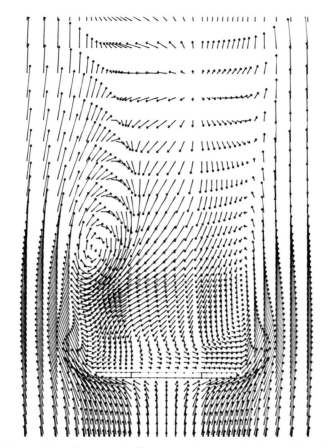

Fig. 8.20 Streamlines for unsteady flow past a hexagonal plate at $N_{Re} = 140$. From Wang and Ji (1997). Reproduced by permission of the American Meteorological Society.

crystal have been done thus far, but numerical solutions for steady flows are available. The computed drag coefficients for flow past hexagonal plates can be fitted into the following empirical formula (Wang and Ji, 1997):

$$C_D = \left(\frac{64}{\pi N_{Re}}\right)\left(1 + 0.142 N_{Re}^{0.887}\right), \qquad (8.75)$$

valid for the range of N_{Re} between 0.2 and 150. The drag coefficient of flow past a broad-branch ice crystal is somewhat higher than that past a hexagonal plate at the same N_{Re}, as illustrated in Fig. 8.21.

The above results are for vertically falling ice crystals with their largest dimension oriented horizontally. As mentioned previously, ice crystals can fall with their largest dimensions oriented obliquely. Recently, Hashino *et al.* (2010) performed a numerical

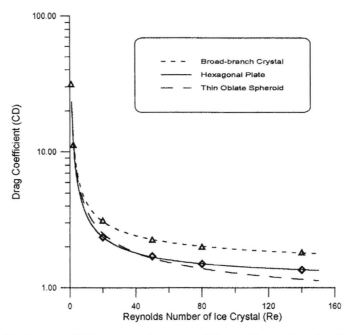

Fig. 8.21 Drag coefficients as a function of Reynolds number for flow past hexagonal plates and broad-branch crystals. From Wang and Ji (1997). Reproduced by permission of the American Meteorological Society.

study for steady flow past obliquely oriented falling ice columns and hexagonal plates. Fig. 8.22 shows an example for flow past an inclined hexagonal plate for $N_{Re} = 20$ and inclination angle $\theta = 30°$. They found that the drag force generally decreases with increasing inclination angle.

Flow fields around falling conical graupel

The flow fields around falling conical graupel has only been studied recently by Kubicek and Wang (2012), who performed numerical calculations for unsteady flow past conical graupel at $N_{Re} = 440$ for various inclination angles. The shape of the graupel is defined by Eq. (2.15). A comparison between the numerical solution and an experimental photograph of the flow field around a falling conical body of similar Reynolds number range is shown in Fig. 8.23. The general features of the numerical solution look similar to those of the experimental flow field.

Fig. 8.24(a) shows the flow field around this graupel when it falls with the apex pointing straight up in the vertical. The flow is unsteady, so that eddy shedding occurs. Many of the flow field features are similar to those of the flow past spheres or ice crystals, e.g. the upstream high-pressure and downstream low-pressure configuration and the high-vorticity region near the edge of the graupel base. The inclined case in

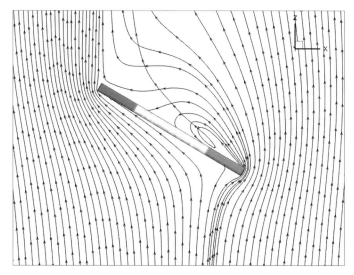

Fig. 8.22 Streamlines for steady flow past an inclined hexagonal plate with $\theta = 30°$ at $N_{Re} = 20$, based on the calculations of Hashino *et al.* (2010).

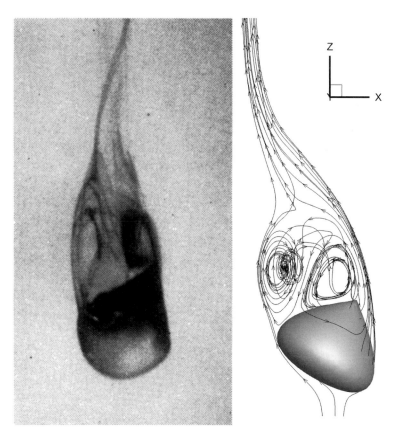

Fig. 8.23 (a) Experimental flow field around a falling cone with hemispherical base at $N_{Re} = 237$. From Jayaweera and Mason (1965). (b) A snapshot of the flow field around a falling inclined conical graupel with inclination angle $\theta = 30°$ at $N_{Re} = 440$, based on the calculations of Kubicek and Wang (2012).

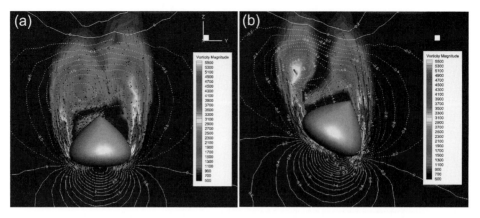

Fig. 8.24 (a) The velocity vector field (black arrows), the pressure field (white contours) and the vorticity field in the central cross-sectional plane around a falling graupel with inclination angle $\theta = 0°$ and $N_{Re} = 440$. (b) The same as (a) except for $\theta = 30°$ based on the calculations of Kubicek and Wang (2012). See also color plates section.

Fig. 8.24(b) shows greater asymmetry, as expected. One of the more special features of the graupel is that, due to its conical shape, the velocity remains relatively high in the wake region near the graupel surface, which is quite unlike the case for ice crystals (especially the plates). This is especially so for the inclined case. This may have important implications for the diffusion growth of the graupel. However, the details remain to be studied.

Thus far, no quantitative studies on the flow fields around falling hailstones have been performed.

8.6 Terminal velocities of falling cloud and precipitation particles

In the previous discussion, we saw that the flow field around falling cloud and precipitation particles is closely related to the Reynolds number N_{Re}, which depends strongly on the fall velocity. Obviously, there is a need for us to know the fall velocities of these particles.

8.6.1 Cloud and raindrops

Let us begin by examining the fall velocity of a small cloud drop. Even if a cloud drop starts out at zero fall velocity (with respect to air, of course), it will soon gain velocity due to the gravitational acceleration and falls faster and faster. At the same time, however, the hydrodynamic drag force acting on the drop will increase because the drag increases with velocity (see Eq. (8.27)). Eventually, the drop will reach a point where the gravitational force is exactly balanced by the drag, and the net force acting on this drop vanishes. This means that there will be no more acceleration and thereafter the drop will fall at a constant velocity – the *terminal velocity*. Because it is the final velocity due to the continual acceleration, the terminal velocity is also the maximum fall velocity of the drop.

For a falling drop, we can determine this terminal velocity easily by examining the equation of motion for the drop:

$$m\frac{d\vec{u}}{dt} = m\vec{g}\left(\frac{\rho_w - \rho_a}{\rho_w}\right) + \vec{F}_D, \tag{8.76}$$

where the first term on the right-hand side is the buoyancy-corrected gravitational force and $\vec{F}_D$ is the drag force. When the drop finally reaches terminal velocity, there will be no acceleration and the left-hand side of (8.76) vanishes, and we then have

$$m\vec{g}\left(\frac{\rho_w - \rho_a}{\rho_w}\right) = -\vec{F}_D. \tag{8.77}$$

For a small cloud drop, we can use the Stokes drag. Hence we have (remember that the drag force acts in the opposite direction to the fall)

$$mg\left(\frac{\rho_w - \rho_a}{\rho_w}\right) = 6\pi\eta a U_\infty \tag{8.78}$$

and

$$U_\infty = \frac{mg(\rho_w - \rho_a)/\rho_w}{6\pi\eta a} = \frac{2a^2 g(\rho_w - \rho_a)}{9\eta} \approx \frac{2a^2 g\rho_w}{9\eta}. \tag{8.79}$$

For very small cloud drops ($a \leq 1$ μm), we need to consider a *Stokes–Cunningham slip correction factor* to the drag because the drop size is close to the mean free path λ_a of air (~ 0.1 μm at sea level, more in the upper atmosphere). This factor is

$$f_{SC} = \left(1 + 1.26\frac{\lambda_a}{a}\right) \tag{8.80}$$

and hence the terminal velocity becomes

$$U_\infty = \left(1 + 1.26\frac{\lambda_a}{a}\right)\frac{2a^2 g(\rho_w - \rho_a)}{9\eta}. \tag{8.81}$$

The correction can be substantial: for example, a 1 μm drop at sea level will have 17% higher terminal velocity by Eq. (8.81) than by (8.79). For drops greater than 10 μm, the correction will be less than 1%.

For larger drops, the determination of terminal velocities becomes more complicated because of the lack of a simple drag expression. Beard (1976) developed the following convenient empirical formulas for determining the terminal velocities of drops of various sizes falling at different pressure levels.

- For $0.5 \leq a \leq 10$ μm, simply use Eq. (8.81).
- For $10 \leq a \leq 535$ μm, several steps are required. First compute

$$C_{\mathrm{D}} N_{\mathrm{Re}}{}^2 = \frac{32a^3 g(\rho_{\mathrm{w}} - \rho_{\mathrm{a}})\rho_{\mathrm{a}} g}{3\eta^2}, \tag{8.82}$$

then set

$$X = \ln(C_{\mathrm{D}} N_{\mathrm{Re}}{}^2) \tag{8.83}$$

and compute

$$Y = B_0 + B_1 X + B_2 X^2 + B_3 X^3 + B_4 X^4 + B_5 X^5 + B_6 X^6, \tag{8.84}$$

where

$$
\begin{aligned}
B_0 &= -0.318657 \times 10^1, & B_1 &= 0.992696, \\
B_2 &= -0.153193 \times 10^{-2}, & B_3 &= -0.987095 \times 10^{-3}, \\
B_4 &= -0.578878 \times 10^{-3}, & B_5 &= 0.855176 \times 10^{-4}, \\
B_6 &= -0.327815 \times 10^{-5}.
\end{aligned}
\tag{8.85}
$$

Then determine the Reynolds number by

$$N_{\mathrm{Re}} = \exp(Y), \tag{8.86}$$

and finally the terminal velocity by

$$U_\infty = \frac{\eta N_{\mathrm{Re}}}{2\rho_{\mathrm{a}} a}. \tag{8.87}$$

- For $535 \leq a \leq 3500$ μm (3.5 mm), several steps are again required. First determine a physical property number N_{P} and a Bond number N_{Bo} by

$$N_{\mathrm{P}} = \frac{\sigma^3 \rho_{\mathrm{a}}^2}{\eta^4 (\rho_{\mathrm{w}} - \rho_{\mathrm{a}})g}, \tag{8.88}$$

$$N_{\mathrm{Bo}} = \frac{4a^2 (\rho_{\mathrm{w}} - \rho_{\mathrm{a}})g}{\sigma}. \tag{8.89}$$

Then set

$$X = \ln\left(\frac{4}{3} N_{\mathrm{Bo}} N_{\mathrm{P}}{}^{16}\right) \tag{8.90}$$

and compute

$$Y = B_0 + B_1 X + B_2 X^2 + B_3 X^3 + B_4 X^4 + B_5 X^5, \tag{8.91}$$

where

$$
\begin{aligned}
B_0 &= -0.500015 \times 10^1, & B_1 &= 0.523778 \times 10^1, \\
B_2 &= -0.204914 \times 10^{-2}, & B_3 &= 0.475294, \\
B_4 &= -0.542819 \times 10^{-1}, & B_5 &= 0.238449 \times 10^{-2}.
\end{aligned}
\tag{8.92}
$$

Then determine the Reynolds number by

$$N_{Re} = N_P^{1/6} \exp(Y),$$ (8.93)

and finally the terminal velocity by

$$U_\infty = \frac{\eta N_{Re}}{2\rho_a a}.$$ (8.94)

Fig. 8.25 shows the variation of drop terminal velocity as a function of size and pressure level.

We see that, at sea level, small cloud drops have fairly small terminal velocities (a few cm s^{-1}) whereas raindrops can fall at ~ 10 m s^{-1}. As expected, at a fixed pressure level, the terminal velocity generally increases with drop size. However, for drops with diameter greater than 4 mm ($a \sim 2$ mm), the terminal velocities remain fairly constant. The reason is that, as drops become larger, they become more flattened and the drag force (acting like a brake) increases. This brake effect compensates for their larger mass and consequently results in near-constant terminal velocities. On the other hand, the terminal velocity of a drop increases with height because the lower density of air results in lower drag.

If the drop is oscillating, its terminal velocity may oscillate also in response to the shape change (and hence change in the drag force). This has been observed experimentally for millimeter size drops.

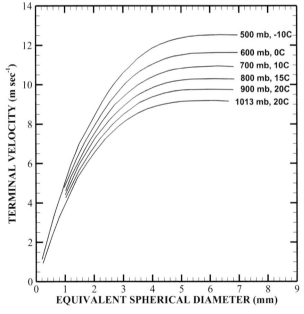

Fig. 8.25 Terminal velocity of water drops vs. equivalent spherical diameter in various pressure and temperature environments. Adapted from Beard (1976).

Distance required for water drops to reach terminal velocity

In most current applications, especially in cloud modeling, cloud and raindrops are always assumed to fall at their terminal velocities. But this cannot be strictly true since it will take a finite amount of time for a drop to reach its terminal velocity. This time is different for different drop sizes, and in general (but not always) a small drop reaches the terminal velocity sooner than a larger drop in the same environment if they both start out from rest. Fig. 8.26 shows experimental results of the distances and times required for raindrops of various sizes to reach their terminal velocities at the surface level.

At sea level, typical cloud drops ($a \sim 10$ μm) only require a few millimeters to reach terminal velocity, and therefore it is justified to assume that they are falling at terminal velocity. But a typical raindrop ($a > 1$ mm) requires 10 m or more to reach terminal velocity. In the upper troposphere, say, 500 mb, this distance would be more than 20 m (see Fig. 8.27). Interestingly, very large drops ($a > 2$ mm) require somewhat smaller distance due to the greater braking effect of the drag.

The required large distances to reach terminal velocities for large drops imply that, in cloud regions where a spectrum of large raindrops is present, it is likely that not all drops are falling at their terminal velocities. For very high-resolution cloud models with grid size down to a few tens of meters, it may be necessary to consider the situation where not all drops are falling at their terminal velocities.

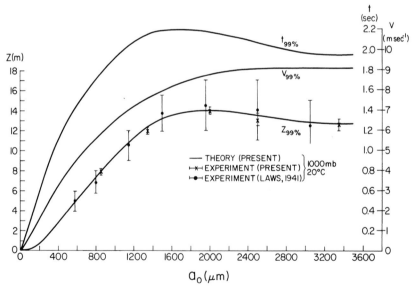

Fig. 8.26 The distance ($Z_{99\%}$) and time ($t_{99\%}$) for a water drop to reach 99% terminal velocity ($V_{99\%}$) vs. drop radius for 1000 mb, 20°C, as experimentally determined by Wang and Pruppacher (1977). Reproduced by permission of the American Meteorological Society.

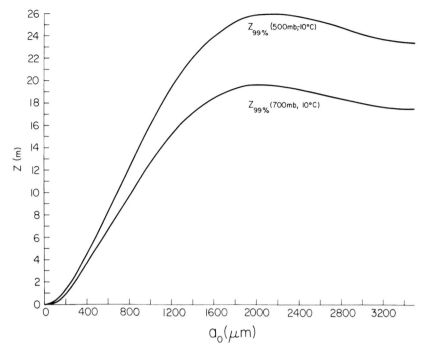

Fig. 8.27 The distance to reach 99% of terminal velocity for a water drop of equivalent radius a_0 under 700 mb, 10°C and 500 mb, −10°C conditions. From Wang and Pruppacher (1977). Reproduced by permission of the American Meteorological Society.

8.6.2 Ice particles

In principle, we can use an expression similar to (8.77) to estimate the terminal velocities of ice particles. Given the complicated shapes and variable densities of ice particles and the lack of adequate information about their drag coefficients, however, it is understandable that there is no unified approach in determining the terminal velocities of ice particles theoretically. Fortunately, there are observational values that can be used. Figs. 8.28 and 8.29 show experimentally measured snow crystal fall velocities. They show, as expected, that the terminal velocity of an ice crystal of a single habit increases with size, and that more compact (higher bulk density) crystals such as thick plates have higher terminal velocities than more open crystals (such as dendrites and stellar crystals) of the same size. Ice crystals of a few millimeters in diameter typically have U_∞ of a few tens of centimeters per second.

Figs. 8.30 and 8.31 show the best fits for the observed fall velocity vs. maximum dimension of various ice aggregates and graupel. Needless to say, it is necessary to remember that there may be substantial uncertainties associated with the velocities, given the wide varieties of ice particles.

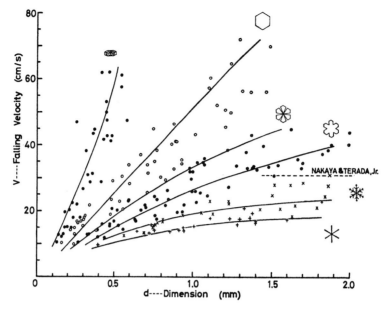

Fig. 8.28 Fall velocity of plane type crystals. From Kajikawa (1972). Reproduced by permission of the Meteorological Society of Japan.

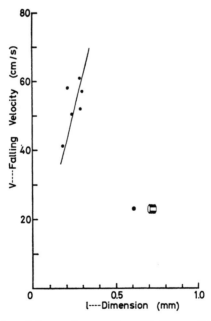

Fig. 8.29 Fall velocity of short columns. From Kajikawa (1972). Reproduced by permission of the Meteorological Society of Japan.

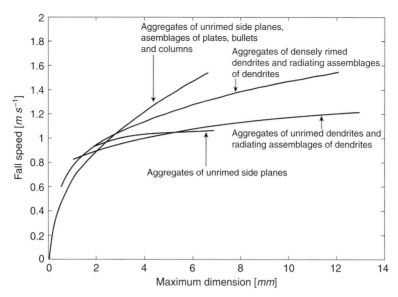

Fig. 8.30 Best-fit curves for fall speed vs. maximum dimension for aggregates of various types. Adapted from Locatelli and Hobbs (1974), with changes.

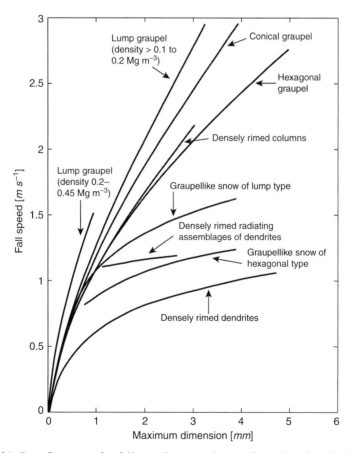

Fig. 8.31 Best-fit curves for fall speed vs. maximum dimension for single solid precipitation particles of various types. Adapted from Locatelli and Hobbs (1974), with changes.

Problems

8.1. Estimate the N_{Re} for the following cases:

(a) a cloud drop of 30 μm radius falling at its terminal velocity at $p = 700$ mb and $T = -10°C$;

(b) a raindrop of 1 mm radius falling at its terminal velocity at $p = 900$ mb and $T = 10°C$;

(c) a hailstone of 5 cm diameter falling at its terminal velocity at $p = 1000$ hPa and $T = 20°C$.

(d) Use the formulas in problem 6.1 to determine the dynamic viscosity of air.

8.2. In treating theoretical problems in fluid dynamics, it is often convenient to work with non-dimensional systems because the exact sizes of the objects involved are often irrelevant. Rather, it is the non-dimensional characteristic numbers (such as the Reynolds number) that are of importance. This problem is an exercise to non-dimensionalize an equation. It uses a technique called scaling. The steady-state Navier–Stokes equation for flow past a rigid sphere can be non-dimensionalized by introducing the following dimensionless quantities:

$$\vec{r}' = \frac{\vec{r}}{a}, \ \vec{u}' = \frac{\vec{u}}{u_\infty}, \ \nabla' = a\nabla, \ p' = \frac{p}{\rho u_\infty^2},$$

where a is the radius of the sphere, ρ is the air density, and u_∞ is the magnitude of the free-stream velocity. The quantities in the denominator in the above equations represent the characteristic scales of those variables.

(a) Show that the Navier–Stokes equation becomes

$$(\vec{u}' \bullet \nabla')\vec{u}' = -\nabla'p' + \frac{2}{N_{Re}}\nabla'^2\vec{u}'.$$

(b) The boundary conditions for this type of flow are:

$$\vec{u} = 0 \quad \text{at } r = a,$$
$$\vec{u} = \vec{u}_\infty \quad \text{at } r \to \infty.$$

How would you express these two conditions in terms of $\vec{u}'$?

(c) How would you non-dimensionalize the unsteady term $\partial\vec{u}/\partial t$?

8.3. The fluid mechanical properties of a falling cloud drop can be approximated by those in the flow past a rigid sphere. Consider the flow past a rigid sphere given in the potential, Stokes, and Oseen approximations.

(a) Determine the velocity components and the vorticity for three points, $(2a, \pi)$, $(2a, 0)$, and $(2a, \pi/2)$.

(b) Discuss the results you have obtained regarding their applicability for the case of a cloud droplet of 10 μm radius.

(c) Determine the maximum and minimum pressures on the surface of the sphere for these three flows and compare their magnitudes. Can you explain why they behave like that?

(d) The drag force acting on a sphere can be determined by the following equation:

$$F_D = 2\pi a^2 \int_0^\pi \left[p \cos\theta + \eta \left(\frac{\partial u_\theta}{\partial r} \right) \sin\theta \right]_{r=a} \sin\theta \, d\theta$$
$$= F_{Dp} + F_{Df},$$

where F_{Dp} is called the pressure (or form) drag and F_{Df} the friction (or skin) drag. Determine the form and the relative magnitude of these two drag components for each of the three flows.

8.4. Show that the terminal velocity of a small cloud drop of radius a (assuming the drag is Stokesian) is given by

$$u_\infty \approx \frac{2}{9} \eta_a a^2 \rho_w g.$$

9

Diffusion growth and evaporation of cloud and precipitation particles

The nucleation we discussed in Chapter 7 is about the initiation of condensed phases of cloud and precipitation particles. Our next question is this: Once these particles are initiated, how do they grow and how fast? This is what we will address in the next two chapters.

There are two broad categories of particle growth modes: (1) diffusion growth and (2) collision growth. The former refers to the mechanism in which water vapor diffuses toward the surface of a particle, resulting in the increase of the particle's mass. The latter refers to the collision between two or more particles, which subsequently coalesce together and hence become a larger particle. The latter can mean collisions among water drops, among water drops and ice particles, or among ice particles.

The reverse of growth is reduction. Both water drops and ice particles can decrease their size by evaporation and fragmentation. Ice particles can also melt to become liquid and thus reduce their size.

9.1 Diffusion of water vapor around a spherical water drop

When a water drop is suspended in supersaturated air (with respect to the drop), then there will be net deposition of water molecules on the drop surface and the drop grows in size. This is the diffusion growth of water drops. Note that the saturation humidity here should take into account both the curvature and solute effects as discussed in Chapter 5.

The diffusion growth of a stationary drop is different from that for a falling drop due to the ventilation effect (to be discussed later). In reality, all water drops in clouds are falling with respect to air, and hence accurate diffusion growth rates of drops should take that motion into account. However, a typical cloud droplet radius is about 10 μm and the ventilation effect is small. In the following, we will derive a

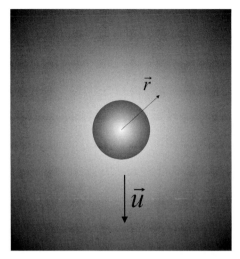

Fig. 9.1 Configuration of the theoretical problem for the diffusion of vapor around a falling spherical drop.

general expression for the diffusion of water vapor around a falling drop first. Then we will simplify it for the case of a stationary drop.

We consider a water drop of radius a falling at a terminal velocity U_∞ in moist air with water vapor density ρ_v (see Fig. 9.1). Like in Chapter 8, we will assume that it is the air that is flowing past a stationary drop with velocity $\vec{u}$. In general, ρ_v is a function of both space and time, i.e. $\rho_v = \rho_v(\vec{r}, t)$. The flux density of water vapor j_v toward the drop surface consists of two components: the diffusion flux density $-D_v\nabla\rho_v$, where D_v is the diffusivity of water vapor in air; and the convective flux density $\rho_v\vec{u}$ due to the motion of the drop. Thus the total vapor flux density is

$$\vec{j}_v = -D_v\nabla\rho_v + \rho_v\vec{u}. \tag{9.1}$$

The water vapor density ρ_v should also satisfy the continuity equation (a statement of the conservation of water vapor mass):

$$\frac{\partial\rho_v}{\partial t} = -\nabla\bullet\vec{j}_v. \tag{9.2}$$

Substituting (9.1) into (9.2), we obtain

$$\frac{\partial\rho_v}{\partial t} = D_v\nabla^2\rho_v - \nabla\bullet(\rho_v\vec{u}) = D_v\nabla^2\rho_v - \vec{u}\bullet\nabla\rho_v, \tag{9.3}$$

where D_v is considered a constant in this circumstance and the incompressibility assumption ensures that $\nabla\bullet\vec{u} = 0$. Eq. (9.3) is the convective diffusion equation that describes the distribution of ρ_v in the vicinity of a falling drop.

Hall and Pruppacher (1976) gave a simple empirical relation for D_v (in cm^2 s^{-1}) in the temperature range -40 to $40°$C:

$$D_v = 0.211 \left(\frac{T}{T_0}\right)^{1.94} \left(\frac{p_0}{p}\right), \tag{9.4}$$

where $T_0 = 273.15$ K and $p_0 = 1013.25$ mb. Thus D_v increases with increasing temperature and decreasing pressure.

9.1.1 Stationary water drop

Now we turn to the special case where the drop is assumed to be stationary with respect to air, i.e. $\vec{u} = 0$. Eq. (9.3) then becomes

$$\frac{\partial \rho_v}{\partial t} = D_v \nabla^2 \rho_v. \tag{9.5}$$

We will further assume that ρ_v is in steady state, i.e. $\partial \rho_v / \partial t = 0$. This means that we consider that the adjustment of the ρ_v distribution in response to environmental change is immediate. This approximation is valid if the time of significant changes in vapor density in the environment is much longer than the diffusion relaxation time $t_c = a^2/\pi D_v$, which usually is the case. Thus under the steady-state assumption, (9.5) becomes the Laplace equation

$$\nabla^2 \rho_v = 0 \tag{9.6}$$

after we divide D_v out.

The symmetry of the present case naturally demands the use of the spherical coordinate system. Inspection of Fig. 9.1 shows that ρ_v can only vary in the radial direction and hence we only need to consider the radial term of the Laplace equation (9.6):

$$\frac{1}{r^2} \frac{\partial}{\partial r} \left(r^2 \frac{\partial \rho_v}{\partial r}\right) = 0. \tag{9.7}$$

We also need to specify the boundary conditions, which are

$$\begin{aligned} \rho_v &= \rho_{v,a} \quad \text{at } r = a, \\ \rho_v &= \rho_{v,\infty} \quad \text{at } r \to \infty. \end{aligned} \tag{9.8}$$

Eq. (9.8) says that the vapor density at the drop surface is a constant $\rho_{v,a}$ and is another constant $\rho_{v,\infty}$ far away from the drop. It is commonly assumed that the vapor density at the drop surface is exactly at saturation, i.e. $\rho_{v,a} = \rho_{v,sat}$. To be exact, the so-called saturation vapor pressure should be that with respect to a drop of radius a, but it is often approximated by the conventional one, i.e. the saturation vapor pressure with respect to a plain water surface. The error is very small in this context.

The solution of (9.7) that also satisfies the conditions (9.8) is

$$\rho_v = \rho_{v,\infty} + (\rho_{v,a} - \rho_{v,\infty})\frac{a}{r} \tag{9.9}$$

and the vapor density gradient is

$$\nabla\rho_v = \frac{\partial\rho_v}{\partial r}\hat{e}_r = -(\rho_{v,a} - \rho_{v,\infty})\frac{a}{r^2}\hat{e}_r, \tag{9.10}$$

where $\hat{e}_r$ is the unit vector in the positive (outward) radial direction.

The growth rate of the drop is the total mass of water vapor added to the drop per unit time, which is obtained by integrating the vapor flux density over the drop surface:

$$\frac{dm}{dt} = \oint_S (-D_v\nabla\rho_v)_{r=a}\cdot d\vec{S}, \tag{9.11}$$

where $d\vec{S}$ represents the increment of drop surface area whose positive direction is also $\hat{e}_r$. From (9.10) we have

$$(\nabla\rho_v)_{r=a} = -\frac{(\rho_{v,a} - \rho_{v,\infty})}{a}\hat{e}_r, \tag{9.12}$$

which is a constant and can be moved out of the integral. Hence

$$\begin{aligned}\frac{dm}{dt} &= D_v\frac{(\rho_{v,a} - \rho_{v,\infty})}{a}\hat{e}_r\oint_S d\vec{S} \\ &= D_v\frac{(\rho_{v,a} - \rho_{v,\infty})}{a}\cdot 4\pi a^2 \\ &= 4\pi D_v a(\rho_{v,a} - \rho_{v,\infty}).\end{aligned} \tag{9.13}$$

The sign in the result (9.13) needs some explanation. In order to have "growth", we need to have a supersaturated environment, i.e. $\rho_{v,\infty} > \rho_{v,a}$. Under this condition, dm/dt in (9.13) is negative. Can this be right?

The answer is yes. Eq. (9.13) represents the total "outward" (in the positive $\hat{e}_r$ direction) flux of water vapor. A negative dm/dt simply means that the flux is in the $-\hat{e}_r$, i.e. inward, direction and hence represents the growth rate. If, instead, we have $\rho_{v,\infty} > \rho_{v,a}$, then dm/dt will be positive and we have evaporation instead of growth.

However, since we are concerned mostly with "growth" here, our life will be easier if we define the "growth rate" as a positive quantity. Hence, from now on, we will reverse the term order in the parentheses and define the growth rate as:

$$\frac{dm}{dt} = 4\pi D_v a (\rho_{v,\infty} - \rho_{v,a}). \tag{9.14}$$

Throughout the derivation, we have assumed that the vapor diffusivity D_v is the same as that measured in free air, in which water vapor behaves as a continuum. But when it is very close to the drop surface, where the distance is smaller than the *vapor jumping length* Δ_v (which is thought to be about the same as the mean free path of air), water vapor behaves more like an ensemble of individual molecules, and some corrections are necessary. A new vapor diffusivity defined as

$$D_v^* = \frac{D_v}{\left[\frac{a}{a+\Delta_v} + \frac{D_v}{a\alpha_c} \left(\frac{2\pi M_w}{RT_a} \right)^{1/2} \right]} \tag{9.15}$$

takes this correction into account, so that (9.14) can be modified as

$$\frac{dm}{dt} = 4\pi D_v^* a (\rho_{v,\infty} - \rho_{v,a}). \tag{9.16}$$

The quantity α_c in (9.15) is the *condensation coefficient*. It can be understood as the fraction of vapor molecules striking the liquid surface that will stick to the liquid state while the rest will bounce back into the vapor state. The experimental value of α_c for vapor condensing into liquid is between 0.01 and 0.07, with an average of about 0.035. There are substantial uncertainties about D_v^*, however, because of uncertainties in the values of D_v, α_c, and Δ. This effect is more important for drops of radius less than 1 μm, and becomes insignificant for drops with radius greater than 10 μm. Everything else being the same, D_v^* or D_v will be greater in the higher troposphere (see (9.4)), and hence according to (9.16) the mass growth rate of a drop with radius a will be greater higher up in a cloud than when near the cloud base.

Eq. (9.16) further indicates that the mass growth rate of a stationary spherical water drop increases with increasing a, that is, the larger the drop, the greater its mass growth rate. On the other hand, the linear growth rate is given by

$$\frac{da}{dt} = \frac{D_v^* (\rho_{v,\infty} - \rho_{v,a})}{a \quad \rho_w}, \tag{9.17}$$

which indicates that, the smaller the drop, the greater its linear growth rate. There are circumstances in which we need to make it clear which growth rate is being discussed to avoid ambiguity.

However, we cannot simply use equations (9.14), (9.16) or (9.17) to determine the diffusion growth rate of water drops. Growth (or evaporation) implies a phase change of water substance and hence involves the release (or consumption) of latent heat. The latent heat will influence the temperature field, which in turn influences the vapor density field. A growing drop will have a warmer surface than its environment because of the latent heat release, and it is this surface temperature that determines the saturation vapor density $\rho_{v,a}$ at the surface. Thus we need to take this temperature effect into account to correctly predict the diffusion growth rate.

9.1.2 Effect of latent heat

In order to take into account the latent heat effect, we will have to understand how heat is transported in the present situation of a stationary drop. It turns out that this is completely analogous to the vapor diffusion, that is, the temperature field also satisfies the diffusion equation

$$\frac{\partial T}{\partial t} = K_a \nabla^2 T, \tag{9.18}$$

where K_a is the thermal conductivity coefficient of the moist air. This equation is sometimes called the *heat conduction equation*. We again assume steady-state conditions, so that (9.18) becomes

$$\nabla^2 T = 0, \tag{9.19}$$

which is again the Laplace equation. The boundary conditions are

$$\begin{aligned} T &= T_a && \text{at } r = a, \\ T &= T_\infty && \text{at } r \to \infty, \end{aligned} \tag{9.20}$$

where T_a and T_∞ are the temperature at the drop surface and at a distance far from the drop, respectively. We see that (9.19) and (9.20) are exactly the same equation and boundary conditions as (9.6) and (9.8) (only the symbols are different). Hence we should obtain the same solution,

$$T = T_\infty + (T_a - T_\infty)\frac{a}{r}, \tag{9.21}$$

and the heat transported is also analogous to (9.14), i.e.

$$\frac{dq}{dt} = -\oint_S K_a \nabla T \cdot d\vec{S} = 4\pi K_a a (T_\infty - T_a), \tag{9.22}$$

where q is the total heat transported. Like D_v, the thermal conductivity coefficient K_a needs some corrections when the drop is very small, i.e.

$$K_a^* = \frac{K_a}{\left[\dfrac{a}{a+\Delta_T} + \dfrac{K_a}{a\alpha_T\rho_a c_{p,a}}\left(\dfrac{2\pi M_a}{RT_a}\right)^{1/2}\right]}, \tag{9.23}$$

where Δ_T is the heat jumping length, which is similar to the vapor jumping length mentioned previously, and α_T is the thermal accommodation coefficient. For a water surface, $\alpha_T \approx 0.96$. Thus

$$\frac{dq}{dt} = 4\pi K_a^* a (T_\infty - T_a).$$ (9.24)

9.2 Diffusion growth of a stationary aqueous solution drop

To determine the diffusion growth rate, we will need to consider the vapor and heat diffusion *simultaneously* because mass and heat transfers are coupled in this case. In addition, we will also consider the effect of solute in the drop, which will only cause a slight modification of the equations.

Eq. (9.17) can be rewritten as

$$a\frac{da}{dt} = \frac{D_v^* M_w}{R\rho_s}\left[\frac{e_\infty}{T_\infty} - \frac{e_a(T_a)}{T_a}\right],$$ (9.25)

where we have utilized the ideal gas law for water vapor and ρ_s is the bulk density of the aqueous solution drop. Note that we did not consider the temperature effect when deriving (9.17). Now we include the temperature effect, and it is important to note that the vapor pressure at the drop surface e_a, which is also the saturation vapor pressure, is to be evaluated at the surface temperature T_a.

In the drop diffusion growth problems, T_∞ is usually given but T_a needs to be determined. To proceed, we note that the latent heat release rate is given by

$$\frac{dq}{dt} = -L_e\left(\frac{dm}{dt}\right).$$ (9.26)

Using (9.24), we easily arrive at

$$T_a = T_\infty + \frac{L_e\rho_s}{K_a^*}\left(a\frac{da}{dt}\right).$$ (9.27)

Substituting T_a in (9.27) into (9.25) and utilizing the Clausius–Clapeyron equation to connect $e_{sat,w}(T_a)$ and $e_{sat,w}(T_\infty)$ (different temperature results in different saturation vapor pressure), we obtain

$$e_{sat,w}(T_a) = e_{sat,w}(T_\infty)\exp\left[\frac{L_e M_w}{R}\left(\frac{T_a - T_\infty}{T_a T_\infty}\right)\right].$$ (9.28)

Substituting (9.28) back into (9.25) and utilizing the Köhler equation (5.60) to take the solute effect into account, we obtain a complex-looking equation

$$a\frac{da}{dt} = \frac{D_v^* M_w e_{sat,w}(T_\infty)}{\rho_s R T_\infty}\left\{ S_v - \frac{1}{(1+\delta)}\exp\left[\frac{L_e M_w}{R T_\infty}\left(\frac{\delta}{1+\delta}\right)\right]\right.$$

$$\left. + \frac{2M_w\sigma}{R T_\infty(1+\delta)\rho_w a} - \frac{v\Phi_s m_s(M_w/M_s)}{\frac{4}{3}\pi a^3\rho_s - m_s}\right\},$$

(9.29)

where

$$\delta = \frac{L_e\rho_s}{T_\infty K_a^*}\left(a\frac{da}{dt}\right).$$

(9.30)

Eq. (9.29) is an implicit equation because both sides contain da/dt, which is the unknown to be determined, and no simple analytical solution has been presented so far. The usual way to solve this type of equation is to use a numerical iteration technique. However, it is possible to obtain approximate solutions, and one of them is given by (see Pruppacher and Klett, 1997):

$$a\frac{da}{dt} \approx \frac{S_v - (1+y)}{\dfrac{\rho_w R T_\infty}{D_v^* M_w e_{sat,w}(T_\infty)} + \dfrac{L_e\rho_w}{K_a^* T_\infty}\left(\dfrac{L_e M_w}{R T_\infty} - 1\right)},$$

(9.31)

where

$$y = \frac{2M_w\sigma}{R T_\infty\rho_w a} - \frac{v\Phi_s m_s(M_w/M_s)}{\frac{4}{3}\pi a^3\rho_s - m_s}.$$

(9.32)

In deriving (9.31), the consideration $\delta \ll 1$ and the approximation $\rho_s \approx \rho_w$ have been used.

Fig. 9.2 shows the growth rates of some drops using (9.31). It is clear from Fig. 9.2 that the linear growth rate of a drop decreases with size. A cloud drop with a radius of a few micrometers initially would grow fairly rapidly by diffusion of vapor. Its growth, however, becomes slower as it becomes larger. As the drop reaches about 40 μm in radius, the growth becomes very slow, so that its size is approaching an asymptote. Thus it is nearly impossible for a drop to reach precipitation size in a reasonable time by diffusion growth alone. We now understand that another growth process, collision and coalescence, will become active even before the drops reach 40 μm. Discussions of collision and coalescence will be given in Chapter 10.

Another interesting implication of Fig. 9.2 is that the diffusion growth mechanism tends to produce a narrow spectrum of drop sizes. We see that smaller drops grow faster in linear dimension than larger drops. Thus even if we start out with an

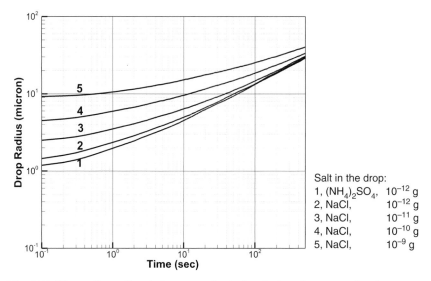

Fig. 9.2 The radius of a stationary spherical solution drop growing by vapor diffusion as a function of time. The slope represents the linear growth rate da/dt.

initially broad spectrum of CCN, the resulting drop spectrum will become narrower as time goes on and approach a monodisperse drop size distribution.

9.3 Ventilation effect

The discussions in the previous section only pertain to the diffusion growth of a stationary drop. In reality, all cloud drops fall relative to air, and hence it is necessary to consider the effect of the falling motion to obtain more accurate estimates of the diffusion drop growth rates.

Once the motion is included, the vapor density distribution will not be spherically symmetric around the drop but will become asymmetric in response to the flow field generated by the fall. The inclusion of motion always results in an *enhancement* of vapor diffusion. If the drop is growing, the motion will enhance the growth rate. Conversely, if a drop is evaporating, it will evaporate faster than when it is stationary. We call this enhancement phenomenon the *ventilation effect*.

The *mean ventilation coefficient for vapor diffusion* is defined as

$$\bar{f}_v = \frac{(dm/dt)}{(dm/dt)_0},\qquad(9.33)$$

where (dm/dt) represents the diffusion growth rate of a falling drop and $(dm/dt)_0$ represents that for a stationary drop (as given by (9.16)). Thus $\bar{f}_v$ is the enhancement factor on the total vapor flux toward the drop due to the effect of the drop motion. As noted above, $\bar{f}_v \geq 1$.

Because $\bar{f}_v$ is caused by the drop motion, its value is purely determined by the motion and is the same for both diffusion growth and evaporation if the pressure, temperature, and Reynolds numbers are the same. Thus, for example, for a drop of radius 300 μm falling at a Reynolds number 100, if it is growing, its diffusion growth rate will be enhanced by $\bar{f}_v$ times. If it is evaporating, its evaporation rate will be enhanced by $\bar{f}_v$ times also. This equivalence is fairly important for experimental measurements of $\bar{f}_v$ because it is very difficult to measure the diffusion growth rates of a falling drop. The ideal equipment to perform such an experiment is again the vertical wind tunnel. In a diffusion growth experiment, we need to produce a supersaturated environment in the tunnel. But as soon as water vapor reaches saturation, it starts to condense on the wall of the tunnel instead of going to the drop, and the experiment cannot proceed. On the other hand, the measurement of drop evaporation rates suffers no such problem.

To take into account the motion effect in diffusion growth, we have to go back to Eq. (9.3). Assuming again steady state, we obtain the time-independent convective diffusion equation for the vapor density:

$$D_v \nabla^2 \rho_v - \vec{u} \cdot \nabla \rho_v = 0. \tag{9.34}$$

Here $\vec{u}$ is the air velocity vector, which is not a constant. To solve (9.34), we will need to solve the Navier–Stokes equation for $\vec{u}$ first, then put the solution into (9.34) to solve for ρ_v and finally the diffusion growth rate, and then use (9.33) to determine $\bar{f}_v$.

An alternative way is to determine the ventilation coefficient experimentally and fit the experimental data to empirical formulas. Beard and Pruppacher (1971) performed an experimental study of the evaporation rates of small water drops falling at terminal velocity using a vertical wind tunnel. From that experiment, they determined the ventilation coefficients for these drops. They fitted their data by the following empirical formulas:

$$\bar{f}_v = 1.00 + 0.108(N_{Sc}^{1/3} N_{Re}^{1/2})^2 \quad \text{for } N_{Sc}^{1/3} N_{Re}^{1/2} < 1.4 \tag{9.35}$$

and

$$\bar{f}_v = 0.78 + 0.308(N_{Sc}^{1/3} N_{Re}^{1/2}) \quad \text{for } 51.4 > N_{Sc}^{1/3} N_{Re}^{1/2} \geq 1.4. \tag{9.36}$$

Here N_{Sc} is the Schmidt number of water vapor, defined as

$$N_{Sc} = \frac{v}{D_v}. \tag{9.37}$$

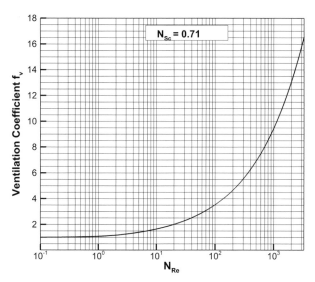

Fig. 9.3 The ventilation coefficient as a function of Reynolds number for a water drop falling in air assuming that the Schmidt number $N_{Sc} = 0.71$.

To get a rough idea about the drop size range in the above two formulas, let us assume $N_{Sc} = 0.71$ (which was the experimental condition). Then (9.35) applies to drops with radius $a < 60$ μm and (9.36) applies to 60 μm $\leq a \leq 1500$ μm.

Fig. 9.3 shows $\overline{f}_v$ vs. N_{Re} assuming that $N_{Sc} = 0.71$ in Eq. (9.35) and (9.36). In general, raindrops have Reynolds numbers in the range from about 1000 and upward. We see here that the ventilation factor can be greater than 10 for raindrops, indicating that both the diffusion growth and evaporation rates of falling raindrops can be more than 10 times the rates for stationary drops of the same size. Such rapid growth and evaporation rates would have great impacts on the thermodynamics and dynamics of clouds. Rapid growth would lead to a large amount of latent heat release and hence vigorous local growth of that cloudy region.

On the other hand, rapid evaporation of large falling raindrops would result in quick cooling of the surrounding air. Such very cold air can sink rapidly and is thought to play a major role in producing the *microburst* phenomenon that is a severe aviation hazard associated with a severe storm (see Fujita, 1985).

As in the treatment for the growth (evaporation) of stationary drops, we need to consider the diffusion of the latent heat released (consumed) in order to determine the actual growth (evaporation) rates. Like the vapor diffusion, the heat diffusion will also be enhanced by the fall motion of the drop. It is usually assumed that the ventilation coefficient for heat diffusion $\overline{f}_h$ is the same as $\overline{f}_v$ for the same drop.

So now, if we consider the ventilation effect, then the heat conservation equation becomes

$$\frac{dq}{dt} = \bar{f}_h \left(\frac{dq}{dt} \right)_0 = -L_e \frac{dm}{dt} = -L_e \bar{f}_v \left(\frac{dm}{dt} \right)_0, \tag{9.38}$$

which leads to the equation for determining the drop surface temperature,

$$T_a = T_\infty - \frac{L_e D_v M_w}{K_a R} \left[\frac{e_{sat}(T_a)}{T_a} - \frac{e_\infty}{T_\infty} \right] \left(\frac{\bar{f}_v}{\bar{f}_h} \right), \tag{9.39}$$

if we use the previous expressions for $(dq/dt)_0$ and $(dm/dt)_0$ in (9.38).

Then the drop diffusion growth rate is given by

$$a \frac{da}{dt} = \frac{D_v M_w}{\rho_w R} \left[\frac{e_{sat}(T_a)}{T_a} - \frac{e_\infty}{T_\infty} \right] \bar{f}_v. \tag{9.40}$$

9.4 Diffusion growth of ice crystals – electrostatic analogy

Like the case of cloud drops, the initial stage of ice particle growth must also be due to the diffusion of water vapor. In principle, we can determine the theoretical diffusion growth rate of ice crystals in the same way that we dealt with the cloud drops. For example, to determine the diffusion growth rates of stationary ice crystals, we can solve the same Laplace equation for ρ_v and T to obtain their distributions, and then proceed to calculate the total flux of water vapor toward the ice surface. But there is one problem here: ice particles are mostly non-spherical, and it becomes difficult to write down the inner boundary conditions mathematically. In the case of spherical particles, the inner boundary surface (i.e. the drop surface) is simply specified by $r = a$ as in (9.8) and (9.20). This simplification is unfortunately not available for more complicated shapes. If we are to write down the governing equation and the boundary conditions of ρ_v around a stationary ice crystal, we will have to write

$$\nabla^2 \rho_v = 0 \tag{9.41}$$

and

$$\begin{align} \rho_v &= \rho_{v,a} \quad \text{at crystal surface,} \\ \rho_v &= \rho_{v,\infty} \quad \text{at } r \to \infty. \end{align} \tag{9.42}$$

A workable way to do this is to use the mathematical formulas presented in Chapter 2 for simulating the shape of ice particle surfaces to specify this inner boundary condition; but this has not yet been done for the purpose of studying the diffusion growth of ice particles. However, it turns out that it is possible to determine the diffusion growth rates for stationary non-spherical ice particles without actually solving the equation set (9.41) and (9.42) to determine the vapor density and temperature distributions. This method is called the *electrostatic analogy*.

This analogy works because the mathematical formulations of the two situations are identical. In electrostatics, the electrostatic potential Φ of a charged perfect conductor (regardless of its shape) in the absence of space charge is given by the Laplace equation:

$$\nabla^2 \Phi = 0. \tag{9.43}$$

To obtain the specific potential distribution around such a conductor, we impose the following boundary conditions:

$$
\begin{aligned}
\Phi &= \Phi_a \quad \text{at the conductor surface,} \\
\Phi &= \Phi_\infty \quad \text{at } r \to \infty.
\end{aligned}
\tag{9.44}
$$

Thus the equation set (9.43) and (9.44) is identical to the set (9.41) and (9.42); only the symbols are different, which is superficial. Thus ρ_v is equivalent to Φ in this analogy. If we know how to treat the electrostatic problem, then we also know how to treat the diffusion growth problem.

A conductor with a total electric charge Q will produce an electric field $\vec{E}$ on its surface (while inside the conductor $\vec{E} = 0$, but this need not concern us here). $\vec{E}$ is related to Φ by

$$\vec{E} = -\nabla \Phi. \tag{9.45}$$

Now in electrostatics there is the *Gauss law*, which states that when we integrate the electric field $\vec{E}$ over a surface enveloping the conductor we get

$$\oint_S \vec{E} \cdot d\vec{S} = \oint_S (-\nabla \Phi) \cdot d\vec{S} = 4\pi Q = 4\pi C (\Phi_\infty - \Phi_a), \tag{9.46}$$

where S is a surface enclosing the conductor. The quantity C is called the *capacitance* of the conductor and is given by

$$C = \frac{Q}{\Phi_\infty - \Phi_a}. \tag{9.47}$$

Capacitance is a quantity not just associated with electricity, as we shall see. Eq. (9.46) is true regardless of the shape of the conductor.

Note that the forms of the Gauss law differ when different unit systems are used. Here we use the cgs unit system. Also, the form we present here is the integral form of the Gauss law. There is also a differential form of this law, but we do not need it here.

Now here comes the analogy. Since ρ_v is analogous to Φ, we replace Φ by ρ_v and $\vec{E}$ by $-\nabla \rho_v$ in (9.46) and get

$$\frac{dm}{dt} = \oint_S (-D_v \nabla \rho_v)_{r=a} \cdot d\vec{S} = 4\pi D_v C (\rho_{v,\infty} - \rho_{v,a}), \tag{9.48}$$

so that we can determine the diffusion growth rate of the ice crystal whose shape is the same as the conductor. Thus if we can determine the capacitance of the ice crystal, we can determine dm/dt without even worrying about the distribution of ρ_v.

Comparing (9.48) with (9.14), we see that $C = a$ for the case of a sphere. Indeed, as long as we view the ice crystal as a perfect conductor, the capacitance is just a function of the ice crystal geometry, whether electric or otherwise. So our next question is: How do we determine the capacitance of an ice crystal?

Experimentally, we can use a conducting metallic material (such as aluminum foil) and cut it out in the shape of an ice crystal, for example, a hexagonal plate. We then charge this plate and measure the charge Q and the potential difference $\Phi_\infty - \Phi_a$. We then use (9.46) to calculate C. This has been done (e.g. McDonald, 1963; Podzimek, 1966) for a number of simpler ice crystal shapes such as hexagonal plates and columns, columns with hourglass-like hollow ends, etc. These experimental measurements have been compared to theoretical values of the capacitances of some simple geometrical shapes that approximately resemble ice crystals. For example, very thin hexagonal ice plates can be approximated by an infinitely thin circular plate, whereas hexagonal plates of various thicknesses can be approximated by thin oblate spheroids. There are formulas readily available for calculating the capacitance of simply shaped conductors. The following provides a few useful examples.

- The capacitance of a circular plate with radius a is

$$C_0 = \frac{2a}{\pi}. \tag{9.49}$$

- The capacitance of an oblate spheroid with semi-major and semi-minor axis lengths a and b is

$$C_0 = \frac{a\varepsilon}{\sin^{-1}\varepsilon}, \quad \varepsilon = \left(1 - \frac{a^2}{b^2}\right). \tag{9.50}$$

- The capacitance of a hexagonal ice column can be approximated by a prolate spheroid of semi-major and semi-minor axis lengths a and b, i.e.

$$C_0 = \frac{A}{\ln[(A+a)/b]}, \quad A = (a^2 - b^2)^{1/2}, \tag{9.51}$$

which in the case $b \ll a$ becomes

$$C_0 = \frac{a}{\ln(2a/b)},$$ (9.52)

which can be used to approximate the capacitance of an ice needle.

For the simple ice crystal shapes as mentioned above, these approximations seem to agree with experimental values to within a few percent. However, for more complicated shapes, such as more open dendrites, these formulas give values that deviate substantially from measurements. Clearly, there is still a need to develop new formulas for the capacitances of more complicated ice crystal shapes.

One of the more complicated but important shapes of ice particles is rosettes, and many observational studies indicate that rosettes are common in many high cirrus. Chiruta and Wang (2003) performed numerical computations to determine the capacitance of rosettes consisting of different lobes. Fig. 9.4 shows their

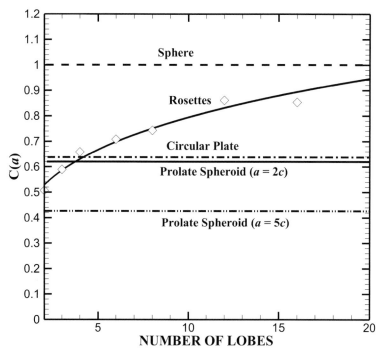

Fig. 9.4 Computed capacitance (diamonds) of rosettes as a function of number of lobes. The solid curve represents a power-law fit by Eq. (9.53). Also shown are the capacitances of a sphere, a circular plate, and two prolate spheroids with semi-axis ratio $a/b = 2$ and 5, respectively. The rosettes, the sphere, and the circular plate all have a radius a. Adapted from Chiruta and Wang (2003).

computed rosette capacitance as a function of the number of lobes. The results show that the capacitance varies from ~ 0.5 to ~ 0.92 as the number of lobes increases from two to 20. When there are just two lobes, the capacitance is close to that of a prolate spheroid. When there are many lobes, the capacitance is close to that of a sphere. This behavior would be expected, as the rosette shape indeed changes from prolate spheroid-like to sphere-like when the number of lobes increases from two to 20.

Chiruta and Wang (2003) fitted their results to relate the capacitance with the number of lobes N, the surface area S, and the volume V by the following empirical formulas:

$$C = 0.434N^{0.257}, \tag{9.53}$$

$$C = 0.3636S^{0.3476}, \tag{9.54}$$

$$C = 0.7472V^{0.4401}. \tag{9.55}$$

Many ice crystals in cirrus clouds are hollow, as we mentioned in Chapter 2, and diffusion growth is the dominant growth mode of ice particles there. Hence there is a clear need to understand the capacitance of these hollow crystals. Chiruta and Wang (2005) performed numerical computations to determine the capacitance of solid and hollow hexagonal ice plates and columns, as depicted in Fig. 9.5. The computed capacitances are shown in Fig. 9.6.

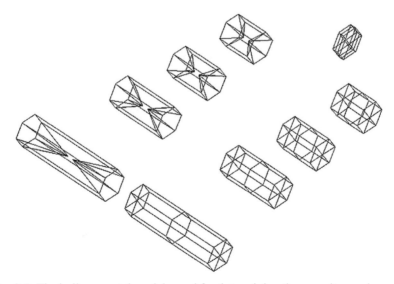

Fig. 9.5 The hollow crystal models used for determining the capacitance shown in Fig. 9.6. From Chiruta and Wang (2005). Reproduced by permission of the American Meteorological Society.

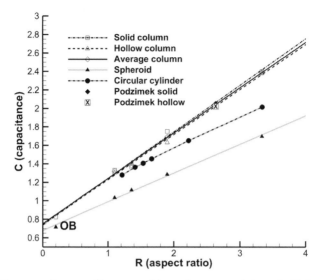

Fig. 9.6 The capacitances of hexagonal ice columns, prolate and oblate (indicated by OB) spheroids, and circular cylinders as a function of the aspect ratio R. Experimental results are taken from Podzimek (1966). From Chiruta and Wang (2005). Reproduced by permission of the American Meteorological Society.

One important finding is that solid and hollow crystals of the same dimension and aspect ratio have the same capacitance. This has an important implication. The same capacitance means the same *dm/dt*, which leads to the conclusion that a hollow crystal will grow faster in linear dimension than a corresponding solid crystal, since a hollow crystal requires less water to grow a certain length. In the next moment, the long hollow crystal will have a greater capacitance, which would make it grow even faster. Fig. 9.7 shows a simple model study of this phenomenon assuming that the growth is only in one direction, for example, along the *c* axis (Chen and Wang, 2009). It is seen that, in a minute, the hollow crystal becomes about 10 times longer than the solid crystal if they start out with the same dimension and aspect ratio. However, this mode of diffusion growth will not last very long, as other effects (such as falling out of the cloud) will set in to change the direction of growth.

Westbrook *et al.* (2008) made refined calculations of the capacitances of hollow and solid hexagonal prisms and confirmed the above finding. They fitted their data by the following equation:

$$C = 0.58(1 + 0.95R^{0.75})a, \tag{9.56}$$

where $R = L/2a$ is the aspect ratio.

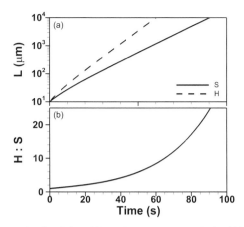

Fig. 9.7 (a) The length of solid (solid line) and hollow (dashed line) ice columns as a function of time in the c axis growth mode. (b) The ratio of length of hollow ice to that of solid ice. From Chen and Wang (2009). Reproduced by permission of the European Physical Society.

Finally, like the diffusion growth of liquid drops, we need to consider the issue of latent heat due to the phase change. The formulation is entirely the same, and we only need to replace the drop radius a by the crystal capacitance C. But for ice we can only write down the mass growth rate. We will neglect the curvature and solute effect in this case, so the growth rate equation is

$$\frac{dm}{dt} \approx \frac{4\pi \overline{f}_{\mathrm{v}} C S_{\mathrm{v,i}}}{\dfrac{RT_\infty}{D_{\mathrm{v}}^* M_{\mathrm{w}} e_{\mathrm{sat,i}}(T_\infty)} + \dfrac{L_{\mathrm{s}}}{K_{\mathrm{a}}^* T_\infty}\left(\dfrac{L_{\mathrm{s}} M_{\mathrm{w}}}{RT_\infty} - 1\right)}, \qquad (9.57)$$

where $\overline{f}_{\mathrm{v}}$ is the ventilation coefficient to be discussed next.

9.4.1 Ventilation effect on falling ice crystals

The ventilation effect of falling ice crystals has been studied theoretically and experimentally. Ji and Wang (1998) numerically solved the convective diffusion equation of water vapor around three types of ice crystals: columns (approximated by circular cylinders of finite length), hexagonal plates, and broad-branch crystals. This equation is

$$\frac{\partial \rho_{\mathrm{v}}}{\partial t} = D_{\mathrm{v}} \nabla^2 \rho_{\mathrm{v}} - \vec{u} \cdot \nabla \rho_{\mathrm{v}} \qquad (9.58)$$

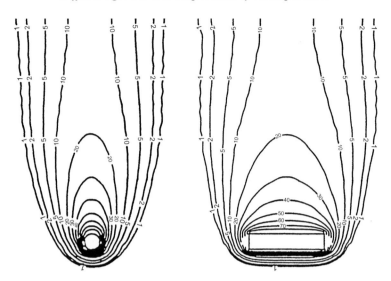

Fig. 9.8 Water vapor density distribution around a falling ice column at $N_{Re} = 10$, in end view (left) and length view (right). The contour levels are (from outside) 1, 2, 5, 10, 20, 30, 40, 50, 60, 70, 80, and 100 (surface). The contour level 90 is not shown to avoid overcrowding of curves. Adapted from Ji and Wang (1998).

subject to the boundary conditions

$$\begin{aligned} \rho_v &= \rho_{v,s} \quad \text{at the surface of the crystal,} \\ \rho_v &= \rho_{v,\infty} \quad \text{far away from the crystal.} \end{aligned} \tag{9.59}$$

Fig. 9.8 shows an example of the computed vapor density field, which illustrates the enhanced vapor density gradient upstream and the reduced gradient in the wake of a falling columnar crystal at $N_{Re} = 10$.

Fig. 9.9 shows the computed ventilation coefficients of the three crystal types together with some previous experimental data. The following are the empirical equations for the three computed curves.

- Columnar crystal ($0.2 \leq N_{Re} \leq 20$):

$$\bar{f}_v = 1.0 - 1.67 \times 10^{-3}X + 1.5 \times 10^{-1}X^2 + 1.15 \times 10^{-2}X^3 - 2.89 \times 10^{-3}X^4. \tag{9.60}$$

- Hexagonal plates ($1.0 \leq N_{Re} \leq 120$):

$$\bar{f}_v = 1.0 - 6.04 \times 10^{-2}X + 2.80 \times 10^{-2}X^2 + 3.19 \times 10^{-4}X^3 - 6.25 \times 10^{-6}X^4. \tag{9.61}$$

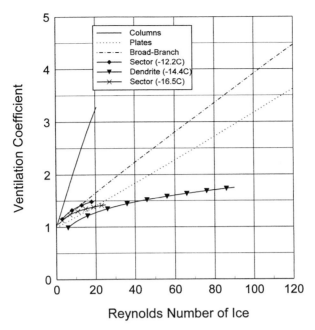

Fig. 9.9 Computed mean ventilation coefficients as a function of the dimensionless parameter X defined in Eq. (9.63). Experimental results of Thorpe and Mason (1966) and numerical results of Pitter *et al.* (1974) are also plotted for comparison. From Ji and Wang (1998). Reproduced by permission of the American Meteorological Society.

- Broad-branch crystals ($1.0 \leq N_{\mathrm{Re}} \leq 120$):

$$\bar{f}_v = 1.0 + 3.55 \times 10^{-2}X + 3.55 \times 10^{-2}X^2. \tag{9.62}$$

In these equations,

$$X = N_{\mathrm{Sc}}^{1/3}N_{\mathrm{Re}}^{1/2}, \tag{9.63}$$

where $N_{\mathrm{Sc}} = v/D_v$ is the Schmidt number.

9.5 Habit change of ice crystals

While convenient, the electrostatic analogy is only an approximation of the true growth process of ice crystals. In the development of the analogy, we assumed that both the temperature and the vapor density are constant at the crystal surface. While the isothermal surface assumption seems to be more valid given the fairly large thermal conductivity, the constant vapor density assumption has more problems. Unlike the relatively smooth surface of a liquid drop, the ice crystal surface consists of steps, as we have mentioned already in Chapter 2, and these steps may be of different sizes and distributed unevenly. We may expect that the vapor–surface equilibrium condition varies from location to location, and hence the constant vapor density assumption seems to be unrealistic.

In addition, it has been observed that ice crystals change habits and possibly also their growth direction in response to the environmental temperature and super-saturation condition. Fig. 9.10 shows this important behavior of the habit change of ice crystals. At a fixed supersaturation, e.g. 0.1%, we see that the crystal habit changes from hexagonal plates (growing along the *a* axis) to columns (growing along the *c* axis) to plates and then to columns as the temperature becomes colder and colder. At a fixed temperature, say −15°C, the crystal habit changes from plates to dendrites as the supersaturation increases. In the colder temperature range, e.g. −30°C, the habit changes from plates to columns. Bullet rosettes preferentially occur at temperatures colder than −40°C and supersaturation higher than 0.3%.

Currently we do not have a detailed understanding of the molecular processes occurring on the crystal surface that are responsible for causing the habit change induced by temperature and supersaturation. Presumably the vapor density distribution around the growing crystal (which is influenced by the shape of the crystal) and the surface processes of how water molecules diffuse and become incorporated into the ice lattice are working together to cause such changes. Libbrecht (2005) provides a review of the growth physics of snow crystals. One of the notable conclusions is that the presence of some minute impurities in ice may have a significant impact on the crystal growth, a subject for which very little research is available. Future observations utilizing new microscopic techniques such as laser confocal microscopy combined with differential interference contrast microscopy (LCM-DIM) mentioned in Chapter 5 are likely to shed more light on the surface process of ice crystal growth.

The changing crystal habit with temperature and humidity obviously will affect the diffusion growth rate of ice crystals. This effect has been observed directly by Takahashi *et al.* (1991) who investigated the diffusion growth rates of free-falling snow crystals in a vertical wind tunnel for up to 30 min at temperatures from −3 to −23°C. Crystals change their habit, which impacts upon their mass and linear growth rates in response to the changing temperature. Figs. 9.11 and 9.12 illustrate their results on semi-log scale.

Fig. 9.11 shows the variation of crystal mass with temperature for various growth times, ranging from 3 to 30 min. The crystal mass shows two maxima at about −5.8 and −14.8°C, although the former maximum is not obvious for growth less than 5 min. These maxima become more pronounced with time, indicating the growth of needles and dendrites. Fig. 9.12 shows the corresponding variations of crystal dimensions with temperature, for the same growth times. The difference in methods applied for measuring the *c* axis length apparently caused the

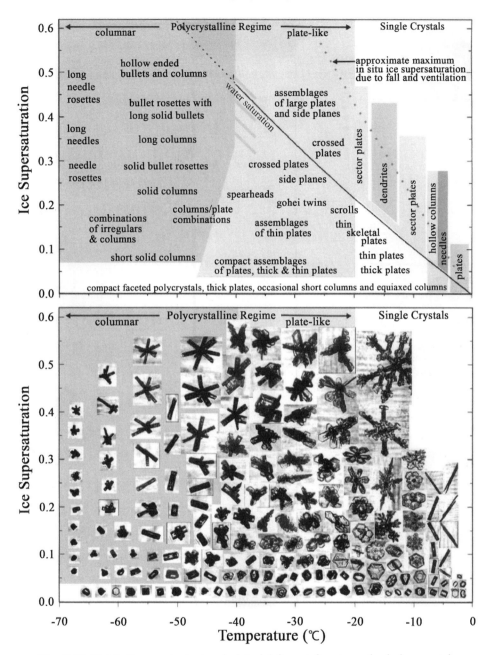

Fig. 9.10 Habit diagram in text and pictorial format for atmospheric ice crystals derived from laboratory results (Bailey and Hallett, 2004) and CPI images gathered during AIRS II and other field studies. Diagonal bars near the middle of the upper diagram are drawn to suggest the possibility of the extension of the bullet rosette habit to temperatures slightly higher than −40°C. Adapted from Bailey and Hallett (2009).

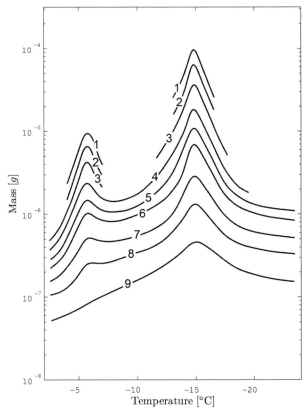

Fig. 9.11 Variation of crystal mass with temperature at different growth times:
(1) 30 min, (2) 25 min, (3) 20 min, (4) 15 min, (5) 12 min, (6) 10 min, (7) 7 min,
(8) 5 min, and (9) 3 min. Adapted from Takahashi *et al.* (1991).

inflections at about −11 and −19°C (boundary temperatures between the thick
plates and the plates). In 30 min of growth, the dendrites and the needles acquired
a diameter of 4 mm and a length of 2 mm, respectively. The growth rates were
maximum along the *c* axis at about −5.8°C (needles) and along the *a* axis at
around −14.8°C (dendrites); whereas minima were observed along the *c* axis at
about −14.8°C and along the *a* axis at about −4.0°C.

Such habit-changing features obviously should play an important role in the
ice crystal growth in real clouds, although the observational data derived from
field studies cannot be as clear-cut as data obtained in laboratory controlled
experiments. This is especially important for the upper tropospheric part of
clouds, where ice crystals are the dominant form of condensates. What kind
and how large these crystals are will impact upon their interaction with incoming
solar radiation and influence the dynamics of clouds and possibly longer-term
climate processes. The remote-sensing research community will need more

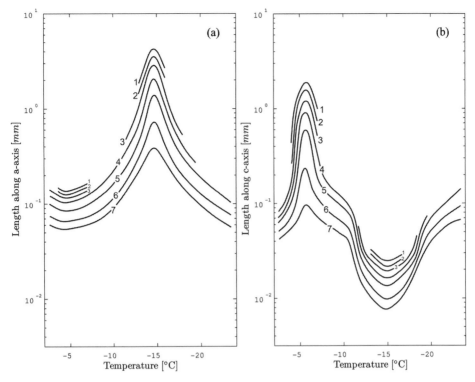

Fig. 9.12 Variations of crystal dimensions with temperature at different growth times (same times as in Fig. 9.11). Adapted from Takahashi *et al.* (1991).

information about ice crystal habits and growth rates in order to interpret their observational data more accurately.

Problems

9.1. We obtained Eq. (9.13) by integrating Eq. (9.11) over the surface of the drop where $r = a$.

(a) What result would you obtain if, instead, you integrated over the spherical surface $r = 2a$?

(b) Can you explain why you get this result?

9.2. (a) Assuming that the surface temperature does not change, calculate the time required for a 10 μm radius stationary water drop to evaporate to 1 μm in an environment of 1013.25 mb, 20°C and 90% relative humidity. [Assume the vapor pressure at the drop surface $e_a = 23.4$ mb.]

(b) Would the evaporation become faster or slower if the surface temperature is allowed to change? Why?

10

Collision, coalescence, breakup, and melting

The diffusion growth discussed in the last chapter undoubtedly dominates the growth of cloud particles in the initial stage of cloud formation. But the diffusion growth rate becomes very slow once a particle reaches a certain size (about 40 μm for a water droplet, and a few hundred micrometers for an ice particle), as we have discussed. In order to form larger particles, especially precipitation-sized particles, we need to look for other, faster growth mechanisms.

One such mechanism is collision and coalescence. While it seems to be intuitive now, the concept took a while to take root. It was first mentioned by Reynolds (1876), who merely suggested its possibility. It was not considered seriously by scientists until Langmuir (1948) proposed the mechanism for rain formation on top of the Bergeron–Findeisen process, and made a quantitative estimate of its efficacy. This mechanism is now accepted as one of the major precipitation formation mechanisms, especially in the warm rain process.

Since cloud droplets are free to move around in air, they can collide with each other and become larger if they also coalesce afterward. If two equal-sized drops collide and coalesce, the resulting drop will have twice the mass of either original drop. Obviously, this is a much faster growth mode than diffusion growth if it occurs. Similarly, ice crystals can collide with supercooled droplets and form rimed crystals. Upon further riming, rimed crystals can eventually grow to become graupel and hail. When graupel and hail fall to lower levels where the temperatures are warm enough, melting may occur and produce raindrops. Similarly, ice crystals can collide and clump together to form large snowflakes, and they can fall as snow or melt into raindrops. Collision and coalescence of large raindrops may be subject to hydrodynamic instability, and the resulting drop may break up into smaller drops and hence change the drop size distributions and the microdynamics of the cloud.

10.1 Definition of collision efficiency

To understand collision growth, we need to consider how such collision can occur in clouds. Two cloud drops, even if moving in each other's path, do not necessarily collide, because the flow field of the air between them may prevent it. Given certain initial conditions (drop speeds, separation, etc.), we will need to know what the likelihood is for the two to collide. First, we need to define a quantity called the *collision efficiency*.

We shall only consider a two-body collision here, since, as is well known in classical mechanics, many-body collision problems are quite complicated. Besides, the distances between two hydrometeors in a cloud are usually large enough (except maybe in very heavy precipitation events) that it is adequate to consider only the two-body collision situation most of the time.

Consider the case of the collision of two spherical drops, one with a larger radius a_1 (called the collector drop) and the other with a smaller radius a_2 (called the collectee drop). Normally, we assume that the collector is located above and falls faster than the collectee, and thus will catch up with the latter given enough time. Just like the flow field problems in Chapter 8, we will consider the equivalent configuration in which a_1 is stationary but a_2 is moving upward. This situation is illustrated in Fig. 10.1.

For such a pair, the collision efficiency is defined as

$$E = \frac{y_c^2}{(a_1 + a_2)^2} = \frac{\pi y_c^2}{\pi (a_1 + a_2)^2} = \frac{A}{A^*}, \qquad (10.1)$$

where y_c is called the *critical initial offset*, which is the initial horizontal distance of the center of the collectee drop from the vertical line joining the two drop centers that results in a grazing collision with the collector drop (see Fig. 10.1). There will be no collision if the initial offset is larger than y_c.

If there is no air, then the collision of two spheres (in the absence any other forces) will be just due to interception – the collectee will move in a straight line and touch the collector when $y_c = a_1 + a_2$ and the collision efficiency will be exactly 1 from Eq. (10.1). However, because of the presence of the air, there will be a flow field around these two drops and that usually prevents the collectee from moving in a straight line. Instead, its trajectory is normally influenced by the hydrodynamic force and the drop will move in a curved trajectory. The quantity A is the collision cross-section, which is the area of the shaded circle with radius y_c. Those drops initially launched within this area will result in a collision. The quantity A^* is the geometrical collision cross-section, whose meaning is self-evident.

Eq. (10.1) can be further written as

$$E = \frac{K}{K^*}, \qquad (10.2)$$

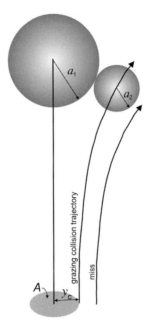

Fig. 10.1 The geometrical configuration of the collision of two spherical drops. The problem of a falling larger drop collecting a smaller drop is replaced by the equivalent configuration in which a smaller drop is going upward and collides with the larger drop.

where

$$K = A(\vec{v}_1 - \vec{v}_2) \quad \text{and} \quad K^* = A^*(\vec{v}_1 - \vec{v}_2). \tag{10.3}$$

Here K and K^* are called the *collision kernel* and *geometrical collision kernel*, and $\vec{v}_1$ and $\vec{v}_2$ are the fall velocity of the collector and the collectee, respectively. In most cases, these velocities are taken as the terminal velocities of the respective drops, but we have to remember that sometimes hydrometeors do not always fall at their terminal velocities. Eq. (10.2) says that K (K^*) is the volume swept out by A (A^*) per unit time. Physically, it means that any drops of size a_2 located inside the volume K in that unit time will eventually collide with a_1.

Wang (1983) pointed out that the definition (10.2) is more general than (10.1) because (10.1) is valid only when the collision is azimuthally symmetric such that y_c is independent of the azimuthal angle. This would be the case for a two-sphere collision in an axisymmetric flow field, but would not be true for general 3D non-spherical hydrometeors, such as ice columns or plates, colliding with water drops. On the other hand, (10.2) is still valid in that case as well as the case where the collectee is not a sphere. When the flow is unsteady, K is also time-dependent, and in that case we will speak of the time-averaged collision kernel.

After the collision, the two particles may or may not coalesce. The probability that they will coalesce is defined as the *coalescence efficiency* E_{coa}. The product of the collision efficiency and the coalescence efficiency is called the *collection efficiency* E_c:

$$E_c = E E_{coa}. \tag{10.4}$$

10.2 Theoretical determination of collision efficiency

In principle, we can determine the two-drop collision efficiency by solving the equation of motion of the collectee drop:

$$m_2 \frac{d\vec{v}_2}{dt} = m_2 \vec{g} + \vec{F}_2, \tag{10.5}$$

where $\vec{F}_2$ represents the sum of all external forces acting on the collectee drop with mass m_2. In the present case, $\vec{F}_2$ is simply the hydrodynamic drag force on the a_2 drop. In order to solve (10.5), it will be necessary to know $\vec{F}_2$ first, which means that we need to know the flow field in this situation.

The flow field we should be considering here is not just the flow field created due to the falling collector drop but the flow field due to the presence of two drops falling simultaneously. The flow fields we discussed in Chapter 8 are all limited to that due to a single falling hydrometeor; hence, strictly speaking, they are not directly applicable to the present case. However, if the collectee is small compared to the collector, then it is probably justified to ignore the effect of it compared to the flow field generated by the falling collector drop. But when the collectee drop is not that small compared to the collector, this assumption becomes unrealistic.

The most accurate way to obtain the flow field in this case is to solve the Navier–Stokes equation in the presence of two spheres, but this has not been done for the determination of the collision efficiency of water drops. One method that has been used in order to partially take into account the two-drop flow field is called the *superposition method*. This is done by assuming that each drop is moving in the flow field generated by the other drop moving in isolation. This has been adopted by some investigators for calculating the collision efficiencies of small cloud droplets, and the results we will summarize below were obtained by this method.

Fig. 10.2 shows the computational results of collision efficiencies between two cloud drops. The numbers on the curves represent the radii of the collectors, and the horizontal axis is the radius ratio a_2/a_1. The general behavior of the collision efficiency can be summarized as below.

First, examine the top curve – the one labeled 74.3 μm (the collector radius a_1). When the a_2 drop is very small, its inertia is also small. Remember that the larger the a_1 drop, the greater its U_∞ in general and hence the greater the drag force. This large

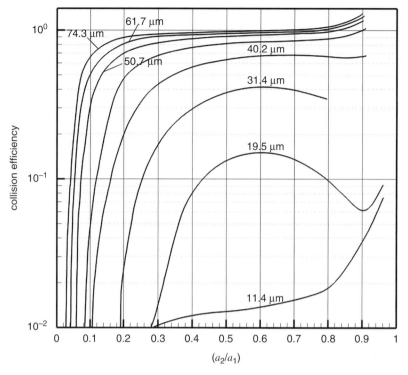

Fig. 10.2 The collision efficiency vs. size ratio a_2/a_1 of a larger drop of radius a_1 colliding with a smaller drop of radius a_2. The number labeled on each curve represents the larger drop radius a_1. Data based on Schlamp *et al.* (1976).

drag will force the a_2 drop to follow more closely the streamline of the flow field. Consequently, the trajectory of the a_2 drop deviates little from the streamline and the collision efficiency is also small as a result. If a_2 strictly follows a streamline, its collision efficiency will be exactly zero (which means that a_2 is just like an air molecule in the flow). But a_2 is of course not as small as an air molecule, and hence its collision efficiency is not zero but very small. When the a_2 size increases, the efficiency increases rapidly due to the increase in its inertia and reaches a plateau of $E \sim 0.9$ for $0.2 \leq a_2/a_1 \leq 0.8$. In this size ratio range, the efficiency increases due to the large deviation of the a_2 trajectory from the streamlines.

The value of E increases to about one, and then becomes somewhat greater than one for $a_2/a_1 > 0.8$. The exact reason for this feature is unclear at present, but may be due to the numerical artifact of the superposition method used for these calculations, which overestimates the drag when two drops are of comparable size and tends to produce too strong "wake capture", namely, a_2 gets caught in the standing eddies of a_1 and collides with it from the rear. In general, it is reasonable to say that the collision efficiency is essentially one for this size range.

When a_1 decreases, the collision efficiency also decreases for the same a_2/a_1 ratio. For $a_1 = 31.4\,\mu\text{m}$, for example, the maximum E is about 0.7 at $a_2/a_1 \sim 0.6$; whereas when $a_1 = 19.5\,\mu\text{m}$, the maximum E is only about 0.15. This indicates that collision growth of small drops is inefficient, as noted before. The collision growth becomes more efficient only when the drops are large enough.

Thus far the theoretical collision efficiency calculations are restricted to drop radius smaller than $\sim 600\,\mu\text{m}$, as drops larger than this size will develop unsteady motion such as eddy shedding and turbulence, and the unsteady flow fields of large drops necessary for the collision efficiency calculations have not yet been developed.

10.3 Impact of turbulence on collision efficiency

The collision efficiencies described in the previous section are obtained by considering only the hydrodynamic forces and gravity. There are other forces in the atmosphere that may influence the collision efficiency. One of these is the electric force, which we will discuss in Chapter 14. Another important factor that may have a significant impact is turbulence. Turbulence on various temporal and spatial scales exists in the atmosphere and certainly in cumulus clouds as well.

How does turbulence impact upon the collision efficiency? In our analysis in the last section, the relative velocity between the two drops is considered to be constant as long as their sizes are fixed. In the presence of turbulence, however, the relative velocity can be changed randomly, so that not only its magnitude but also its direction become highly variable, which results in inter-drop shears. Such shears will surely lead to changes in the collision efficiency. The general approaches to study such problems are usually statistical in nature.

One of the more recent studies on the impact of turbulence on the collision efficiency of cloud droplets was performed by Pinsky *et al.* (1999) and Pinsky and Khain (2004), who suggested that turbulence will cause an increase in the collision efficiency of cloud droplets. The increase can be up to 25% for weak turbulence, with a dissipation rate of 200 cm^{-2} s^{-3}, which is typical for small cumulus clouds, but the increase can be up to a factor of 2.5 for the dissipation rate of 1000 cm^{-2} s^{-3} typical for well-developed deep convective clouds. Turbulence is a highly complicated phenomenon and further studies will be necessary to fully elucidate its impact on the collision of hydrometeors.

10.4 Coalescence of water drops

When two water drops collide, they do not necessarily merge into one, as experiments have shown. Rather, one of the following outcomes may occur. Only the second mode fits into what we call collision growth here.

(1) *Rebound* – the two drops will bounce apart as if there is something preventing them from sticking together (Fig. 10.3). The rebound is mainly due to the air film that exists between the two drops when they collide. If the air film cannot drain away in time, then rebound results.

(2) *Coalescence* – the two drops will merge into one larger drop, which remains as a single entity long after the interaction is over (Fig. 10.4). The merged drop often oscillates vigorously due to the momenta of the original drops.

(3) *Coalescence and breakup* – the two drops coalesce for a very short time but break apart to become two or more drops afterwards (Fig. 10.5). In the case of breakup into two drops, the resulting drops may or may not be the same size as the original ones.

Figs. 10.3–10.5 show experimental photographs of these three processes taken from Park (1970). The experiments were not performed under the condition that the drops are moving at their respective terminal velocities, and the two drops approach each other at various angles, so the collision geometries are not quite the same as that illustrated in Fig. 10.1. However, Park's results demonstrate that the collision between two drops can yield rather complicated outcomes that depend on factors such as collision angle, drop speed, and size. In real clouds, other factors such as drop surface impurities, electric charges, and ambient conditions (pressure, temperature, and humidity) may also influence the outcome as well. For example, Czys (1994) performed a laboratory study of the temperature effect and showed that the coalescence efficiency decreases at higher temperature for drops in the radius range 306–350 μm. Ochs *et al.* (1995) demonstrated that relative humidity has an impact on the outcome of the collision when permanent coalescence does not occur.

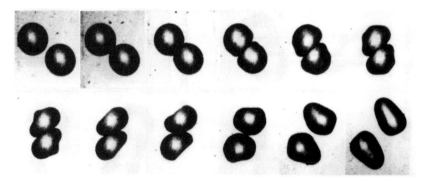

Fig. 10.3 Collision and subsequent rebound of two water drops of 200 μm diameter colliding at a velocity of 160 cm s^{-1} and angle 50°. From Park (1970), courtesy of Dr Robert W. Park.

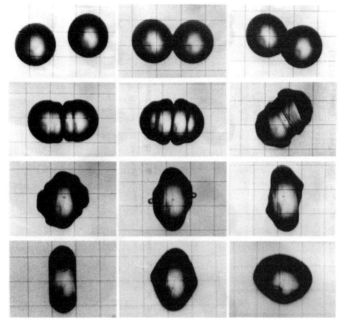

Fig. 10.4 Collision and subsequent coalescence of two water drops of 200 μm diameter colliding at a velocity of 63 cm s^{-1} and angle 79°. From Park (1970), courtesy of Dr Robert W. Park.

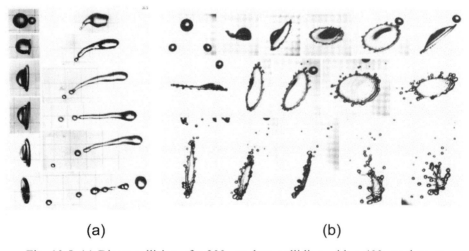

(a)   (b)

Fig. 10.5 (a) Direct collision of a 200 μm drop colliding with a 400 μm drop at 720 cm s^{-1} and the subsequent breakup. (b) Collision and subsequent spatter of two drops of 1630 and 1080 mm colliding at 558 cm s^{-1}. From Park (1970), courtesy of Dr Robert W. Park.

Our current knowledge of the coalescence of two drops is inadequate, and for simplicity many theoretical collision growth studies simply assume that $E_{coa} = 1$. Under this assumption, the collection efficiency E_c is the same as the collision efficiency E.

Experimental measurements of drop–drop collision are difficult to perform, and consequently there are only a few experimental data available for comparison with theoretical values. Beard and Ochs (1983) performed an experimental study to measure the collection efficiencies of a_1 drops of radius 63–100 µm collecting a_2 drops of radius 10–26 µm. They found that the efficiencies fall in a narrow range of 0.6–0.7. These values are consistently lower than the theoretical values, suggesting that the coalescence efficiencies are between 0.6 and 0.8. Fig. 10.6 shows the empirical curves for E_{coa} based on the experiments of Ochs *et al.* (1986). It shows that the coalescence efficiency is low for large colliding pairs but is higher for large drops collecting small drops.

More extensive experiments utilizing a wind tunnel were performed by Low and List (1982a,b), who investigated the coalescence and breakup of drops after collision under the surface pressure (1000 hPa). Fig. 10.7 shows the results of Low and List (1982a). More recently, List *et al.* (2009a,b) performed similar experiments

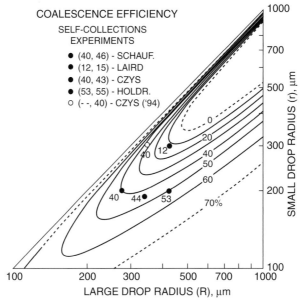

Fig. 10.6 Contours of the coalescence efficiency ε_0 (%) from the self-collection correlation compared with ε_0 data for freely falling drops. Numbers in the key are ε_0 and ε, where ε_0 is the average of the critical offsets $(\varepsilon_C, \varepsilon_B)$ and ε is the average of the minimal-charge efficiencies. The dashed curves indicate extrapolation from an empirical formula well outside the range of the experiments. From Beard and Ochs (1995).

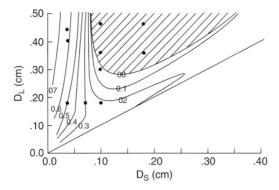

Fig. 10.7 Values of the parameterized coalescence efficiency E_{coa} of all drop pairs of sizes D_L and D_S. The studied drop pairs are indicated by dots. The straight line $D_L = D_S$ corresponds to a value of $E_{coa} = 0.19$. From Low and List (1982a).

under reduced pressure conditions (500 hPa) to simulate cloud conditions in the middle troposphere. Both were performed at a laboratory temperature of about 20°C. A few salient conclusions obtained by List *et al.* (2009a) are summarized below.

(1) Lowering the pressure from 1000 to 500 hPa increases fragment numbers, but only for high and intermediate *collision kinetic energies* (CKE). At low CKE, the pressure effect is negligible or even reversed.
(2) The breakup types are as follows:
   • *Filament breakup* – a small drop hits a large one near or at its equator but is soon carried away and separated by its own momentum from the large drop while temporarily connected to it by a filament of fluid. This filament will eventually disintegrate into a string of tiny fragments much smaller than the small drop.
   • *Sheet breakup* – if the small drop hits closer to the center, then it rips a whole sheet out of the large drop. The sheet develops instability and disintegrates into many fragments. The small drop loses its identity into the fragments of the sheet.
   • *Disk breakup* – when the small drop hits the large one in or close to the center, the resulting conglomerate forms a horizontal disk and disintegrates.
   • *Bag breakup* – a bag results when a disk is inflated by air and forms a bag, which then disintegrates into many, many fragments. This is the same as that shown in Fig. 8.16.
(3) Only 20% of the collisions that they examined resulted in coalescence. This clearly demonstrates that coalescence is rare and breakup plays a fairly important role in determining the drop size distribution.

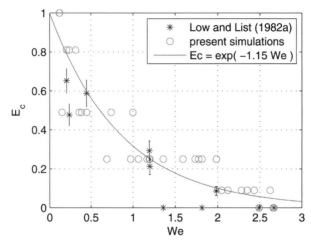

Fig. 10.8 Coalescence efficiency E_{coa} as a function of the Weber number We. From Straub *et al.* (2010). Reproduced by permission of the American Meteorological Society.

Schlottke *et al.* (2010) and Straub *et al.* (2010) performed direct numerical simulations of the collision breakup of raindrops and produced breakups very similar to Figs. 10.3–10.5. Fig. 10.8 shows the coalescence efficiencies computed by Straub *et al.* (2010). It demonstrates that the numerical simulation is able to reproduce the experimental results of Low and List (1982a) successfully. The Weber number in the present context is defined as

$$We = \frac{CKE}{\sigma_w}, \qquad (10.6)$$

where σ_w is the surface tension of water and CKE is the collision kinetic energy defined as

$$CKE = \frac{\pi}{12} \rho_w \frac{d_L^3 d_S^3}{d_L^3 + d_S^3} v_r, \qquad (10.7)$$

where d_L and d_S are the diameters of the large and small drop, and v_r is their relative velocity. The Weber number is a dimensionless number characterizing the relative importance of the drop inertia vs. the surface tension.

The impact of the coalescence and breakup of drops in a population of droplets will be discussed in the next chapter.

10.5 Collision between ice particles and supercooled water drops

In mixed-phase clouds, ice particles and supercooled water drops may coexist in regions with temperature colder than 0°C. If the drops are of cloud drop size, they usually stick to and freeze on the surface of the ice particles to become rime because liquid water is a metastable phase in a subfreezing environment. The process is called riming. Riming is a mode of accretion growth of ice particles.

10.5.1 Dry growth, wet growth, and the Schumann–Ludlam limit

When an ice crystal collects supercooled droplets and turns them into rime, it becomes a rimed crystal. Presumably, this process can also occur if the original ice particle is a frozen drop instead of a pristine ice crystal. When riming proceeds to such a degree that the original ice particle is completely masked by the rimes, it becomes a graupel. When a graupel grows larger than 5 mm in size, it is called hail.

Depending on the environmental conditions, ice accretion growth can occur either in the *dry growth* or *wet growth* regime. Dry growth refers to the process when supercooled drops collide with an ice particle and freeze on it. However, we have to remember that the freezing of water also releases latent heat, which will warm the ice particle surface to a temperature above the environmental air. However, as long as the heat can be dissipated efficiently so that the crystal surface temperature remains colder than 0°C, freezing will occur and the accreted frozen drops form a layer of highly polycrystalline and finer grain structure. This is the *dry growth regime*.

But if the collection of supercooled water droplets is going on rapidly enough such that the accumulated heat cannot be dissipated quickly, the ice surface may approach 0°C and the above spontaneous freezing may or may not occur. This occurs where the liquid water content of the supercooled droplets is high. The amount of ice formed depends on how fast the heat is dissipated, and not all the water droplets accreted turn into ice. The particle is now in the *wet growth regime*. The boundary between the dry and wet growth regimes marks the critical condition for which all the accreted supercooled water freezes on ice and the ice surface acquires a 0°C temperature. This condition is called the *Schumann–Ludlam limit*, as such conditions were studied first by Schumann (1938) and Ludlam (1958).

In their studies of the wet growth regime, Schumann and Ludlam assumed that the growing ice particles remain solid and that all liquid film on them would be shed away. However, later wind tunnel studies (e.g. Macklin, 1961) using rotating ice cylinders to simulate the hail growth process showed that this is not the case. Usually little or no shedding of water occurred. Instead, water and air bubbles would be soaked into the capillaries of the complicated ice meshes to form the so-called *spongy ice* (List, 1965). The shedding and sponginess depend strongly on the detailed motion of the ice particles.

10.5.2 A general description of the riming process

Wind tunnel riming experiments using a simulated ice plate that collects super-cooled droplets (Pflaum *et al.*, 1978) show that riming starts at the lower face of the plate, and drops that come later often collide with the earlier drops that have become rime instead of with the original crystal surface. The result is a tassel-like rime hanging like beads on the downward face of the crystal whose densities can be $0.1–0.3\,\mathrm{g\,cm^{-3}}$. This is also observed in natural rimed crystals. Fig. 10.9 shows an example of rimed sector plate crystal.

At a certain riming stage, whose exact nature is still unclear but most likely due to hydrodynamic torque, the rimed ice would flip over upside down and the rimed face will be turned upward. This behavior has been observed repeatedly in vertical wind tunnel experiments. The new riming process will now take place on the downward-facing surface. Presumably the rimed particle will now continue to rime all over its surface. After a while, the original ice crystal shape can no longer be distinguished.

Whether or not the riming sequence described above actually happens in clouds is unknown presently. Some of the rimed crystals may grow to become graupel, which are frequently conical in shape (see Chapter 2). Rimed crystals

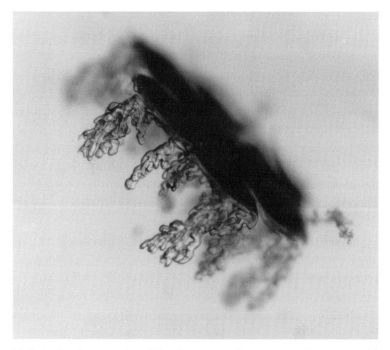

Fig. 10.9 A rimed sector plate. The rime grows just on one side, presumably the underside during the fall of the plate. Photo courtesy of Dr Charles A. Knight.

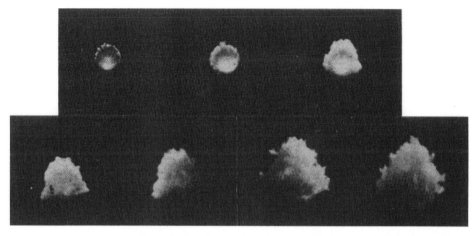

Fig. 10.10 The growth evolution from a drop into a conical graupel. The diameter of the frozen drop is about 450 μm; the diameter of the graupel at the last stage of development is about 2 mm. From Pflaum *et al.* (1978). Reproduced by permission of the Royal Meteorological Society (UK).

can have rather complicated motions, as we described in Chapter 8, but one of the modes – the bell swing, where the rimed particle swings left and right like a bell – is possibly responsible for the formation of the conical graupel. Fig. 10.10 shows the formation of the initial stage of a conical graupel starting with a frozen drop.

Further riming growth of graupel can lead to the formation of hail. Theoretically, both graupel and hail can go through the dry and wet growth regimes mentioned previously but most observations are related to hail. It is possible that hail may go through two or more dry and wet growth cycles and hence consists of two or more layers of alternating opaque and clear ice, like an ice onion. Such dry and wet growth layers can be easily distinguished when a thin slice of the hailstone is illuminated by polarized light from below. A wet growth layer consists of larger single crystalline patches (a single ice crystal structure with uniform molecular orientations) whereas a dry growth layer consists of small polycrystalline grains with random molecular orientations due to many small separate frozen drops. Fig. 10.11 shows three examples. The layers with larger grain size (with stronger dark and light contrast) are the wet growth layers, while those with small grains are the dry growth layers. The layers are quasi-concentric in Fig. 10.11(b) but not so in Fig. 10.11(a). The latter probably indicates that the stone fell in a particular orientation and hence the growth occurred preferentially on one side. The stone in Fig. 10.11(c) appears to be dominated by wet growth.

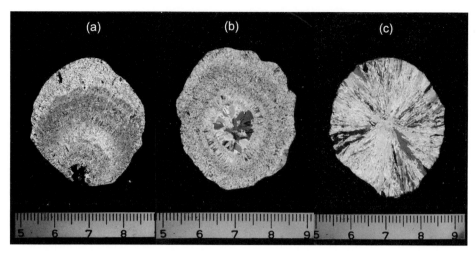

Fig. 10.11 A thin section of a hailstone illuminated by polarized light showing the layered structure of dry and wet growth regimes. Photos courtesy of Prof. Albert Waldvogel.

10.6 Collision efficiency between ice crystals and supercooled drops

Thus far, owing to the difficulty of computing the flow fields around larger falling ice particles such as heavily rimed crystals, graupel, and hail, the theoretical calculations of collision efficiencies between ice particles and supercooled droplets have been limited to that between small clean ice crystals up to a few hundred micrometers in diameter collecting small cloud droplets up to about 40 µm. Consequently, the results apply to the initial stage of riming but not to the later stage.

Using similar formulations as described for drop–drop collisions, we can determine the collision efficiencies of ice crystals colliding with supercooled water droplets. Details of the computational processes and results are given by Wang (2002). In the following we select some typical cases to show the characteristics of the ice–drop collision behavior. Fig. 10.12 shows a typical ice–drop collision trajectory for a broad-branch crystal colliding with a 2 µm droplet.

It is commonly assumed that, once a supercooled drop hits the crystal's surface, it will freeze and stick to the crystal, and hence the coalescence efficiency is one. So far there appear to have been no studies disputing this assumption. Thus, in the following discussion, we use "collision efficiency" and "collection efficiency" interchangeably.

Wang and Ji (2000) calculated the collision efficiencies of ice crystals collecting small supercooled water drops. Their results are shown in Figs. 10.13–10.16. Figs. 10.13–10.15 shows the computed collision efficiencies of three types of ice

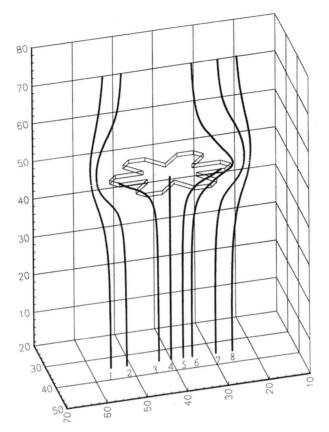

Fig. 10.12 Trajectories of a droplet of 2 μm radius moving in the vicinity of a falling broad-branch ice crystal at $N_{Re} = 10$. Trajectories 1, 2, 6, 7, and 8 are misses, and trajectories 3, 4, and 5 are hits. From Wang and Ji (2000). Reproduced by permission of the American Meteorological Society.

crystals – columns, hexagonal plates, and broad-branch crystals – colliding with supercooled water droplets of 1–100 μm radius (Wang and Ji, 2000). Tables 10.1–10.3 show the dimensions of the ice crystals considered in these calculations.

We see that the collision efficiencies of ice crystals collecting small cloud droplets behave in a similar way to drops collecting drops in Fig. 10.2, and thus can be explained similarly. The most obvious feature is that, in general, larger crystals have larger collision efficiencies for the same drop size. Both columns and plates have a fairly wide plateau of E with values between 0.9 and 1.0 for drop size between ~20 and ~90 μm. The efficiencies quickly drop to zero for larger drop size because drops fall faster than the crystal in that range and hence no collision can occur. The plateau of E for broad-branch crystals is smaller than the other two because the broad-branch crystals have smaller fall velocities than the other two and are "outrun" sooner by the falling drop.

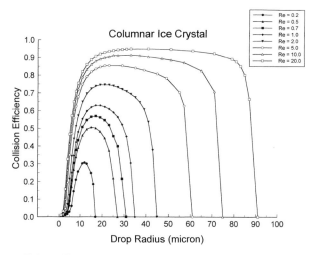

Fig. 10.13 Collision efficiencies of columnar ice crystals colliding with supercooled droplets. From Wang and Ji (2000). Reproduced by permission of the American Meteorological Society.

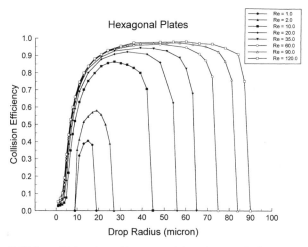

Fig. 10.14 Collision efficiencies of hexagonal ice plates colliding with supercooled droplets. The last data points (at large drop size end) for $N_{Re} = 20$–120 are extrapolated. From Wang and Ji (2000). Reproduced by permission of the American Meteorological Society.

An important question about riming is: How large should a crystal be for riming to begin? Table 10.4 shows some field observations of the cutoff size for ice crystals of various types to commence riming.

Fig. 10.16 shows a plot of the "maximum" collision efficiencies of each crystal type as a function of their size. The size where E drops to zero represents the cutoff size predicted by the numerical results. The agreement between the theory and

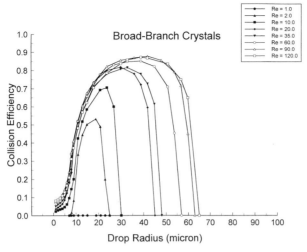

Fig. 10.15 Collision efficiencies of broad-branch crystals colliding with super-cooled droplets. The last data points (at large drop size end) for $N_{Re} = 20–120$ are extrapolated. From Wang and Ji (2000). Reproduced by permission of the American Meteorological Society.

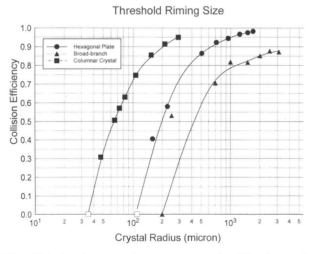

Fig. 10.16 Cutoff riming ice crystal sizes as extrapolated by the results shown in Figs. 10.13–10.15. For broad-branch crystals, the data point for crystal radius 2.5 mm was ignored when obtaining the best fit. From Wang and Ji (2000). Reproduced by permission of the American Meteorological Society.

observations seems reasonable, although numerical results predict a cutoff somewhat larger than observations. Discrepancies may be due to the difference in exact shapes and dimensions, and also because the theoretical predictions are based on extrapolations that may introduce some errors.

Table 10.1 *Reynolds numbers, dimensions, and capacitances of columnar ice crystals in the present study.*

N_{Re}	Radius (μm)	Length (μm)	Capacitance
0.2	23.5	67.1	1.3628
0.5	32.7	93.3	1.3628
0.7	36.6	112.6	1.4054
1.0	41.5	138.3	1.4535
2.0	53.4	237.4	1.6511
5.0	77.2	514.9	2.0151
10.0	106.7	1067	2.5067
20.0	146.4	2440	3.3959

Table 10.2 *Reynolds numbers, dimensions, and capacitances of hexagonal ice plates in the present study.*

N_{Re}	Radius (μm)	Thickness (μm)	Capacitance
1.0	80.0	18.0	0.7298
2.0	113.3	20.0	0.6977
10.0	253.3	32.0	0.6639
20.0	358.2	37.0	0.6485
35.0	473.8	41.0	0.6371
60.0	620.0	45.0	0.6278
90.0	750.0	48.0	0.6221
120.0	850.0	49.0	0.6179

Table 10.3 *Reynolds numbers and dimensions of broad-branch ice crystals in the present study.*

N_{Re}	Radius(μm)	Thickness(μm)
1.0	100.0	15.0
2.0	125.0	18.0
10.0	350.0	32.0
20.0	500.0	40.0
35.0	750.0	50.0
60.0	1000.0	60.0
90.0	1250.0	65.0
120.0	1550.0	73.0

Table 10.4 *Observed critical riming size. The code after the crystal habit is the Magono and Lee (1966) classification of natural snow crystals. From Wang and Ji (2000).*

Crystal habit	Wilkins and Auer (1970)	Reinking (1979)	Bruntjes *et al.* (1987)
Hexagonal plate (Pla)	–	–	$d = 150\,\mu m$
Broad-branch crystal (Plc)	–	$d = 275\,\mu m$	$d = 240\,\mu m$
Columnar crystal (Cle)	$l = 100\,\mu m$, $d = 40\,\mu m$	–	$l = 125\,\mu m$, $d = 40\,\mu m$
Long solid column (Nle)	$l = 100\,\mu m$, $d = 30\,\mu m$	–	–

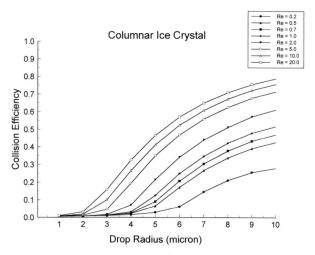

Fig. 10.17 Collision efficiencies of columnar ice crystals colliding with supercooled droplets windowed to drop size from 0 to $10\,\mu m$ to reveal the detailed features. From Wang and Ji (2000). Reproduced by permission of the American Meteorological Society.

At the smaller drop size end, the collision efficiencies are small, but, in contrast to the case of drop–drop collision, the E curves do not drop to zero sharply; rather, they taper off more gradually toward the small drop size end. An example of this more gradual decrease of E is shown in Fig. 10.17. We see that it is possible for ice crystals to collect droplets as small as $1\,\mu m$, albeit the efficiency is very low. The capability of collecting very small drops is the highest for broad-branch crystals and smallest for columns. There were some field observations showing the lack of rime drops smaller than 5 μm, and earlier theoretical studies suggested that E is zero for such drops. In view of Figs. 10.16 and 10.17, this lack of small rime drops is probably due to the local clouds that lacked such small drops but not due to the intrinsic zero collision efficiency.

10.6.1 Preferential riming near the rim of the crystal

Examination of the trajectories of the droplets around plate-like ice crystals indicates that drops collide with the plates preferentially near the rim of the plate. This seems to agree with the observation that rimes are usually the heaviest around the edge of the plate and the center part is usually relatively void of rime. Fig. 10.18(a) shows such an example. Fig. 10.18(b) is a side view, showing that the rimes form

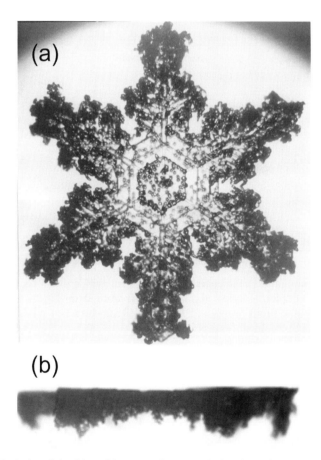

Fig. 10.18 A rimed dendrite with a center hexagonal plate in (a) face view and (b) side view. From Iwai (1983). Photos courtesy of Prof. Kunimoto Iwai. Reproduced by permission of the Meteorological Society of Japan.

almost completely on one side of the crystal, presumably the down-facing side, as also shown earlier in Fig. 10.9.

Pitter and Pruppacher (1974) explained this phenomenon as due to the wide stagnation region near the center of the downward-facing surface of the plate, which is a high-pressure region that tends to prevent drops from approaching. Consequently, most drops have difficulty penetrating this central high-pressure region and are swept to near the edge and, if collected, stick to the plate within an annular region close to the rim.

10.6.2 Riming on snowflakes

Riming certainly occurs to snowflakes as well. Unlike simple ice crystals, snowflakes consist of many ice crystals clumped together and hence have various porosities. Lew et al. (1986a, b) performed wind tunnel experiments to study the riming on both natural and simulated disk snowflakes of porosities ranging from 15 to 50%. They found that the riming growth rates are fairly insensitive to the snow porosity larger than 15%, but that the rates are about one order of magnitude greater than those for non-porous disks of the same size and rime at the same velocity.

10.7 Growth rates of rimed crystals to form graupel and hail

Continued riming of ice crystals would eventually lead to the formation of graupel. In principle, graupel can also grow in dry or wet growth regimes and that results in a wide range of graupel density, from 0.05 to 0.9 g cm^{-3}. Our current understanding of the growth rates of graupel mainly comes from laboratory experimental studies. Various investigators have performed wind tunnel experiments to determine the growth rates of graupel using either spherical ice particles suspended in a vertical wind tunnel (Pflaum and Pruppacher, 1979; Blohn et al., 2009) or suspended conical particles in an icing tunnel (Cober and List, 1993). Blohn et al. (2009) produced the following empirical relations for the collection kernels of graupel:

$$K_{ice6} = 10.06(mv)^{0.847}, \tag{10.8}$$

$$K_{ice10} = 10.22(mv)^{0.738}, \tag{10.9}$$

$$K_{ice15} = 10.72(mv)^{0.728}, \tag{10.10}$$

where K_{ice6}, K_{ice10}, and K_{ice15} are the collection kernels of graupel collecting drops of 6, 10 and 15 μm, respectively, and mv is the momentum of the graupel of mass m falling at velocity v. Here the unit of m is g, v is in cm s^{-1} and K is in cm^3 s.

There is currently no systematic theoretical study of the collision growth of rimed crystals to form graupel and hail, due to the previous difficulty of simulating unsteady flow past ice particles of more complicated shapes, but recent work on flow past conical graupel by Kubicek and Wang (2012) can be used to perform such calculations.

There are no wind tunnel measurements of hail growth rates using freely floating particles so far, due to the difficulty of confining such a particle stably in the tunnel. The available growth rate data are mainly obtained by using a fixed target in an icing tunnel, which can simulate the dry and wet growth environmental conditions (see e.g. Cober and List, 1993).

10.8 Collision between ice particles

Certainly ice particles can collide with each other to form larger ice particles. Not only can ice crystals collide with each other to form snowflakes, but they can collide with graupel and hail as well. However, our main concern in this section is the formation of snowflakes by crystal–crystal collision.

This is probably the area about which we have the least understanding among all the accretion growth scenarios of hydrometeors. There are only a few experimental studies that have measured the collision efficiencies of small ice crystals colliding with a fixed ice or other artificial target (e.g. Hallgren and Hosler, 1960; Hosler and Hallgren, 1960; Latham and Saunders, 1970; Keith and Saunders, 1989). Most of these experiments are more relevant to the problem of graupel collecting ice crystals instead of the snow aggregation process. When ice particles collide, they can either stick together (called sintering, a process of fusing the surfaces), simply interlock due to the branches (such as the case for dendrites), or may bounce apart. Fig. 10.19 shows a summary plot of previous experimental results. Given the complicated motions of the ice crystals in air, it is not surprising that the scatter of the data is fairly substantial. But there appears to be a trend that the collision efficiency is higher at higher temperature.

This increasing aggregation efficiency with temperature may also explain the observation of the maximum dimension of aggregates, which also shows a general increase with increasing temperature, with the maximum close to 0°C (Fig. 10.20). One possible explanation is that near 0°C surface melting (see Chapter 3) may commence and the sintering of ice crystals to form aggregates thus proceeds at a rate higher than at lower temperature. However, there is no detailed quantitative study on this mechanism.

Other investigators have estimated the aggregation efficiency (essentially cloud-averaged collision efficiency) based on aircraft observations of ice particles in cirrus clouds. They performed the so-called "Lagrangian spiral descent" by lowering the altitude of the aircraft at a rate approximately equal to the terminal fall speed of the ice particles and, at the same time, sampling the ice crystals including both monomers and aggregates. Then the aggregation efficiency is deduced by applying a bulk cloud microphysical model. In this way, Mitchell *et al.* (1996) deduced a cloud-averaged aggregation efficiency of 0.5. Field and Heymsfield (2003) deduced the

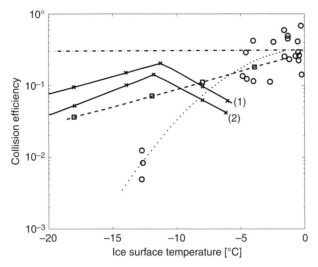

Fig. 10.19 Summary of experimental results of ice–ice collision efficiency: dashed-dotted, Latham and Saunders (1970); dashed, Rogers (1974); dotted, Hallgren and Hosler (1960); full, Hosler and Hallgren (1960). Adapted from Keith and Saunders (1989).

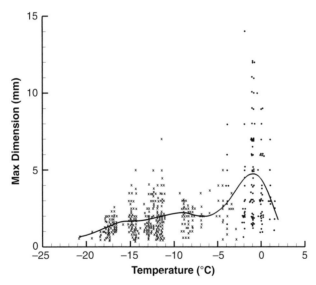

Fig. 10.20 Maximum dimensions of aggregates as a function of the temperature at which they were collected. Circles represent ground data, and crosses represent aircraft data. The solid line gives the most probable maximum dimension of the aggregates. Data derived from Hobbs *et al.* (1974).

aggregation efficiencies from 13 Lagrangian descents to be around 0.3, but they also stated that values of 0.1 also fitted the observations well. Mitchell *et al.* (2006) found that the aggregation efficiency is about 0.55 for a case dominated by dendrites at the cloud top. Naturally, the uncertainties of the estimates from this type of deduction must be substantial.

Recently, Connolly *et al.* (2012) performed a laboratory study where ice crystals are nucleated in a cloud chamber of 10 m height and allowed to fall and collide with each other to form aggregates. The temperature range is −30°C to 5°C. Aggregate samples are taken in the middle and at the bottom of the chamber. A cloud model with various schemes is used to interpret the data and to determine the aggregation efficiency. They found that they can rule out aggregation efficiencies larger than 0.5 at temperatures other than −15°C at the 75th percentile, whereas at −15°C the efficiency is somewhere between 0.35 and 0.85. The results are applicable to the initial stage of aggregation growth with particle size up to ~ 500 μm.

Currently there is no theoretical study of the aggregation efficiency of snowflakes due to the difficulty in simulating the realistic motions of snow crystals.

10.9 Melting of graupel, hail, and snowflakes

Melting of ice particles contributes to the production of water drops and impacts the drop size distributions, and hence is an important mechanism of the precipitation process. In principle, melting can start when ice particles fall to a level (or move into a region) where the temperature is warmer than 0°C. However, whether or not melting will occur also depends on the evaporative cooling rate at the surface of the falling particle. If the cooling rate is fast enough, melting may not start at 0°C but at warmer temperatures. For example, an ice particle falling in an environment of 50% relative humidity will start to melt when the environmental temperature is about 4°C.

Depending on the particle size and velocity and the heating rate of the environmental air, some particles, such as snowflakes, melt in a very short distance (a few hundred meters) into the warmer layer, while others, such as large hailstones, may melt just slightly even after falling for a few kilometers and end up with the bulk of the solid part reaching the ground.

There are both theoretical and experimental studies of the melting problem of ice particles. We shall take a brief look at some of these studies.

10.9.1 Melting of small frozen drops

The melting of small frozen drops has been observed in a vertical wind tunnel by first injecting a liquid drop into the observation section and allowing it to be freely suspended in a subfreezing air stream. Next some nuclei are introduced into the air

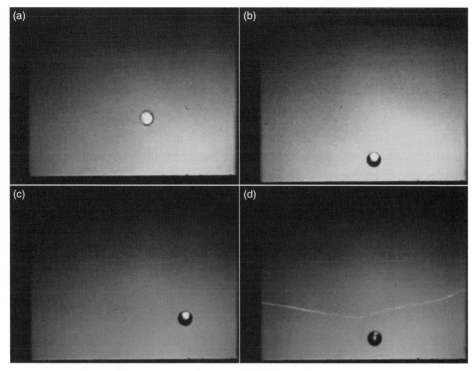

Fig. 10.21 Sequence of snapshots in a vertical wind tunnel experiment showing a drop of 350 mm radius, $RH = 0.85$, and heating rate 1.8°C min^{-1} or 1°C/100 m. Photos courtesy of Prof. Hans R. Pruppacher.

stream from below so as to nucleate the drop into ice. Then the upstream part of the tunnel is heated at a fixed rate so that the ice can start to melt. Fig. 10.21 shows a sequence of snapshots of a melting frozen drop of 350 μm radius in air of 85% relative humidity and heating rate of 1.8°C min^{-1}.

At first, the whole sphere is ice (white part) as in Fig. 10.21(a). The light shining through the ice sphere is strongly scattered by the crystal structure, making the ice sphere bright. Since the ice surface is not really smooth like a liquid, it often performs more complicated motions such as sailing and precession. As the melting commences, the liquid layer (the darker region) thickens and the ice sphere shrinks (Fig. 10.21b). The motion becomes more stable. But note that the ice sphere floats to the top, as its density is smaller than that of the liquid (Fig. 10.21b,c). Thus the ice sphere is not concentric with the drop but is eccentric with it. Later the ice core may not remain spherical either but rather becomes conical (with apex pointing *downward*). This process continues until the ice completely melts (Fig. 10.21d). Now the drop is completely liquid and the light shining through the drop appears as a point in the center. One important characteristic of the melting of small ice spheres is that the whole ice–water system remains essentially spherical during the melting process. It is clear from this sequence of photos that the melting of falling ice particles in clouds is a moving boundary problem even for a small ice sphere like this one if we want to develop a theory to calculate the melting rate.

10.9.2 *Melting of large ice particles*

The melting of falling large ice particles is much more complex, as the sequence in Fig. 10.22 shows. In this sequence, we first see that the ice starts to melt from below and liquid water appears (Fig. 10.22a). However, unlike the melting of small spheres, the ice core is too large to float to the top; instead, the liquid is pushed up by the vertical wind to form a skirt-like water torus around the ice (Fig. 10.22b). As time goes on, the liquid portion increases in volume and assumes a shape similar to a raindrop, i.e. flatter bottom and round top. The remaining ice core may still stick out at the top and bottom of the

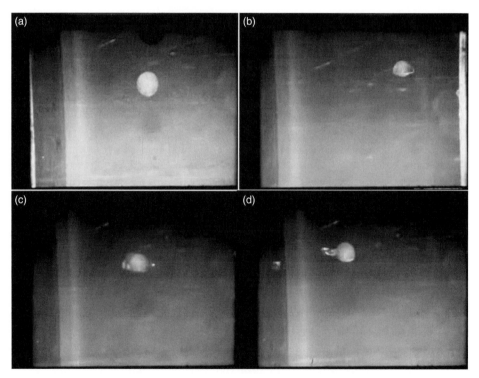

Fig. 10.22 The melting of an ice sphere of diameter 1.08 cm suspended in a vertical wind tunnel with $T = 20°C$ and relative humidity 40%. The sphere is attached to the tunnel by a flexible nylon string, as the tunnel does not generate enough wind speed to support such a large particle. Photos courtesy of Prof. Hans R. Pruppacher

water torus if it is still large enough (Fig. 10.22c). The water torus may not be in a steady state as the ice particle often rotates or vibrates and therefore may protrude more to one side than the other. Intermittent shedding of water drops may occur at this stage if the water oscillation becomes unstable (Fig. 10.22d). Such shedding may contribute significantly to the rain formation in deep convective precipitation. Eventually, the ice core will become so small that its melting behavior will be similar to that of a small ice sphere.

Based on wind tunnel experiments, Rasmussen *et al.* (1984b) summarized the observed melting behavior of large and small ice spheres as shown in Fig. 10.23. Here we see that, as the ice sphere of ~ 20 mm (Fig. 10.23, stage (1)) starts to melt, a liquid torus formed by the melt water from the lower half of the sphere is advected to near the "equator" (Fig. 10.23, stage (2)). As the melting goes on, the liquid mass increases and the water torus becomes unstable. Part of the water sheds away intermittently. The water torus also moves upstream (lower part of the particle) as the ice core becomes smaller (Fig. 10.23, stage (3)). After more shedding of drops occurs, the particle becomes smaller and the water torus becomes indistinct and forms a water cap with the ice core inside (Fig. 10.23, stage (4)). When the ice core becomes sufficiently small, it is completely submerged in the water, which now assumes the shape of a regular raindrop. No shedding occurs at this stage and the ice core floats to the top of the liquid drop (Fig. 10.23, stage (5)). The following two stages (Fig. 10.23, stages (6) and (7)) are the same as the melting of a small ice sphere described before.

10.10 Theoretical models of ice particle melting

Theoretical treatments of the melting of ice particles can be complicated, not only because they involve the complicated geometry and the moving boundary problem mentioned previously, but also because of the multimedia nature – liquid and solid phases – of the problem. However, they can be approximated by simplifications as in Mason (1956) and Drake and Mason (1966). They assumed that the ice core and the liquid layer remain concentric throughout the melting process, that the heat is exchanged by conduction, and that the internal circulation in the liquid portion (and hence convective heat transfer) can be neglected. As time goes on, the liquid layer thickens whereas the ice sphere shrinks until the ice is completely gone. Fig. 10.24 illustrates this assumption graphically.

Following Rasmussen and Pruppacher (1982), the above system can be treated by a semi-empirical theory as follows. Assuming that the melting process proceeds in a steady state and the system is stationary, the temperature distribution should satisfy the Laplace equation

$$\nabla^2 T = 0, \qquad (10.11)$$

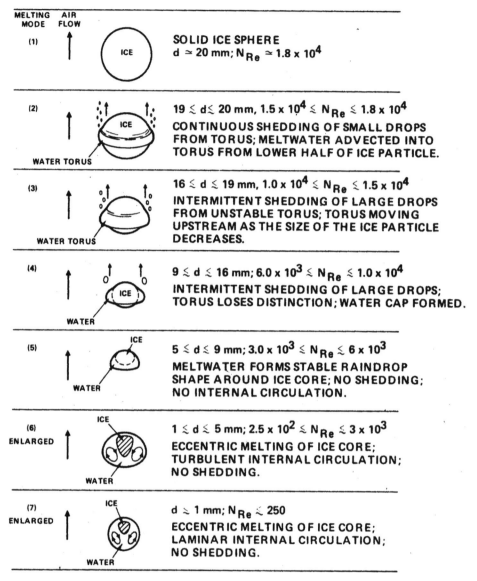

MELTING MODE	AIR FLOW		
(1)		ICE	SOLID ICE SPHERE $d \simeq 20$ mm; $N_{Re} \simeq 1.8 \times 10^4$
(2)		ICE / WATER TORUS	$19 \leq d \leq 20$ mm, $1.5 \times 10^4 \leq N_{Re} \leq 1.8 \times 10^4$ CONTINUOUS SHEDDING OF SMALL DROPS FROM TORUS; MELTWATER ADVECTED INTO TORUS FROM LOWER HALF OF ICE PARTICLE.
(3)		WATER TORUS	$16 \leq d \leq 19$ mm, $1.0 \times 10^4 \leq N_{Re} \leq 1.5 \times 10^4$ INTERMITTENT SHEDDING OF LARGE DROPS FROM UNSTABLE TORUS; TORUS MOVING UPSTREAM AS THE SIZE OF THE ICE PARTICLE DECREASES.
(4)		ICE / WATER	$9 \leq d \leq 16$ mm; $6.0 \times 10^3 \leq N_{Re} \leq 1.0 \times 10^4$ INTERMITTENT SHEDDING OF LARGE DROPS; TORUS LOSES DISTINCTION; WATER CAP FORMED.
(5)		ICE / WATER	$5 \leq d \leq 9$ mm; $3.0 \times 10^3 \leq N_{Re} \leq 6 \times 10^3$ MELTWATER FORMS STABLE RAINDROP SHAPE AROUND ICE CORE; NO SHEDDING; NO INTERNAL CIRCULATION.
(6) ENLARGED		ICE / WATER	$1 \leq d \leq 5$ mm; $2.5 \times 10^2 \leq N_{Re} \leq 3 \times 10^3$ ECCENTRIC MELTING OF ICE CORE; TURBULENT INTERNAL CIRCULATION; NO SHEDDING.
(7) ENLARGED		ICE / WATER	$d \simeq 1$ mm; $N_{Re} \leq 250$ ECCENTRIC MELTING OF ICE CORE; LAMINAR INTERNAL CIRCULATION; NO SHEDDING.

Fig. 10.23 Schematic of melting modes of an ice sphere of about 20 mm diameter as seen in vertical wind tunnel experiments. From Rasmussen *et al.* (1984b). Reproduced by permission of the American Meteorological Society.

with the boundary conditions

$$
\begin{cases}
T = T_0 & \text{at } r = a_i, \\
T = T_a & \text{at } r = a_d,
\end{cases}
\tag{10.12}
$$

where a_i and a_d are the radius of the ice core and the drop, respectively, and $T_0 = 273.15$ K $= 0°$C. Mathematically, this is the same kind of problem as the

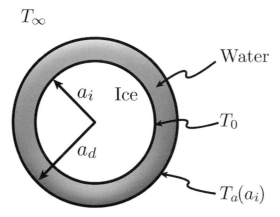

Fig. 10.24 Configuration of the problem of the idealized melting of an ice sphere. Adapted from Rasmussen and Pruppacher (1982).

system of (9.19) and (9.20) except for the slightly different outer boundary condition, and so we get the same kind of solution:

$$T = \frac{(T_0 - T_a)a_d a_i}{r(a_d - a_i)} + \frac{a_i T_0 - a_d T_a}{(a_i - a_d)}, \qquad (10.13)$$

and therefore

$$\left(\frac{dT}{dr}\right)_{r=a_i} = \frac{(T_0 - T_a)a_d a_i}{a_i^2(a_d - a_i)}. \qquad (10.14)$$

In steady state, the conservation of energy requires that the heat flux to the ice core is the same as the heat flux consumed by melting. Thus

$$4\pi a_i^2 k_w \left(\frac{dT}{dr}\right)_{r=a_i} = L_m \frac{dm_i}{dt} = 4\pi \rho_i L_m a_i^2 \left(\frac{da_i}{dt}\right), \qquad (10.15)$$

where m_i is the mass of the ice sphere and k_w is the thermal conductivity of liquid water. Finally, we get

$$\frac{da_i}{dt} = \frac{k_w[T_0 - T_a(a_i)]a_d}{\rho_i L_m(a_d - a_i)a_i}. \qquad (10.16)$$

Eq. (10.16) can be used to calculate the time required for the ice sphere to melt completely:

$$t_m = \int_0^{t_m} dt = -\frac{L_m \rho_i}{k_w a_d} \int_{a_i=0}^{a_i=a_d} \frac{(a_d - a_i)a_i}{[T_0 - T_a(a_i)]} da_i. \qquad (10.17)$$

Up to this stage, the quantity $T_a(a_i)$ is still an unknown quantity. It is related to the heat transfer occurring at the outer surface of the liquid. Now, for steady state, the rate of heat transfer through the water layer is balanced by the rate of heat dissipation by forced convection and evaporation at the surface of the falling, melting ice particle. In reality, both ice and water are not stationary but in motion, but this effect can be approximated by using the empirical ventilation coefficients. Thus we have

$$\frac{4\pi k_{\mathrm{w}}[T_0 - T_a(a_{\mathrm{i}})]a_{\mathrm{d}}a_{\mathrm{i}}}{(a_{\mathrm{d}} - a_{\mathrm{i}})} = -4\pi a_{\mathrm{d}}\{k_{\mathrm{a}}[T_\infty - T_a(a_{\mathrm{i}})]\bar{f}_{\mathrm{h}} + L_{\mathrm{e}}D_{\mathrm{v}}\left(\rho_{\mathrm{v},\infty} - \rho_{\mathrm{v},a_{\mathrm{d}}}\right)\bar{f}_{\mathrm{v}}\},$$

(10.18)

where T_∞ and $\rho_{\mathrm{v},\infty}$ are the temperature and vapor density far away from the drop surface, and $\bar{f}_{\mathrm{h}}$ and $\bar{f}_{\mathrm{v}}$ are the mean ventilation coefficients for heat and vapor, respectively (see Chapter 9 for a discussion of the ventilation effect). Eq. (10.18) allows us to determine $T_a(a_{\mathrm{i}})$ once the environmental conditions (temperature, relative humidity, pressure, and initial frozen drop size) are fixed, and from there t_{m} can be determined from (10.17).

Curve 1 in Fig. 10.25 shows the radius of the ice core as a function of melting time as determined by (10.17) and (10.18). The complete melting requires ~ 60 s for an ice sphere of 350 μm, which is $\sim 30\%$ longer than measured experimentally. Rasmussen *et al.* (1984a) found that, even after making improvements by considering an eccentric

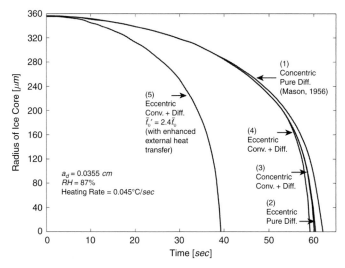

Fig. 10.25 Diagram giving the radius of the ice core as a function of melting time for various theoretical models. Curve 1, theory for concentric arrangement with pure diffusion as given by Mason (1956); curve 2, theory for eccentric arrangement with pure diffusion; curve 3, theory for concentric arrangement with internal circulation in the melt water; curve 4, theory for eccentric arrangement with internal circulation in the melt water; and curve 5, theory as for curve 4 but with $\bar{f}'_{\mathrm{v}} = 2.4\bar{f}_{\mathrm{v}}$. Adapted from Rasmussen *et al.* (1984b).

arrangement due to the ice core floating to the top of the liquid drop (curve 2), concentric arrangement with internal circulation in liquid (curve 3), and eccentric with internal circulation in liquid (curve 4), the improvements are limited to just a few percent. Only by adopting a new ventilation coefficient $\overline{f}'_v = 2.4\overline{f}_v$ do the theoretical results agree with experiments (curve 5). In general, the experimentally measured melting time is 30–40% less than predicted by (10.17) and (10.18). Rasmussen *et al.* (1984a) suggested that the combination of several effects, including (i) the increased surface area due to the frozen drop's oblateness and surface irregularities, (ii) the shedding of eddies at the rear of a melting particle falling in air, and (iii) the unsteady heat transfer due to random sailing motions of the particle during melting, is responsible for the enhancement in heat transfer that acts to shorten the melting time.

The melting of larger particles can be treated by similar considerations. Macklin (1963) performed such studies earlier upon which other later works are based. For details, see Rasmussen *et al.* (1984b).

Owing to the complicated melting and shedding, the falling behavior of large ice particles is even more complicated. The detailed evolution of hailstones is sensitive to many parameters, such the initial size, density, and humidity. Rasmussen and Heymsfield (1987) calculated various properties of a few typical large ice particles during their fall in an idealized environment (no wind and no liquid water), whose vertical temperature and humidity profiles are shown in Fig. 10.26. These profiles are modified based on the sounding of a severe storm that occurred in Montana on 1 August 1981, and hence are relevant to the hail production environment. As an example, Fig. 10.27 shows the sensitivity of the hailstone history to the initial density.

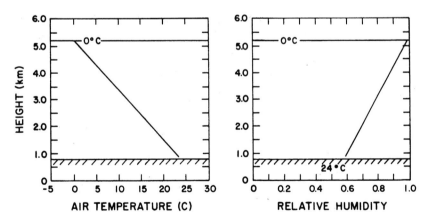

Fig. 10.26 Constant air temperature (°C) and relative humidity profiles used in 1D sensitivity runs by Rasmussen and Heymsfield (1987). The 0°C level is at 5.2 km above mean sea level (MSL) at 525 mb pressure, while the ground level is 0.8 km above MSL at 920 mb pressure. From Rasmussen and Heymsfield (1987). Reproduced by permission of the American Meteorological Society.

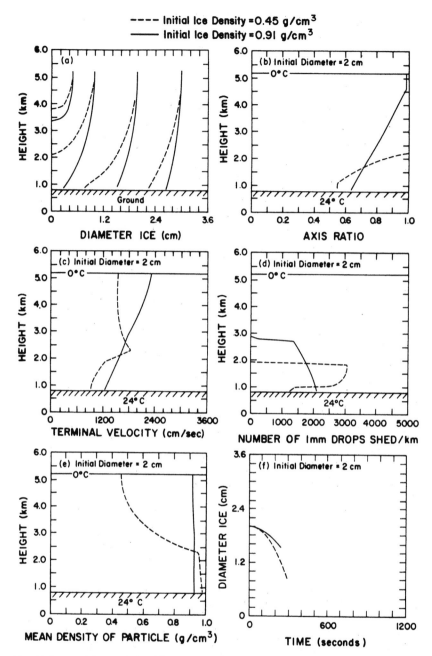

Fig. 10.27 Melting and shedding behavior of spherical ice particles with initial ice densities of 0.45 g cm^{-3} (dashed line) and 0.91 g cm^{-3} (solid line) falling through the temperature and relative humidity profiles shown in Fig. 10.26, with updrafts/ downdrafts and liquid water content set to zero. (a) Results for particles with initial diameters of 0.5, 1.0, 2.0, and 3.0 cm; (b)–(f) results for particles with 2.0 cm initial diameter. (a)–(e) Variation with height of (a) ice diameter (cm), (b) axis ratio, minor axis/major axis, (c) terminal velocity (cm s^{-1}), (d) number of 1 mm drops shed per kilometer during melting, and (e) mean density of the particle (g cm^{-3}). (f) Diameter of ice (cm) vs. time (s) from the start of melting. From Rasmussen and Heymsfield (1987). Reproduced by permission of the American Meteorological Society.

We see that, for a hailstone of initial diameter 2 cm and density $0.91\,\mathrm{g\,cm}^{-3}$ falling from 5.2 km, the size would become 1.5 cm diameter, whereas the same size hailstone but with density of $0.45\,\mathrm{g\,cm}^{-3}$ would have become a 0.7 cm particle (Fig. 10.27a). The higher-density hail would fall with a velocity decreasing more or less monotonically due to the decreasing size, but the lower-density hail would experience the increase and decrease of terminal velocity due to the melting and shedding behavior, which changes not only its size but its density as well (Fig. 10.27c,e). Different melting and shedding also result in a difference in the number of 1 mm drops shed per kilometer (Fig. 10.27d).

10.11 The melting of snowflakes and the bright band

The melting of snowflakes also contributes to the production of rain. The initial Bergeron–Findeisen precipitation theory actually proposed that this is the main source of raindrops. While this turns out not to be the case for convective rainfall, it is still important for many stratiform precipitations in higher latitudes.

The radar range height indicator (RHI) scan of precipitation systems, especially stratiform precipitation, often shows a quasi-horizontal band of enhanced reflectivity several hundred meters thick around the 0°C level (Fig. 10.28). This is called the *bright band*. The radar reflectivity of the bright band can be 6–15 dB higher than

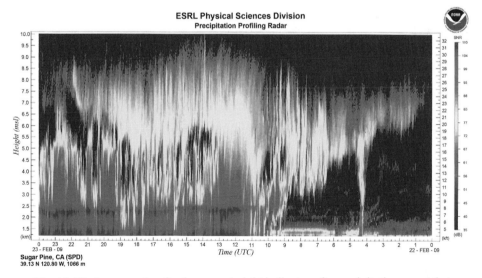

Fig. 10.28 An example of radar range height indicator of a precipitation event that occurred on 22 February 2009 at Sugar Pine, California, showing a bright band (the darker gray (red) color) between 2 and 2.5 km altitude. Image courtesy of NOAA. See also color plates section.

the rain reflectivity (Zawadzki *et al.*, 2005). There are many factors that contribute to this enhanced reflectivity, and one of the most important is the melting of large snowflakes. Upon melting, a snowflake would contain liquid water whose dielectric constant is much higher than that of dry snow. Because of the relatively large size of snowflakes (as compared to the size of the raindrops they eventually turn into) they appear to be fairly large raindrops to radar, hence the brightness of backscattered waves. When these snowflakes melt completely and collapse into raindrops, there is a sudden increase in fall velocities and decrease in particle concentration, reflectivity decreases dramatically, and the bright-band phenomenon disappears.

Theoretical and experimental studies of the snow melting process have been performed by several investigators, including Knight (1979), Matsuo and Sasyo (1981), and Mitra *et al.* (1990). The following are taken from Mitra *et al.* (1990), who observed the melting behavior of freely suspended natural snowflakes (mostly dendrites obtained during snowfall) in the Mainz vertical wind tunnel.

10.11.1 Fall attitudes of falling melting snowflakes

The fall attitudes of dry snowflakes have been described by Kajikawa (1982). Dry snowflakes may perform spinning, helical, shaking, and swing motions at certain frequencies depending on their size, shape, and mesh texture. As soon as melting starts, there is a sudden change in their fall attitudes, for example, the frequencies of rotation and oscillation, that do not seem to correlate with any geometrical or dynamical variables. They may perform sideways sailing in straight or curved paths but interrupted irregularly by helical, rotational, and swing motions until the ice framework collapses totally. At that point, there is an abrupt erratic downward acceleration. One notable observation is that they never perform tumbling.

10.11.2 Morphological changes of snowflakes during melting

Snowflake morphology is observed to go through four stages of changes:

(1) Melting starts at the tip of the branches where small droplets of tens of micro-meters diameter appear. Melting is the most intense at the periphery and the lower side of the flake.
(2) The melt water is directed to flow toward the linkages of the flake branches due to capillary forces and surface tension effect. The flake surfaces are not covered by water and hence appear ragged.
(3) Small branches of the flake interior melt. The melt water flows from these branches to the linkages of the main branches. This causes some structural rearrangements to take place in the flake. Some of the main branches bulge out

or suddenly flip inward. The crystal mesh of the flake changes from one with many small openings to one with a few larger openings.

(4) The main ice frame collapses suddenly. The melt water covers the remaining ice frame completely and pulls itself together into a drop shape.

Similar behavior has been reported by Knight (1979) and Matsuo and Sasyo (1981).

Problems

10.1. The presence of hydrodynamic drag greatly complicates collision problems (and, in fact, all problems of motion in air). Let us look at a simpler case where two spherical drops of radius a_1 and a_2, respectively, are colliding in an environment where air is absent (so no drag is involved). The drops are electrically charged and initially separated by a vertical distance $10a_1$. Determine the collision efficiencies for the following situations. You should determine the relative velocity between the two drops using Beard's formulas given in Chapter 8 assuming that $p = 900$ mb, $T = 10°C$, and $RH = 100\%$. Assume that the electric force is simply that between two point charges q_1 and q_2.

(a) Determine the collision efficiency for $a_1 = 74.3$ μm, $a_2/a_1 = 0.01, 0.05, 0.1, 0.2, \ldots, 0.9$. The electric charge on a_1 is $q_1 = +0.2a_1^2$ and the charge on a_2 is $q_2 = -0.2a_2^2$. The charge unit is esu (electrostatic unit) and a_1 should be in the unit of cm when calculating the electric force here.

(b) Do the same as (a) but for $a_1 = 19.5$ μm.

(c) Do the same as (a) and (b) but for $q_1 = +2a_1^2$ and $q_2 = -2a_2^2$ (this is the mean thunderstorm condition).

(d) Compare the results you obtain in (a), (b), and (c) with the corresponding results in Fig. 10.2. What conclusions can you get from such a comparison?

(e) The above cases are for attractive electric forces. Do the same for (a)–(d) above but for repulsive electric forces.

10.2. Based on Eq. (10.9), determine the time it takes for a spherical graupel of 3 mm diameter and density 0.5 g cm^{-3} falling through a cloud of uniform supercooled droplets of 10 μm radius to grow to a 3 cm hailstone by riming (assuming the environment is 800 hPa and $-10°C$).

11

Cloud drop population dynamics in the warm rain process

In Chapter 10, we discussed the collision, coalescence, breakup, and melting of individual water drops and ice particles. There we treated only the very short-time-interval event where only two particles are interacting. In a real cloud, there are a huge number of particles. A hydrometeor right after its inception will grow first by diffusion and, if it survives, will soon reach a size at which collision and coalescence may occur. It will interact with many other particles as time goes on. The time for such interaction to occur is very short and we can make an order-of-magnitude estimate of this time interval τ. For a small cumulus cloud with narrow drop size spectrum centered around $10\,\mu$m, the mean distance between cloud droplets is $\sim 1000\,\mu$m (1 mm). For cloud drops to collide, they have to have different velocities. The terminal fall speed of a $10\,\mu$m radius drop is about $1\,\mathrm{cm\,s}^{-1}$. Therefore, if we assume that the typical velocity difference is $\sim 1\,\mathrm{mm\,s}^{-1}$, then it takes about 1 s for a larger drop to interact with another drop. In other words, within $\tau \sim 1$ s, we only need to consider the collision between two drops. However, for a time interval longer than τ, it will be necessary to consider the time evolution of such events.

If coalescence occurs, this hydrometeor will lose its identity and the cloud will have a new population of hydrometeors of different types, sizes, and number concentrations. Thus the population of hydrometeors evolves as a function of time and space. In this chapter we will discuss the time evolution of the hydrometeor population, mainly the evolution of the cloud particle size distribution as a function of time. Our focus will be on the evolution of the cloud drop size distribution because it is the simplest to understand. The evolution of other hydrometeors is much more complicated.

At first glance, the collision growth of cloud drops looks like a simple problem. As long as the cloud environment (pressure, temperature, humidity) is fixed, and assuming that the coalescence efficiency is constant for the time being, the collision growth rate is just a function of drop size a_1 and a_2 of the collision pair. One would think that it would be a straightforward matter to determine the time evolution of the cloud drop size distribution. It turns out to be more complicated, as we shall see. In this chapter, we shall use the drop–drop collision as an example to illustrate the concept. The formation of precipitation that does not involve the ice phase is called the *warm rain process*.

11.1 Continuous growth model

The most straightforward conceptual model is the *continuous growth model*, which assumes that all drops of radius a_1 will have the same growth rate if they fall through a cloud region and collect a group of drops of uniform size a_2 so that after the collection process all a_1 drops will grow to a new size a_3. This idea can be traced back to Langmuir (1948), who used this model to estimate the rain formation rates. In this conceptual model, the a_2 droplets are assumed to be continuously distributed in space so that a_1 drops will grow in a continuous manner as they fall through such a cloud. This process is illustrated schematically in Fig. 11.1.

The number concentration (number of drops per unit volume of air) $n(a_2)$ of a_2 drops is related to the liquid water content of the cloud by $w_L = n(a_2)m(a_2) = n(4\pi/3)\rho_w a_2^3$. After an a_1 drop falls through this volume

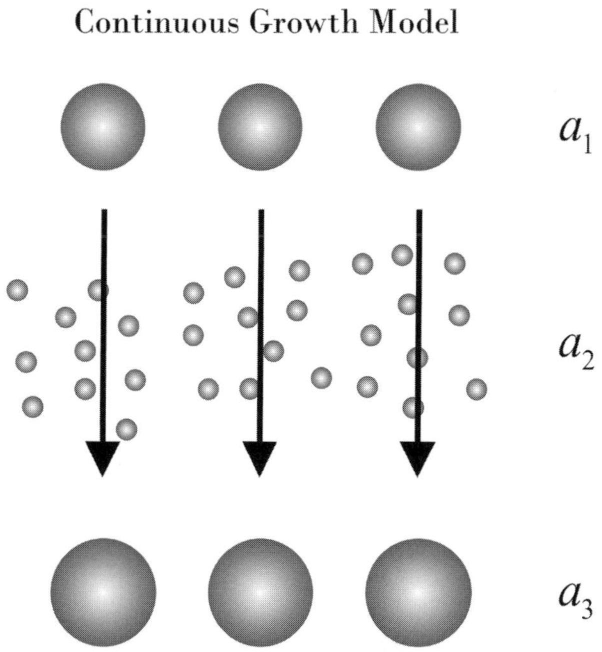

Continuous Growth Model

Fig. 11.1 Schematic of the continuous growth model.

of cloud in a time interval Δt, its mass will increase by an amount $\Delta m = K(a_1, a_2)n(a_2)m_2$ because the collection kernel K represents the volume swept out by an a_1 drop per unit time in which all drops will be collected (see Eq. (10.3)). Thus the mass growth rate of the a_1 drop is

$$\frac{dm_1}{dt} = K(a_1, a_2)n(a_2)m_2 = K(a_1, a_2)w_L = E_c\pi(a_1 + a_2)^2(U_{\infty,1} - U_{\infty,2})w_L,$$

(11.1)

where E_c is the collection efficiency and $U_{\infty,1}$ and $U_{\infty,2}$ are the terminal speed of drops a_1 and a_2, respectively. We have also used the relations $w_L = n(a_2)m_2$ and Eq. (10.4).

The above derivation assumes that the cloud region consists of evenly distributed uniform-sized (i.e. monodisperse) a_2 drops. For a polydisperse cloud consisting of a distribution of drop sizes, (11.1) can be generalized as

$$\frac{dm_1}{dt} = \int K(a_1, a_2)n(a_2)m_2 da_2 = \frac{4\pi\rho_w}{3} \int K(a_1, a_2)n(a_2)a_2^3 da_2,$$

(11.2)

where $n(a_2)$ is the concentration of a_2 drops per unit size increment. Eq. (11.2) indicates that the growth rate of an a_1 drop is a function of the collection kernel K and the size distribution. Once these two quantities are known, (11.2) can, in principle, be integrated.

11.1.1 Influence of updraft on drop size

The idea of continuous growth based on Eq. (11.1) can be used to obtain a quick (though qualitative) estimate of the time required for a drop to grow by collision and coalescence to a certain raindrop size in a cloud with constant updraft. This was done by Bowen (1950). First, we assume that the growing drop $a_1 \gg a_2$ so that the velocity of the latter can be ignored. Then (11.1) can be rewritten as

$$\frac{d}{dt}\left(\frac{4\pi\rho_w a_1^3}{3}\right) = E_c\pi a_1^2 U_{\infty,1}w_L,$$

and hence

$$\frac{da_1}{dt} = \frac{E_c U_{\infty,1}w_L}{4\rho_w}.$$

(11.3)

Now suppose that the updraft in the cloud is $W > U_{\infty,1}$ so that the a_1 drop will be carried upward by the velocity $(W - U_{\infty,1})$ and collide with smaller drops of radius a_2. After a while, this drop may grow large enough that its fall velocity is greater than W and thus may fall back to the cloud base level. With this in mind, (11.3) can be rewritten as

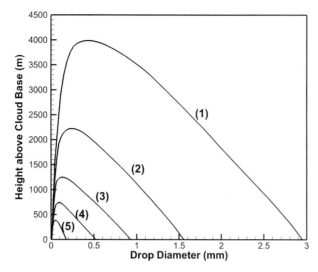

Fig. 11.2 The change in drop diameter with height for a range of vertical air velocities: (1) 200 cm s^{-1}; (2) 100 cm s^{-1}; (3) 50 cm s^{-1}; (4) 20 cm s^{-1}; and (5) 10 cm s^{-1}. Adapted from Bowen (1950).

$$\frac{da_1}{dz}\frac{dz}{dt} = (W - U_{\infty,1})\left(\frac{da_1}{dz}\right) = \frac{E_{\mathrm{c}}U_{\infty,1}w_{\mathrm{L}}}{4\rho_{\mathrm{w}}},$$

and thus

$$\frac{da_1}{dz} = \frac{E_{\mathrm{c}}U_{\infty,1}w_{\mathrm{L}}}{4\rho_{\mathrm{w}}(W - U_{\infty,1})}. \tag{11.4}$$

Solving (11.4) for a given set of updraft W and liquid water content w_{L}, we can obtain a_1 as a function of z. Fig. 11.2 shows an example (Bowen, 1950). We can see that the magnitude of the updraft would influence the drop size: drops will be larger when the updraft is larger. A cloud would need to have substantial updraft to cause the growth of large raindrops! This conclusion is generally supported by observations and is also applicable to graupel and hail, which also grow mainly by collision and coalescence. Conversely, it implies that the observation of large hydrometeors in a cloud must indicate that there are large updrafts in the cloud.

While Bowen's idea is conceptually plausible, the actual growth of large particles in clouds, especially those in thunderstorms, is not as straightforward as described above. The distribution of particles in a thunderstorm is highly heterogeneous in type, size, and concentration, and the growth history of any category of hydrometeors is complicated. The production of raindrops is not simply due to collision and coalescence of cloud drops, but can come from the melting of large snowflakes, graupel, and hail. Nevertheless, the idea that large particles can only be grown in a cloud with large updraft stands.

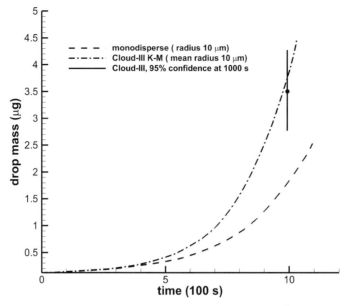

Fig. 11.3 Comparison of mass growth, $a_0 = 30\,\mu m$, $w_L = 1\,g\,m^{-3}$ in a monodisperse cloud of drop radius $10\,\mu m$ (dashed curve) and a cloud with a KM spectrum with mean drop radius $10\,\mu m$ (dashed-dotted curve). The point indicated with the black dot shows the mean mass from random sampling for the latter cloud with $n = 32$ at 95% confidence interval at $t = 1000\,s$. Adapted from Chin and Neiburger (1972).

Using the continuous growth model, Chin and Neiburger (1972) performed a study to compare the mass growth of a drop of initial radius $30\,\mu m$ collecting a monodisperse cloud of $10\,\mu m$ droplets and $w_L = 1.0\,g\,m^{-3}$ with another cloud with the same w_L but with a Khrgian–Mazin (KM) drop size distribution whose volume mean drop radius is $10\,\mu m$. Fig. 11.3 shows their results. Obviously, the growth in the polydisperse cloud is much faster than in the monodisperse cloud even though the two have the same mean drop radius. Thus the cloud drop size distribution certainly has an impact on the growth rate, and a polydisperse cloud allows faster growth of drops than a monodisperse one. The reason for the faster growth in the polydisperse cloud is due to the presence of large drops. Even though about 75% of the droplets are less than $10\,\mu m$ in the polydisperse cloud, droplets $>10\,\mu m$ made a major contribution to the growth that outperformed the monodisperse cloud. This observation underlines the importance of large drops in rain formation.

11.2 Stochastic growth model

The above discussions are all based on the concept of the continuous growth model – all drops of the same size will grow at exactly the same rate if they fall through a cloud of uniform-sized drops. While intuitively appealing, this model predicts the formation of precipitation-sized drops in a time twice or more than observed. So what was wrong?

There were suggestions that perhaps electric charges on droplets or the presence of giant condensation nuclei are necessary to achieve quick formation of raindrops. However, it is unlikely that electric charges should play a significant role in smaller clouds as the electrification is usually weak in this case. Giant salt nuclei turned out to exist in the precipitation events by Hawaiian trade wind cumuli (Beard and Ochs, 1993), but their role in the warm rain process in other locations remains unclear.

Surprisingly, the most convincing answer turns out to be that one should allow for the possibility of unequal collision growth rates for drops of the same size. Only then does the predicted time of precipitation formation become close to that observed. This collision growth model is called the *stochastic growth model* (called "statistical growth model" in the earlier literature). The importance of the stochastic nature of the collision and coalescence process was first recognized by Telford (1955), who pointed out that the continuous model only computes the average growth rate but ignores the possibility that some drops are luckier than others in collision. This seemingly minor point turned out to have great impact on how fast raindrops form.

Fig. 11.4 illustrates the main concept of this model. Suppose there are 100 drops of size a_1 at $t = 0$. Now let the collision growth start, but, instead of equal growth rate, only 10% of this population has the chance to grow. The simplest way to configure such a possibility is to consider the scenario (and quite realistic) that the small droplets are not uniformly distributed so that only a portion of large drops will collide with others in a time interval Δt. Thus Fig. 11.4 says that at $t = 1$, only 10 drops become larger with size a_2 while the remaining 90 drops stay at a_1 size. Using the same reasoning, at $t = 2$, only one drop out of the 10 a_2 drops can grow to a_3 while the remaining nine stay at a_2. Meanwhile, nine out of the 90 a_1 drops can also grow to a_2 size. Thus the total number of a_2 drops is 18 at this time, while there are 81 drops that remain at a_1 size.

One obvious feature here is that, unlike the continuous model, the stochastic model produces a broad size spectrum and allows the formation of large drops quickly. In contrast, by restricting to equal growth rate, the continuous model forces the drops to share the same average quantity of smaller droplets and consequently all of them grow but at a slower pace. As we have seen in Fig. 10.2, the collection efficiency only becomes significant when the collector drop size is large enough. The slowness of the continuous model to produce large drops results in the overall slow cloud growth.

The basis of the stochastic growth model is the *stochastic collection equation* (SCE)

Stochastic Growth Model

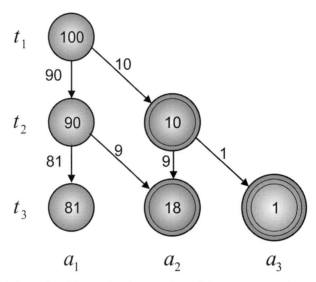

Fig. 11.4 Schematic of the stochastic growth model. Here t_1, t_2, and t_3 represent the time sequence whereas a_1, a_2, and a_3 represent the drop radius. The numbers next to the arrows represent the number of drops transferred to a certain size category.

$$\frac{dN_k}{dt} = \frac{1}{2}\sum_{i=1}^{k-1} A_{i,k-i} N_i N_{k-i} - N_k \sum_{i=1}^{\infty} A_{ik} N_i, \tag{11.5}$$

where N_k is the number concentration of a drop in size category k, and $A_{i,k}$ is the probability per unit time of coalescence between a drop of size i and a drop of size k. The first term on the right-hand side of (11.5) determines the number of drops of size k formed due to the coalescence of two smaller drops. The factor 1/2 on the right-hand side of (11.5) is to eliminate double counting because an i drop colliding with a $k - i$ drop is the same as a $k - i$ drop colliding with an i drop. The second term determines the number of drops that move out of the size k category because a k size drop collides with another drop. Eq. (11.5) can also be written in continuous form as

$$\frac{dN(k,t)}{dt} = \frac{1}{2}\int_0^k A(i, k-i)N(i,t)N(k-i,t)di - \int_0^{\infty} A(i,k)N(k,t)N(i,t)di, \tag{11.6}$$

where $N(k, t)dk$ represents the number concentration of drops with size between k and $k + dk$.

Comparing (11.5) and (11.6) with (11.1) and (11.2), we see that the probability $A_{i,k}$ or $A(i, k - i)$ plays the role of the collection kernel (see Gillespie, 1975, for a discussion of this term).

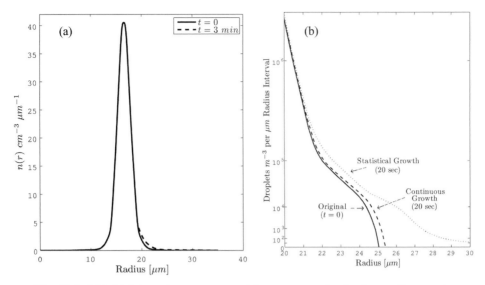

Fig. 11.5 (a) Initial drop size spectrum and the spectrum after 3 min on a linear scale computed based on a stochastic model. (b) The initial spectrum and the spectra after 20 s from continuous and stochastic models. Adapted from Twomey (1964).

Eq. (11.6) is usually solved numerically. One of the earliest numerical solutions of (11.6) was performed by Twomey (1964), who used the coalescence probabilities given by Hocking (1959), and Fig. 11.5 shows his results. The droplet size varies between 10 and 22 μm with a mode at 16 μm, $N = 135\,\text{cm}^{-3}$, and $w_{\text{L}} = 2.56\,\text{g m}^{-3}$. Fig. 11.5(a) shows the initial drop size spectrum (solid) and the spectrum after 3 min (dashed). The two spectra differ little in a linear plot like this except for the tail in the larger size end. To see the difference, we need to use the log scale in the concentration and focus on the tail, as in Fig. 11.5(b). Here we see that, after 20 s, the continuous model results in a spectrum that differs very little from the initial. The stochastic model, on the other hand, yields a wider spectrum and the largest drop size goes beyond 30 μm. Certainly, the stochastic collection process is a more efficient way to produce precipitation-sized particles.

11.3 Impact of the initial cloud drop size distribution

The stochastic model produces more reasonable predictions of the rate of rain formation and can be used in studying the population dynamics of cloud particles. One of the central problems of cloud physics is the colloidal stability of a cloud, as we have mentioned in Chapter 2. Many clouds have similar liquid water contents but the water is distributed differently, i.e. different drop size distributions. How would these distributions influence the stability of the cloud?

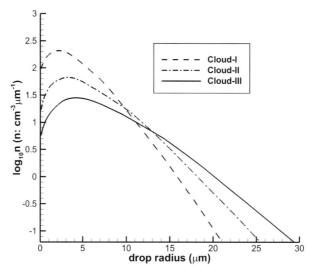

Fig. 11.6 Three clouds with KM drop size distribution with mean drop radius (I) 4.5 μm (dashed), (II) 6.0 μm (dashed-dotted), and (III) 7.5 μm (full). Adapted from Chin and Neiburger (1972).

Chin and Neiburger (1972) performed a Monte Carlo simulation of a stochastic growth model to address this question. They studied three clouds with different Khrgian–Mazin type cloud drop size distributions, as shown in Fig. 11.6. All three spectra have the same liquid water content $w_L = 1.0 \, \text{g m}^{-3}$ but different mean drop radius of (I) 4.5 μm, (II) 6.0 μm, and (III) 7.5 μm, as indicated. Cloud I has the narrowest spectrum, while cloud III has the broadest spectrum.

The evolution of the drop size spectra of these three clouds is shown in Fig. 11.7. The spectrum of cloud I shows very slow change. At $t = 3200$ s, the spectrum does not differ much from the initial spectrum and remains unimodal. Only 0.2% of the mass has moved beyond 100 μm. No sign of precipitation can be discerned in this cloud and it has high colloidal stability. Cloud II with a broader initial spectrum evolves faster than cloud I. At $t = 1800$ s, a second maximum appears at about 40 μm so that the spectrum is now bimodal. The second maximum moves to the right and increases in amplitude as time goes on. Cloud III with the broadest initial spectrum evolves the fastest. The second maximum appears shortly after $t = 1200$ s. Its amplitude increases rapidly and, at the same time, the amplitude of the original maximum decreases. By $t = 1800$ s, the number of drops larger than 100 μm is substantial enough to initiate precipitation albeit the precipitation is more like the drizzle level. This timing of precipitation initiation appears to agree reasonably well with observations.

The above results provide a good explanation of the observed cloud stability in relation to the drop size spectrum we discussed in Chapter 2. There we said that clouds with narrow drop size spectra are usually more stable whereas clouds with broader spectra are less stable and more likely to produce rain. They also tend to

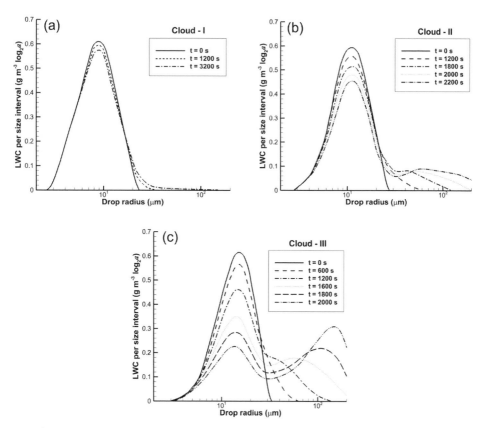

Fig. 11.7 Time evolution of the cloud drop size spectra of the three clouds in Fig. 11.6: (a) cloud I; (b) cloud II; and (c) cloud III. Adapted from Chin and Neiburger (1972).

have bimodal size distributions. We now see that the broader spectrum implies larger mean drop size and a higher number of large drops. These large drops have higher collection efficiencies and they cause rapid growth of even larger drops that enhances the growth process even more.

As one further demonstration of the sensitivity of cloud evolution to the initial drop size distribution, Chin and Neiburger (1972) also simulated a case using an initial cloud drop spectrum that is Gaussian. The Gaussian spectrum has the same liquid water content, mean drop size, and relative dispersion of droplets (~0.54) as the cloud III above, but unlike the KM distribution, the Gaussian distribution is not skew but symmetric with respect to the peak concentration. The results show that the Gaussian cloud develops at a much slower pace than cloud III even though the two have the same w_L, mean drop size, and spread. Thus a slight difference in similar-looking spectra can result in very different outcomes.

More elaborate calculations based on the stochastic collection model were performed by others, and the following examples are taken from Berry and Reinhardt

(1974). They used the collection kernels determined by Hocking and Jonas (1970) for drop radius $a \leq 40\,\mu$m and by Shafrir and Neiburger (1963) for $a > 40\,\mu$m. Unlike cloud droplet sizes in Chin and Neiburger (1972) that reach drizzle drops after ~30 min, Berry and Reinhardt (1974) yield millimeter-sized raindrops after 30 min.

These figures again illustrate the sensitivity of rain formation to the initial drop size distribution in both the mean drop size and the spread of the spectrum. Comparing Fig. 11.8(a) and (c), where the mean drop radius is 10 and 12 μm, respectively, we can see that, by changing the mean radius a little, the latter evolves into a bimodal distribution with most drops in drizzle drop size (~500 μm) after 30 min, whereas the former only evolves into mostly large cloud drop size (~30 μm). Comparing Fig. 11.8(b) and (d), we can see that, for both cases with the same mean drop radius (14 μm), the one with

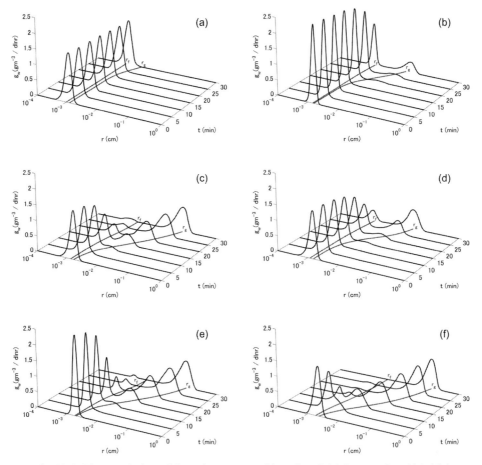

Fig. 11.8 Time evolution of drop size spectra with various initial properties: (a) initial mean radius $r_f^0 = 10\,\mu$m and relative dispersion var $x = 1$; (b) $r_f^0 = 14\,\mu$m, var $x = 0.25$; (c) $r_f^0 = 12\,\mu$m, var $x = 1$; (d) $r_f^0 = 14\,\mu$m, var $x = 1$; (e) $r_f^0 = 18\,\mu$m, var $x = 0.25$; and (f) $r_f^0 = 18\,\mu$m, var $x = 1$. Adapted from Berry and Rheinhardt (1974).

larger spread (Fig. 11.8d) evolves into mostly small raindrops (~700 μm) whereas the one with small spread produces only small drizzle drops (~150 μm) in 30 min. Finally, Fig. 11.8(e) and (f) show that, with initial mean drop size 18 μm, the cloud produces 1 mm raindrops in about 30 min, which agrees reasonably well with observation.

11.4 The condensation broadening of the coalescence growth size spectrum

When drops are growing due to collision and coalescence, diffusion growth does not stop but can go on at the same time as long as the conditions are right. Although the diffusion growth rate is generally smaller than the collection growth rate, it has an amplification effect on the collection growth due to the above-mentioned size sensitivity of the spectrum. This effect was studied by Kovetz and Olund (1969) for supersaturation and shown clearly by Ryan (1974) as illustrated in Fig. 11.9.

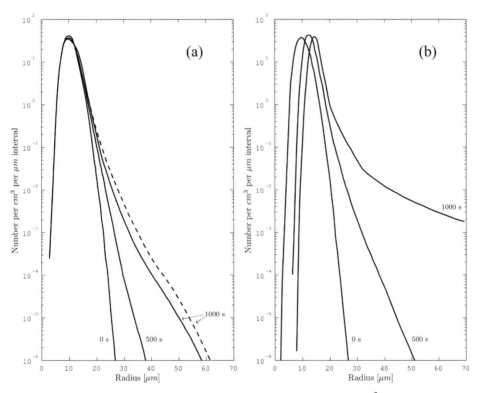

Fig. 11.9 (a) The change in an initial drop distribution of 200 cm^{-3} after 500 and 1000 s when the Klett and Davis (1973) kernels are used. The dashed line shows the change after 1000 s when the shear kernel is used. (b) The effect of including condensation controlled by the wet adiabatic lapse rate and an updraft of 1 m s^{-1} with the coalescence equation. Adapted from Ryan (1974).

Fig. 11.9 demonstrates that, with the inclusion of diffusion growth at the same time as coalescence growth, the drop size spectrum broadens faster than without. Without considering the condensation, the largest drop size after 1000 s is only about 60 μm. When the condensation effect is included, the largest drop size produced after 1000 s is way beyond 70 μm and the spectrum is much broader. This is because the diffusion growth increases the drop size in general, which also makes the collection kernel larger in general, and hence results in quick broadening. We shall call this effect *condensation broadening* of the coalescence size spectrum.

11.5 The impact of giant and ultragiant condensation nuclei

We have seen that a few large drops in the cloud can have a great impact on the coalescence growth of the whole drop size spectrum. In Chapter 2, we see that maritime clouds typically have fewer CCN than continental clouds and hence usually contain larger drops. It is commonly acknowledged that a maritime cloud with a similar liquid water content to a continental cloud would be more unstable colloidally and more ready to rain. On the other hand, continental clouds should normally be quite stable given the observed usually high CCN concentrations that lead to many small drops. Hence, in all likelihood, the warm rain process should usually not occur in continental clouds. Indeed, many continental clouds do dissipate without producing rain, but there are also many that produce rain rather rapidly without any involvement of the ice phase. How can this happen?

In the growth models presented in previous sections, the initial drop size spectra were prescribed and the distribution is based on the idea that the initial droplets form by nucleation. Small nuclei form small drops, whereas large nuclei form large drops. But these are just the "drops" due to condensation. There might also be particles that are not drops, such as the giant and ultragiant nuclei. As far as collision goes, it does not really matter whether the collision is between two drops or between a drop and a dry particle. Certainly the collision between a giant dry particle and a large drop and the subsequent coalescence would produce an even larger drop that serves to accelerate the coalescence growth of the cloud. Therefore, if giant nuclei are present in a cloud, they should be taken into account when calculating the coalescence growth even if they are not good CCN and cannot form cloud drops by condensation.

Are there giant nuclei in continental clouds? The answer is "yes" and researchers have been aware of them for a long time. For example, Squires (1958a) found that the colloidal stability of continental clouds was caused by higher concentrations of large nuclei rather than by insufficient concentrations of giant nuclei. Not only do giant nuclei exist but some of them are very large and are called *ultragiant nuclei* (those with radius greater than 10 μm). Ultragiant nuclei exist in both maritime and continental clouds. In maritime clouds they are most likely sea salt particles, whereas in continental clouds they are mostly of soil origin, relatively insoluble, and generally not effective CCN. But some of these land nuclei can be quite large. For example, Hobbs *et al.* (1985) found variable concentrations of 50 μm diameter particles in the range of 100–1000 m^{-3} from measurements taken during many flights around the US High Plains in the mixed layer (1–3 km above ground level).

Ochs (1978) performed a numerical study of the coalescence growth of maritime and continental clouds using different CCN distributions, but included in both a very small number of ultragiant particles of 50 μm diameter (assumed to be NaCl for the maritime cloud and soil particles for the continental cloud) at a concentration 1000 m^{-3}, which is also the typical concentration of raindrops. He was able to initiate the formation of raindrops quickly in both cases. While it is not surprising to see the formation of rain in the maritime cloud, it is somewhat surprising to see that the continental cloud also develops rain rather quickly (see Fig. 11.10). This underlines

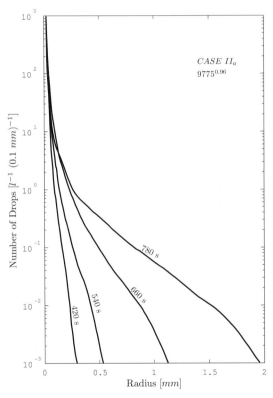

Fig. 11.10 Evolution of large drop concentration of a continental cloud using the average of data gathered by Braham (1974) in March and April 1973 in upwind areas near St. Louis, Missouri, as part of project METROMEX and simulated using moment-conserving numerical techniques. See the original paper for details. Adapted from Ochs (1978).

the plausibility of the idea that the presence of these ultragiant nuclei in clouds may promote the efficiency of rain formation even in continental air masses.

There are also suggestions that mixing different cloudy air parcels may lead to more broadening of the drop size distribution than for the original spectra in the unmixed parcels. Such broadening may then help to initiate large drops and hence accelerate the coalescence growth. However, there is no conclusive observation to substantiate this theory.

11.6 Drop breakup effect on the drop spectrum

Once large raindrops form, they are subject to instability induced by either collisions or environmental turbulence. We have discussed the mechanism of drop breakup in Chapter 10. Clearly, the drop breakup will influence the drop size spectrum. Theoretical studies by various researchers, such as Valdez and Young (1985), List *et al.* (1987), Brown (1988), and List and McFarquhar (1990), predict that the inclusion of drop breakup in the warm rain process would result in

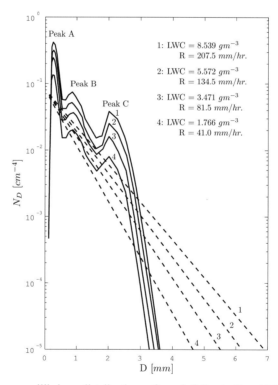

Fig. 11.11 Four equilibrium distributions, for rainfall rate $R = 41.0, 81.5, 134.5$, and 207.5 mm h^{-1}, and their MP counterparts for the same liquid water contents. Adapted from Valdez and Young (1985).

a three-peaked equilibrium distribution (3PED), and Fig. 11.11, adapted from Valdez and Young (1985), is an example.

Valdez and Young (1985) used a Markov chain model to treat the collision and coalescence process, and the drop breakup parameterization by Low and List (1982a,b) to obtain the four equilibrium drop size distributions. The corresponding Marshall–Palmer (MP) distributions of the same amount of liquid water content are also shown as dashed lines. There are three distinct peaks: A at 0.24 mm, B at 0.87 mm, and C at 2.0 mm. They also performed the flux analysis to determine the origin of these peaks, and concluded that satellite droplets created due to filament breakup, sheet and disk breakup fragments, and the combined solitary large fragments of disk breakup are the three mechanisms responsible for these peaks. Brown (1988) performed a similar study and showed clearly that these three mechanisms do produce the three peaks (Fig. 11.12).

The above trimodal distributions are derived from theoretical studies. To this day, however, no such distributions have been observed. Several possible reasons for this absence have been offered, among them, instrumental errors in disdrometers, turbulence and evaporation factors, and time sorting of raindrop sizes

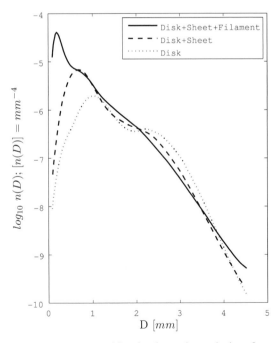

Fig. 11.12 Drop spectrum at $t = 12$ min through evolution from an initial MP distribution due to coalescence and disk breakup (dotted curve), coalescence, disk and sheet breakup (dashed curve) and coalescence and all three types of breakup (solid curve). Adapted from Brown (1988).

(Pruppacher and Klett, 1997). Clearly, more studies are needed to understand this problem.

The discussions in this chapter pertain only to the warm rain process, i.e. no ice particles are involved. Thus the clouds involved in this type of rainfall event cannot be too deep. In the precipitation events by deep convective clouds, raindrops are also produced by the melting of ice particles and the shedding of liquid water from them in addition to the collision and coalescence of drops. Therefore, the resulting spectra would have to take into account these additional factors, and the above arguments need to be completely modified.

The production of rain in a typical severe thunderstorm using a cloud model will be discussed in Chapter 13.

Problem

11.1. Fig. 11.2 is for drop diameter vs. height. Recalculate and plot a similar chart but for drop diameter vs. time. What conclusions can you make from your chart?

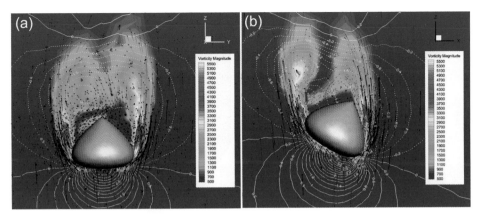

Fig. 8.24 (a) The velocity vector field (black arrows), the pressure field (white contours) and the vorticity field in the central cross-sectional plane around a falling graupel with inclination angle $\theta = 0°$ and $N_{\text{Re}} = 440$. (b) The same as (a) except for $\theta = 30°$ based on the calculations of Kubicek and Wang (2012).

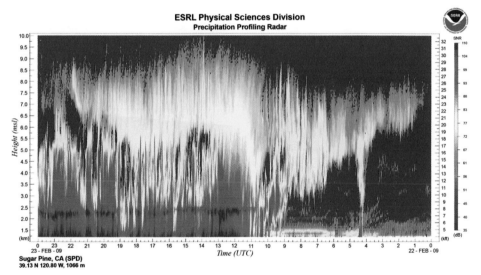

Fig. 10.28 An example of radar range height indicator of a precipitation event that occurred on 22 February 2009 at Sugar Pine, California, showing a bright band (the darker gray (red) color) between 2 and 2.5 km altitude. Image courtesy of NOAA.

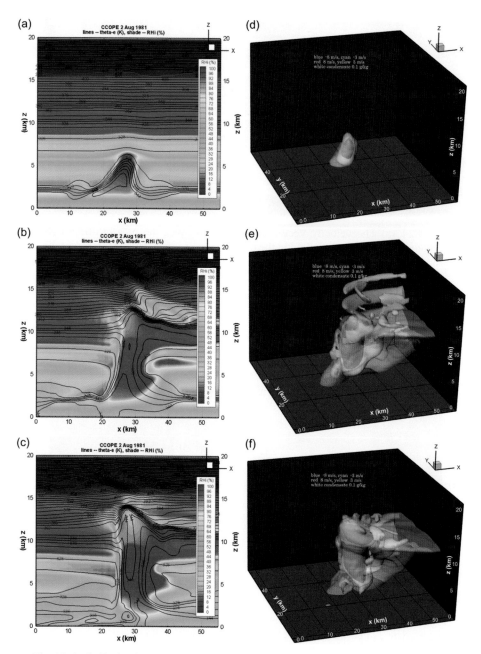

Fig. 13.4 (left) The θ_e field (black contours) in the central east–west cross-section of the storm at (a) $t = 10$ min, (b) $t = 30$ min, and (c) $t = 60$ min. The color shades represent the relative humidity with respect to ice (RH_i) field. (right) The 3D structure of the vertical velocity field at (d) $t = 10$ min, (e) $t = 30$ min, and (f) $t = 60$ min. Isosurface colors: blue, -8 m s^{-1}; cyan, -3 m s^{-1}; yellow, 3 m s^{-1}; red, 8 m s^{-1}. Faint white isosurface: total condensate mixing ratio 0.1 g kg^{-1}, which is used to approximately represent the visual cloud boundary.

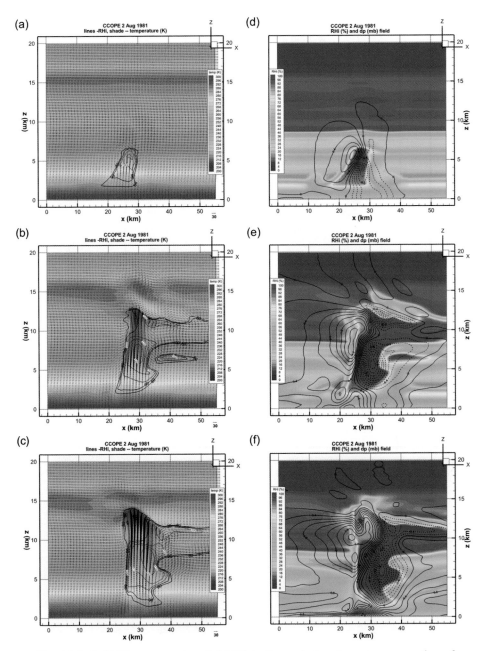

Fig. 13.5 (left) The temperature field (K) in the central east–west cross-section of the storm at (a) $t = 10$ min, (b) $t = 30$ min, and (c) $t = 60$ min. Black contours are for $RH_i = 70$–90% while purple contours are for $RH_i = 100\%$ or greater. Black arrows show the wind vectors projected onto this x–z plane. A reference vector of 30 m s^{-1} is shown in the lower right corner. (right) The perturbation pressure (p') in the central east–west cross-section of the storm at (d) $t = 10$ min, (e) $t = 30$ min, and (f) $t = 60$ min. Solid (dashed) curves represent positive (negative) pressure perturbations. The color shades represent the RH_i field.

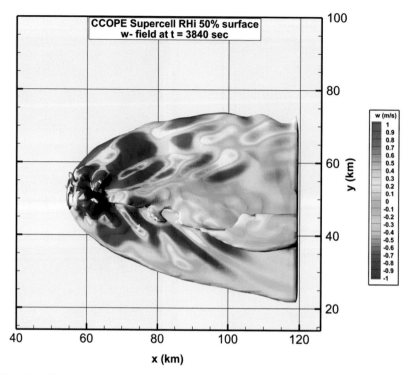

Fig. 13.6 Vertical velocity field at the storm top (represented by the $RH_i = 50\%$ isosurface) showing the ship wave signature.

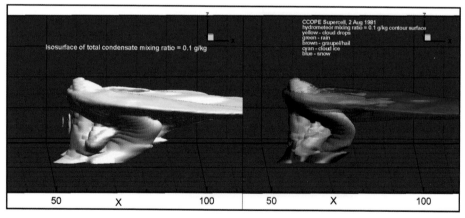

Fig. 13.11 (a) The isosurface of total condensate mixing ratio $= 0.1 \, \mathrm{g \, kg^{-1}}$. (b) The corresponding distribution of hydrometeors represented by the $0.1 \, \mathrm{g \, kg^{-1}}$ isosurface of the respective hydrometeor. Isosurface colors: yellow, cloud water; green, rain; red, graupel/hail; cyan, cloud ice; blue, snow.

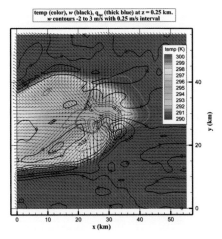

Fig. 13.7 Temperature field (color) and vertical velocity field (black contours: solid = positive, dashed = negative) at the surface of the simulated CCOPE supercell. The thick blue curve is the cloud boundary represented by the $0.1\,\mathrm{g\,kg^{-1}}$ total condensate mixing ratio (sum of all condensed forms of water).

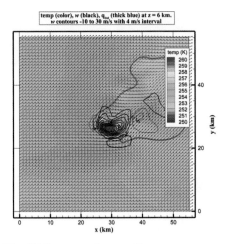

Fig. 13.9 Same as Fig. 13.7, except for $z = 6$ km.

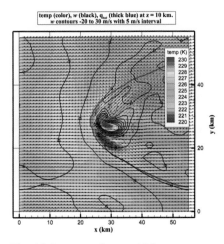

Fig. 13.10 Same as Fig. 13.7, except for $z = 10$ km.

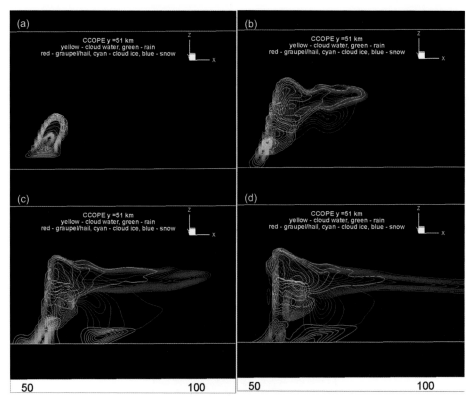

Fig. 13.12 Snapshots of hydrometeor distribution in the central east–west cross-section (y = 51 km) at (a) 10 min, (b) 20 min, (c) 30 min, and (d) 60 min. Color and range of mixing ratio contours ($g\,kg^{-1}$): yellow, cloud water (0.1–2.8); green, rain (0.1–4.2); red, graupel/hail (0.1–8); cyan, cloud ice (0.1–2); blue, snow (0.1–1.3).

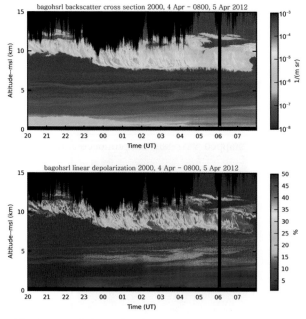

Fig. 13.17 Lidar image of cirrus development from 20:00 UTC 4 April to 08:00 UTC 5 April 2012 at Madison, Wisconsin (UTC = Coordinated Universal Time). (a) Aerosol backscattering cross-section σ. The color scale is $\log_{10}\sigma$ in unit of $m^{-1}\,sr^{-1}$ (sr = steradian). (b) Aerosol linear depolarization (%). Courtesy of Dr Edwin Eloranta.

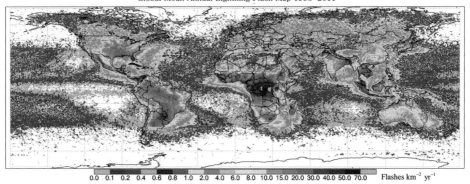

Flashes km^{-2} yr^{-1}

Fig. 14.8 The mean annual flash rate from combined space-borne Optical Transient Detector (OTD) and Lightning Imaging Sensor (LIS) data at 0.5° grid resolution from May 1995 to December 2011. Data from NASA, courtesy of Ms Sherry Harrison.

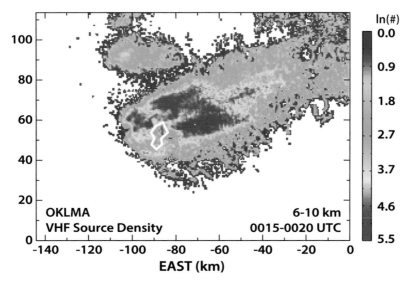

Fig. 14.10 Density of mapped VHF sources produced at 6–10 km from 00:15 to 00:20 UTC on 30 May 2004, during the Thunderstorm Electrification and Lightning Experiment (TELEX). The density color scale is a $\log_e = \ln$ scale. The white outline delineates the BWER. Note that the minimum lightning density in the hole is within the BWER and the core of maximum updraft speed. Coordinates are relative to the location of the KOUN radar. From MacGorman *et al.* (2008). Reproduced by permission of the American Meteorological Society.

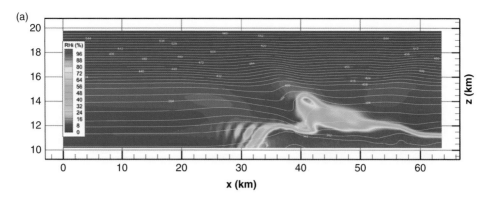

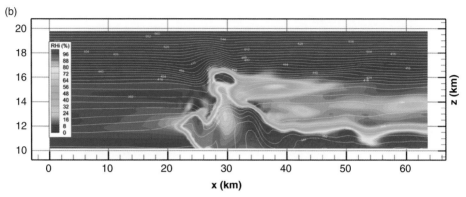

Fig. 15.16 Two snapshots of the gravity wave breaking process at the cloud top of the simulated CCOPE supercell (see Chapter 13). (a) Wave breaking in the anvil sheet. (b) Wave breaking in the overshooting area. Grayscale/color: RH_i. White contours: θ_e.

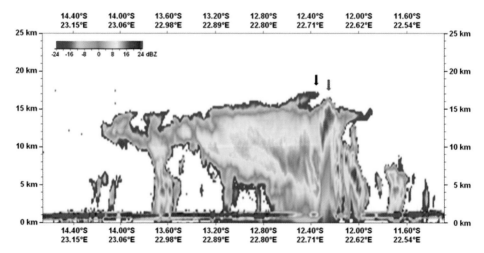

Fig. 15.18 CloudSat image, of 14 November 2009 at 12:20 UTC, showing a jumping cirrus occurrence (black arrow) at the top of a deep convective storm system in the Southern Hemisphere over eastern Angola in Southern/Central Africa. The overshooting top of the storm is indicated by the red arrow. The jumping cirrus was downwind of the overshooting top. Adapted from Wang *et al.* (2011). Data source: NASA/CloudSat.

12

Fundamental cloud dynamics

In the discussions of cloud microphysical processes, we have implicitly assumed that the dynamical and thermodynamic environments remain unchanged. In reality, cloud microphysical processes are closely coupled with dynamics and thermo-dynamics, such that changes in cloud microphysics will lead to changes in cloud dynamics and thermodynamics, and vice versa. For example, a phase change of water substance will cause the release or consumption of latent heat, which will heat or cool the surrounding air and eventually lead to air motion. The formation of large hydrometeors such as raindrops or hail increases the downward drag of the air parcel they are in and thus may cause the air to descend. The ascent or descent of air causes cooling or heating due to adiabatic expansion or compression, and may lead to the formation or dissipation of hydrometeors.

To really take these interactions into account, we need to develop a cloud model that includes the coupled dynamical, thermodynamic, and microphysical processes. We will discuss the cloud models in the next chapter, but it is useful to review some key dynamical processes of cloud development in a simplified setting so as to understand the fundamental processes involved. This is the subject of this chapter.

12.1 Cloud motions

Before we discuss the basic cloud dynamics, let us try to clear up a common misunderstanding that many casual cloud watchers may have. It is our common experience to see a small puffy cumulus floating in air against the blue sky and the cloud sometimes moves in a certain direction. Our intuition would be that the cloud moves with the wind. Is that true? The answer is "no"! It is the air that is moving, and the cloud forms in the air. The moving speed of the cloud may or may not be the same as the air speed. In some cases, such as the wave clouds that form in the lee of a mountain, the clouds often appear to be stationary but the air moves rapidly! Fig. 12.1 shows two frames of wave clouds taken from a video dated 16 August 2007

Fig. 12.1 Mountain wave cloud formation at Rauris Valley (Raurisertal) in Hohe Tauern National Park, Salzburgerland, Austria, on 16 August 2007. Frames (a) and (b) were separated by about 30 min. The wind was blowing, but the cloud appears to be stationary. The video is available at http://www.setvak.cz/timelapse/2007.html. Courtesy of Dr Martin Setvák.

(available at http://www.setvak.cz/timelapse/2007.html). The two frames were taken 30 min apart. By watching the video, it is clear that there were winds but the cloud in the center left of the picture appears to stay in the same place, although its shape has changed. What actually happened was that the wind blew across the mountain and produced mountain waves. The wave motion at the location where this cloud appeared is such that it is favorable for cloud formation. When water vapor carried by winds arrives here, condensation occurs and cloud appears. But when cloud particles (also carried by winds) pass this position, they evaporate because of unfavorable conditions and the cloud disappears. But new air with water vapor keeps coming and condensation keeps occurring at the same location, and that gives the illusion that the "cloud" is always there.

When we watch these time-lapse cloud videos, we sometimes see that the clouds literally "disappear before our eyes", and sometimes "appear from nowhere". It is clear that we are not seeing the motion of an object, such as a solid ball that does not change its content during its motion. Rather, a cloud is an ensemble of gases (air, water vapor, etc.) and condensates (liquid drops, ice particles). These condensates, and hence the cloud, may appear or disappear depending on the environmental conditions, and what we are seeing as the "cloud" entity is the motion of the "pattern" of the cloud, not the cloud particles.

The so-called "touch-down" of a tornado funnel is of similar nature. The funnel is made of water drops due to the rapid expansion of air in the funnel, so it is essentially a cloud. The air in the funnel is going up rapidly, as can sometimes be visualized from the motions of debris in it, yet the funnel seems to be lowering from above during the touch-down. In reality, it is the "pattern" of the funnel that is coming down but the particles in it are going up! The pattern seems to come down because the condensation at higher level forms first and that in the lower levels forms later. The "time sequence of condensation" produces the illusion that the funnel cloud lowers from above.

In a sense, cloud motion is akin to (though not really equivalent to) the propagation of waves. The wave pattern (phase) moves with the phase speed but the medium where the waves occur may not be moving at all.

12.2 Adiabatic ascent of an unsaturated air parcel

The most common process of cloud formation in our atmosphere is adiabatic cooling, as we have derived in Sec. 4.4. There we started out by assuming that an unsaturated air parcel that is rising adiabatically will cool down at the dry adiabatic lapse rate of $\sim 9.76\,\text{K}\,\text{km}^{-1}$, or roughly $10\,\text{K}\,\text{km}^{-1}$. The rate would change slightly if there were water vapor in the air parcel, but the magnitude is small enough to be negligible for the possible amount of water vapor in air as long as the air is unsaturated.

Note that, since $dQ = TdS = 0$ during an adiabatic process, the entropy S of the parcel is conserved. Hence an adiabatic process is also called an isentropic process. In atmospheric science, it is common to use the *potential temperature* θ to represent the conserved quantity during this process instead of entropy, θ being defined as

$$\theta = T\left(\frac{P_0}{P}\right)^{\kappa}, \quad \kappa = \left(\frac{R_d}{c_p}\right) = 0.286, \tag{12.1}$$

and the reference pressure P_0 is taken as 1000 mb (or 100 hPa). Thus the potential temperature is the temperature the air parcel would have if it were displaced from anywhere to 1000 mb via an adiabatic process, and it remains constant during this process. In simplified theoretical studies, atmospheric processes are often assumed to be adiabatic and the potential temperature can be used as a "tracer" to keep track of the movements of air parcels. This is especially useful to atmospheric chemistry, where the source origin of certain species is of central importance.

12.2.1 Lifting condensation level

We see that, as the parcel rises adiabatically, it expands and cools at a rate of $9.76\,\mathrm{K\,km^{-1}}$. As the air parcel cools, the water vapor in the air parcel becomes closer to saturation because the saturation vapor pressure decreases with decreasing temperature. As the air parcel rises high enough, it will eventually reach a level at which its temperature becomes cold enough to saturate the vapor, and any further ascent (hence cooling) will cause excess water vapor to condense – and we will start to see cloud appearing! The level where the air parcel becomes saturated due to adiabatic ascent is called the *lifting condensation level* (LCL). The LCL can be determined graphically from adiabatic charts (such as the so-called skew-T log-P chart) for a specific sounding. Many elementary textbooks of meteorology contain discussions of these adiabatic charts.

Further ascent of the air parcel above the LCL would cause more condensation as long as there is still excess water vapor in the parcel and hence further cloud growth. But the thermodynamic process in this stage is no longer dry adiabatic because the release of latent heat due to the phase change of water begins to influence the temperature of the air parcel. We need to examine this more quantitatively.

12.3 Moist adiabatic process

The formation of condensed water in clouds introduces complications into the thermodynamics. To a first approximation, we will consider a moisture-containing air parcel that reaches saturation but retains all the condensed water and the latent heat released within the parcel. If the parcel descends, then we assume that water drops or ice particles will evaporate and latent heat is consumed, but again all materials and energy remain in the parcel. In other words, this system is thermodynamically reversible and also isentropic. This thermodynamic process is called the *reversible moist adiabatic process*.

While it is conceptually simple, this reversible process is cumbersome to apply because the air parcel must now carry the condensed water along during its ascent or descent, which makes the calculation of energy balance very complicated. A common practice to simplify the situation is to assume that all the condensed water falls out of the parcel as soon as it is formed but carries no energy away. Since the amount of energy carried by condensed water is mostly negligible compared to that of the air parcel, at least for the case of small cloud formation, this assumption is acceptable as a first approximation, and we can conduct our calculations as if the air parcel were still undergoing an adiabatic process. This process is called the *pseudo-adiabatic process* and is the basis of the following discussion.

Now let us examine how the temperature lapse rate of this pseudo-adiabatic system differs from that of a dry adiabatic system. For simplicity, we will just consider the condensation of water drops and ignore the ice phase, as the ice process is analogous. Thus, when the air parcel reaches the LCL, the water vapor is saturated, latent heat is released, and adiabatic energy conservation requires that, instead of (4.2), we should have

$$c_{p,m}dT + L_e dq_{v,sat} - \alpha_m dp = 0, \tag{12.2}$$

where $q_{v,sat}$ is the saturation mixing ratio of water vapor, $c_{p,m}$ is the specific heat, and α_m is the specific volume of the moist air. Unlike the case of dry air, the latter two quantities are not constant because they depend on how much water vapor there is in the parcel, i.e. they are a function of $q_{v,sat}$. Expressing these quantities explicitly and using the ideal gas law for the third term, we rewrite (12.2) as

$$(c_{p,a} + q_{v,sat}c_{p,v})dT + L_e dq_{v,sat} - (R_a + q_{v,sat}R_v)T\frac{dp}{p} = 0. \tag{12.3}$$

The terms with "v" subscript represent that quantity for water vapor and those with "a" subscript those for dry air. Here the first term represents the energy due to the change of parcel temperature dT (measured at constant pressure), the second term the latent heat released, and the third term the change due to the change in pressure. Note that we have completely ignored the heat carried away by the water drops that fall out of the air parcel, which is very small anyway.

We further simplify (12.3) by noting that the vapor-related quantities in the first and third terms are small compared to the dry air quantities and hence can be ignored. Thus, after rearrangement, we have a new approximation:

$$c_{p,a}\frac{dT}{T} + L_e\frac{dq_{v,sat}}{T} - R_a\frac{dp}{p} = 0. \tag{12.4}$$

Following the same procedure as in Sec. 4.4, we obtain the temperature lapse rate of the moist air parcel as

$$\Gamma_s = -\frac{dT}{dz} = \frac{g}{c_{p,a} + L_e(dq_{v,sat}dT)}. \tag{12.5}$$

This is called the *pseudo-adiabatic lapse rate*. Since all quantities on the right-hand side of (12.5) are positive, we immediately see that this lapse rate is always smaller than the dry adiabatic lapse rate (4.8), i.e. $\Gamma_s < \Gamma_d$. Secondly, the pseudo-adiabatic lapse rate is not constant but depends on $dq_{v,sat}/dt$, the amount of water vapor condensed due to cooling. Typical values of Γ_s range from $\sim 4\,\mathrm{K\,km^{-1}}$ (lower troposphere) to $\sim 7\,\mathrm{K\,km^{-1}}$ (upper troposphere).

During a pseudo-adiabatic process, the potential temperature θ is no longer conserved because the latent heat effect is not included in the definition of the potential temperature (12.1). However, it is possible to include this effect by defining another quantity, the *equivalent potential temperature* θ_e, which will be conserved during a pseudo-adiabatic process:

$$\theta_e = \left(T + \frac{L_e q_v}{c_p}\right)\left(\frac{P_0}{P}\right)^\kappa = T_e\left(\frac{P_0}{P}\right)^\kappa, \tag{12.6}$$

where q_v is the mixing ratio of water vapor and T_e as defined above is called the *equivalent temperature*. Thus the equivalent potential temperature can be used as a tracer during a pseudo-adiabatic process.

12.4 Buoyancy and static stability

It is usually assumed that, as soon as an air parcel rises to a certain level, its pressure will immediately be balanced with that of its surrounding environment. In reality, of course, it always takes a finite amount of time to do so, but this is a good approximation, as there is not really a rigid boundary between them. At the same pressure, an air parcel will be lighter than its environment if it is warmer, and the parcel will be subject to a buoyant force defined by

$$F_B = mg\left(\frac{\rho - \rho'}{\rho'}\right) = mg\left(\frac{T - T'}{T'}\right) = mg\left(\frac{\theta - \theta'}{\theta'}\right), \tag{12.7}$$

where the primed quantities are that of the environment. In this case, the parcel will rise due to the buoyancy. If, on the other hand, the parcel is colder than its environment, then it will be subject to a negative buoyant force and sink.

Since the formation of clouds is usually related to the ascending motion of air, the static stability of air in which clouds form plays an important role. When an air parcel, which was originally in thermal equilibrium with its environment (so that the two have the same temperature), is displaced vertically by a perturbation from its initial position (either upward or downward), it is the static stability of its environment that decides the parcel's reaction. If the parcel continues to accelerate away from its initial position, then its environment is unstable. If, instead, the parcel moves back to its initial position, then its environment is stable. Obviously, there is a third condition that sits right between the stable and unstable conditions – the neutral condition, where the parcel moves at a constant speed (no acceleration) and in the direction caused by the perturbation. If the motion of the air parcel is adiabatic, then the static stability condition of its environment is determined by the relative magnitude between the environmental temperature lapse rate and the adiabatic lapse rate according to the following relations:

$$\left.\begin{array}{l} \Gamma < \Gamma_d \quad \text{stable} \\ \Gamma = \Gamma_d \quad \text{neutral} \\ \Gamma > \Gamma_d \quad \text{unstable} \end{array}\right\} \text{unsaturated} \quad \text{or} \quad \left.\begin{array}{l} \Gamma < \Gamma_s \quad \text{stable} \\ \Gamma = \Gamma_s \quad \text{neutral} \\ \Gamma > \Gamma_s \quad \text{unstable} \end{array}\right\} \text{saturated.} \quad (12.8)$$

Remember, of course, that Γ_s is not a constant and $\Gamma_s < \Gamma_d$. Now it may happen that the environmental lapse rate Γ is such that $\Gamma_s < \Gamma < \Gamma_d$, then clearly the stability depends on whether the air parcel is saturated or subsaturated. This situation is called *conditional instability*.

The stability conditions can also be expressed in terms of the potential temperatures. From (12.1) we have

$$\ln\theta = \ln T + \kappa \ln p_0 - \kappa \ln p$$

and therefore

$$\begin{aligned} \frac{1}{\theta}\frac{d\theta}{dz} &= \frac{1}{T}\frac{dT}{dz} - \frac{\kappa}{p}\frac{dp}{dz} = \frac{1}{T}\frac{dT}{dz} + \frac{(R_d/c_p)}{\rho_a R_d T}(\rho_a g) \\ &= \frac{1}{T}\left(\frac{g}{c_p} + \frac{dT}{dz}\right) = \frac{1}{T}(\Gamma_d - \Gamma), \end{aligned} \quad (12.9)$$

after applying the hydrostatic balance equation and the ideal gas law. Thus

$$\frac{d\theta}{dz} = \frac{\theta}{T}(\Gamma_d - \Gamma). \quad (12.10)$$

Using (12.8) and (12.10), we can express the stability condition in terms of the potential temperature as

$$\left.\begin{array}{l} \dfrac{d\theta}{dz} > 0 \quad \text{stable} \\[2mm] \dfrac{d\theta}{dz} = 0 \quad \text{neutral} \\[2mm] \dfrac{d\theta}{dz} < 0 \quad \text{unstable} \end{array}\right\} \text{unsaturated} \quad \text{or} \quad \left.\begin{array}{l} \dfrac{d\theta_e}{dz} > 0 \quad \text{stable} \\[2mm] \dfrac{d\theta_e}{dz} = 0 \quad \text{neutral} \\[2mm] \dfrac{d\theta_e}{dz} < 0 \quad \text{unstable} \end{array}\right\} \text{saturated.} \quad (12.11)$$

12.5 The adiabatic parcel model of cloud formation

We can now form a conceptual model of cloud development based on the assumption that all processes are dry adiabatic before saturation or pseudo-adiabatic after saturation of the air parcel. We further assume that the parcel motion does not disturb its environment. This is called the *parcel model*. When such an air parcel rises, it expands and initially cools at a rate equal to Γ_d. Eventually, the parcel rises to the height of the LCL. Here, the air inside the parcel becomes saturated, condensation starts, and a cloud begins to appear. Now latent heat is released, the parcel continues to rise (since it still has positive buoyancy), and cools according to Γ_s, and more condensation (hence a taller cloud) occurs if excess water vapor is still available. This continues as long as the parcel is still warmer, hence lighter, than its environment. Eventually, the parcel may reach a level where its temperature is the same as that of the environment. The level at which the parcel and its environment have the same temperature is called the *level of neutral buoyancy* (LNB) or the *equilibrium level* (EL) because the parcel loses its buoyancy here.

However, the parcel will not come to a complete halt at the LNB because it still has kinetic energy. Instead, it will continue to rise until it expends all the kinetic energy. The height that the parcel rises above the LNB is called the *overshooting*. In the case of severe thunderstorms, where the LNB is often located at the tropopause, the overshooting represents a part of the cloud that protrudes into the stratospheric level (but see Chapter 13 for a clarification of what this implies) and is often called the *overshooting top* or *overshooting dome* because of its dome shape (Fig. 12.2).

Fig. 12.2 The overshooting top of a thundercloud over Utah on 18 July 2008 as seen from an aircraft flying at the tropopause level. Photo by Pao K. Wang.

For smaller cumulus congestus or cumulonimbus, however, the overshooting may occur entirely in the troposphere.

12.5.1 Convection condensation level

Using the adiabatic ascent idea described above, we can form a simple conceptual model of convective cloud formation. Fig. 12.3 shows a possible scenario of the evolution of a sounding on a typical summer day. In the early morning, the

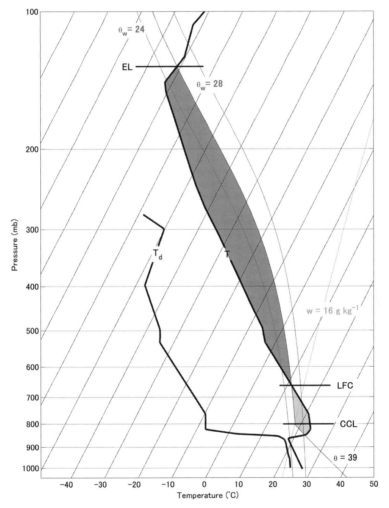

Fig. 12.3 A typical sounding favorable for the formation of a severe convective storm in the Great Plains region of the USA. The dark shaded area above ~ 650 mb represents the convective available potential energy (CAPE) and the lighter shaded area between ~ 950 and ~ 650 mb represents the convective inhibition (CIN). Adapted from Doswell (2001).

temperature sounding curve shows a shallow inversion near the ground due to the overnight radiative cooling. When the Sun is out, the surface starts to receive heat and the temperature of the lower-level air increases by conduction from the surface and by mixing in the air. The mixing homogenizes the water mixing ratio of the low-level air and essentially brings the moist boundary layer air upward, forming the mixing layer. The mixing layer continues to grow as the morning becomes later and eventually a temperature T_c on the surface is reached some time in mid-morning. Using the adiabatic chart, we see that (assuming that the mixing ratio of water vapor remains constant), at this time, a parcel on the surface that rises adiabatically would reach a level at which the T curve intersects the mixing ratio (w_v) curve, where condensation should occur and we can expect to see the formation of a cumulus. This level is called the *convective condensation level* (CCL). The cumulus will continue to grow if the layer above the CCL is moist unstable.

The above scenario provides a way to forecast the cumulus cloud base height if the convective surface temperature can be predicted. In many typical convective cloud cases, the CCL is very close to the LCL mentioned previously.

12.6 Corrections to the parcel model

12.6.1 Burden of condensed water

In the above discussions, we assumed that the parcel rises according to the pseudo-adiabatic theory and we did not consider the impact of the weight of the condensed water on the motion of the parcel. After the parcel reaches the LCL, condensed water forms and its weight becomes a burden (downward force) or drag on the parcel that will impede its buoyancy. Therefore, instead of (12.7), the net force acting on the parcel should be

$$F_B = mg\left(\frac{T - T'}{T'}\right) - m_{con}g = mg\left(\frac{T - T'}{T'} - \frac{m_{con}}{m}\right) = mg\left(\frac{T - T'}{T'} - q_{con}\right),$$

$$(12.12)$$

where m_{con} and q_{con} represent the mass and mixing ratio of the condensed water in the parcel. Eq. (12.12) implies that the parcel will not rise as fast as in the no-burden case.

12.6.2 Adiabatic slice model – a first correction to the test parcel model

While the test parcel model gives us useful insights into the cloud formation process, it is admittedly a crude first approximation. Several important processes are not considered by this model, the two most important of which, namely dynamic coupling and mixing with the environment, both impact significantly on cloud development. These processes will be discussed briefly in the following sections. More detailed discussions of them can be found in Scorer (1997) and Iribarne and Godson (1973).

12.6.3 *Dynamic coupling of the parcel and the environment*

Since the air is a continuous fluid, when there are updrafts in a region of air that was initially at rest, there must be compensating downdrafts somewhere else in that region. This is the case when a thermal rises. The rising motion of the thermal would therefore cause the air around it to make a compensating descent. This is proven by aircraft observation of vertical air velocity around a developing cumulus, as shown in Fig. 12.4.

Here we see that the positive liquid water content region in the lower panel corresponds to the cloudy air formed by the thermal. The vertical velocity profile in the upper panel shows that there are updrafts in the cloud, whereas the regions immediately outside of the cloud edge are dominated by downdrafts, which are the compensating downward motions of the updrafts.

Now, since the environmental air is subsaturated, its downward motion will cause it to heat up by dry adiabatic compression. Its initial temperature profile

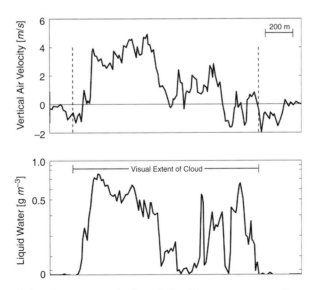

Fig. 12.4 Relation between updraft and liquid water content from an aircraft penetration of a cumulus cloud. The horizontal axis represents the horizontal distance. The two vertical dashed lines in the upper panel denote the cloud boundary. Adapted from Warner (1969).

and hence the static stability will be altered. This change can be estimated by the *slice method* (also called *layer method*). We will call it the *slice model* here. The main idea is that, as a result of continuity, the total mass flux due to the compensating downdrafts should be the same as the mass flux of the air parcel:

$$\rho A w = \rho' A' w', \tag{12.13}$$

where ρ, A, and w are the density, horizontal cross-sectional area, and vertical velocity of the parcel, and the prime indicates the corresponding quantities of the environment. Thus the compensating downdraft can be estimated by

$$\frac{w'}{w} = \frac{\rho A}{\rho' A'} \approx \frac{A}{A'}. \tag{12.14}$$

Thus the impact depends on the ratio of the cross-sectional areas. If the parcel area is very small, i.e. $A \to 0$, then the parcel motion has nearly no impact on the environment. If instead A is substantial, then the induced downdraft w' is also substantial.

Now suppose that the parcel ascends and cools at the pseudo-adiabatic lapse rate, whereas the surrounding environment descends and warms up at the dry adiabatic lapse rate. Thus, in a short time dt, the parcel with initial temperature T_0 that ascends to a new level will have a temperature $T = (T_0 + \Gamma w \, dt) - \Gamma_s w \, dt$, whereas the environmental air that descends to this level will have a temperature $T' = (T_0 - \Gamma w' dt) + \Gamma_d w' dt$, where the quantities in the parentheses are the temperatures of the parcel and the environmental air at their respective initial level. The difference between the two temperatures is

$$\Delta T = T - T' = (\Gamma - \Gamma_s)w - (\Gamma_d - \Gamma)w'. \tag{12.15}$$

The air will be unstable (stable) if ΔT is positive (negative). Using (12.14), the stability criteria are

$$\begin{aligned}
\frac{(\Gamma - \Gamma_s)}{(\Gamma_d - \Gamma)} &< \frac{A}{A'} \quad \text{stable,} \\
\frac{(\Gamma - \Gamma_s)}{(\Gamma_d - \Gamma)} &= \frac{A}{A'} \quad \text{neutral,} \\
\frac{(\Gamma - \Gamma_s)}{(\Gamma_d - \Gamma)} &> \frac{A}{A'} \quad \text{unstable.}
\end{aligned} \tag{12.16}$$

It is of interest to examine the neutral condition. Since $A/A' > 0$, this condition requires that $\Gamma > \Gamma_s$. In the case of the parcel model, it is only necessary to have $\Gamma = \Gamma_s$ to be neutral. The new requirement implies that the compensating downdraft causes stabilization of the layer, and it requires a steeper lapse rate to make it unstable.

12.7 Brunt–Väisälä frequency

Oscillation is a general phenomenon of motion in a stably stratified fluid region such as at the surface of water. The layer around the LNB is also stably stratified and hence we expect to see oscillation here if the air at this level is disturbed. To see this, we note that the vertical equation of motion of the parcel under the influence of buoyant force (12.7) is

$$m\left(\frac{d^2z}{dt^2}\right) = mg\left(\frac{\theta - \theta'}{\theta'}\right) \quad \text{or} \quad \frac{d^2z}{dt^2} = g\left(\frac{\theta - \theta'}{\theta'}\right). \tag{12.17}$$

The quantity $(\theta - \theta')/\theta'$, which is a function of z, is much smaller than unity and can be expanded as

$$\left(\frac{\theta - \theta'}{\theta'}\right) = \left(\frac{\theta - \theta'}{\theta'}\right)_0 + \left[\frac{\partial}{\partial z}\left(\frac{\theta - \theta'}{\theta'}\right)\right]_0 z + \cdots. \tag{12.18}$$

The quantities with "0" subscript mean the values at the LNB. The first term on the right-hand side is zero because $\theta = \theta'$ at the LNB. The second term is

$$\left[\frac{\partial}{\partial z}\left(\frac{\theta - \theta'}{\theta'}\right)\right]_0 z = \left[\frac{\partial}{\partial z}\left(\frac{\theta}{\theta'} - 1\right)\right]_0 z = -\left(\frac{\theta}{\theta'^2}\frac{\partial\theta'}{\partial z}\right)z \approx -\left(\frac{1}{\theta'}\frac{\partial\theta'}{\partial z}\right)z, \tag{12.19}$$

where we have used the fact that θ is constant because the parcel motion is adiabatic whereas θ' is not. Thus (12.17) becomes

$$\frac{d^2z}{dt^2} = -\left(\frac{g}{\theta'}\frac{\partial\theta'}{\partial z}\right)z = -N^2z, \tag{12.20}$$

where $N = \sqrt{(g/\theta')(\partial\theta'/\partial z)}$ (with unit s^{-1}) is called the *Brunt–Väisälä frequency* or *buoyant frequency*. Eq. (12.20) has the form of a wave equation and has general solutions of the form

$$z = A\exp(-iNt), \tag{12.21}$$

where A represents the wave amplitude. When N is positive, the solution (12.21) indicates that the parcel's vertical position z oscillates with frequency N. Since a positive N implies that $(\partial\theta/\partial z) > 0$, which indicates a stable stratification from (12.11), we naturally expect such oscillations to occur. On the other hand, if $(\partial\theta/\partial z) < 0$, the stratification is unstable and N becomes imaginary. Eq. (12.21) then indicates that z will increase exponentially with time, i.e. the parcel accelerates away from the LNB, which is indeed what would occur in an unstable layer.

Such oscillatory motions generate *internal gravity waves* that can transport energy vertically. Even small cumulus clouds can induce internal gravity waves, and a cumulonimbus will certainly generate such waves with larger amplitudes. Oscillatory motions near the overshooting dome of severe thunderstorms have been observed by radar and satellites, and have been successfully simulated by numerical thunderstorm models. We will discuss this in Chapter 15.

12.8 Convection process

The above discussions of air parcel thermodynamics and cloud formation are all based on the assumption that the motions are adiabatic. Whereas the adiabatic assumption is useful, it is admittedly only a crude approximation of the real cloud formation process. One of the most serious omissions is the mixing with environmental air because, after all, there is not really an impenetrable boundary between an air parcel and its environment, and velocity shears between them can easily cause mixing. Mixing has a great impact on cloud formation, especially convective clouds, and needs to be carefully examined.

12.8.1 Thermals and plumes

We used the concept of an air parcel in the above discussions as an abstract object without defining its size and shape. Since we will now discuss the interaction between the air parcel and the environment that eventually forms a cumulus, which does have finite size and shape, it is better for us to consider what that original air parcel looks like. There are three models for such air parcels (Fig. 12.5).

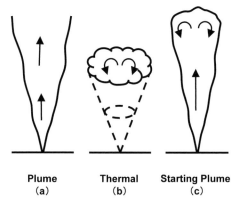

Plume   Thermal   Starting Plume
 (a)   (b)   (c)

Fig. 12.5 Three convection models in the atmosphere: (a) plume; (b) thermal (bubble); and (c) starting plume. Adapted from Turner (1969).

(a) *Plume* – a plume is a continuous column of ascending buoyant air that starts with relatively small diameter near the surface and expands as it goes up higher. The simplest plume model is one that expands upward at a constant angle. It has no top because it is assumed to be continuous.

(b) *Thermal* – in this model, air parcels that eventually form cumulus are thought to be buoyant bubbles more or less spherical in shape. Glider pilots are familiar with the existence of such warm bubbles, which they call thermals. The thermals rise and interact with the environment and lead to the formation of Cu.

(c) *Starting plume* – this model looks like a combination of the plume and thermal models. One can imagine that a thermal ascends first, and then successive thermals follow the same track as the first one, and eventually merge together to become a continuous plume but with a top that looks like a thermal.

A true continuous plume in the atmosphere is quite unlikely, so either the thermal or starting plume model is more realistic. However, the plume model is still mentioned from time to time due to its simplicity.

12.8.2 Lidar observations of thermals and small cumulus clouds

Before the formation of clouds, the thermals or plumes would be located in the boundary layer. Without the formation of clouds, they are basically invisible to the naked eye. However, the atmospheric boundary layer is usually full of aerosol particles, which, although invisible to the human eye, can be detected by lidar, as the particles backscatter strongly the incident laser beam. By recording the back-scattered signal, lidar can "see" the structure of the boundary layer. The presence of thermals and plumes will influence the motions and concentrations of the aerosol. Thus, by examining the time evolution and space distribution of the aerosol backscattered profile, we can learn about the structure of the thermals and plumes.

Figs. 12.6–12.9 show four frames of a movie of the boundary layer evolution made by a volume imaging lidar on 27 July 1989 at Manhattan, Kansas. The top panel of each figure shows a vertical cross-section of the profile along the direction marked by the dashed line in the lower six panels. The lower six panels show the horizontal cross-sections of the backscatter field of $z = 250$ m, 500 m, 750 m, 1000 m, 1250 m, and 1500 m, respectively. Gray shades indicate aerosol, while brighter white shades indicate clouds.

Fig. 12.6 shows the backscatter fields at 9:37:46 a.m. Although it is relatively early in the morning, the boundary layer begins to build up as the ground begins to be heated. Convection cells start to be organized, although it is difficult to tell whether the cells are like bubbles or starting plumes from these images. The top panel shows

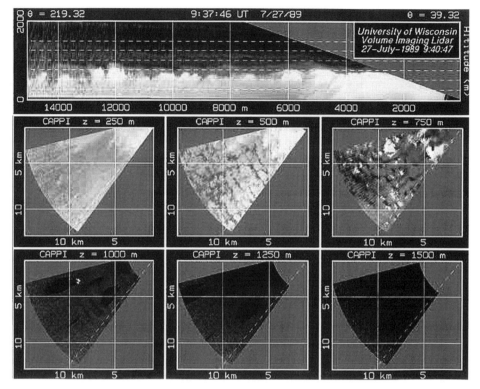

Fig. 12.6 Lidar observation of convection in the boundary layer at 9:37:46 a.m. on 27 July 1989 in Manhattan, Kansas. The top panel shows the vertical scan image. The lower six panels show horizontal scan images at z = 250 m, 500 m, 750 m, 1000 m, 1250 m, and 1500 m level. Gray objects are aerosol, while brighter white color represents clouds. Courtesy of Dr Edwin Eloranta.

the cell-like structure at the top of the boundary layer at z between 400 and 600 m. The middle figure in the middle panel ($z = 500$ m) shows the horizontal structure of this top layer, clearly indicating the cellular features with horizontal cell size ~ 500 m. At high level, for example, $z = 1000$ m, the atmosphere is relatively clean.

As time goes on, at 10:15:51 a.m. (Fig. 12.7), the boundary layer becomes thicker and the top of the layer pushes to a level between 800 and 1000 m. Some CCN at the top of the thermals are activated and small cumulus clouds begin to appear, as can be seen in the $z = 1000$ m panel (lower left). The fact that the clouds form directly on top of the thermals demonstrates that the thermals are indeed the "roots" of the clouds. The cellular structure below 500 m disappears but is still visible at $z = 750$ m. The cell size increases somewhat.

At 10:40:05 a.m. (Fig. 12.8), the boundary layer further thickens and the cloud top reaches a little higher than before. Cloud widths increase somewhat but cloud number becomes fewer, which is a quite common phenomenon in the development

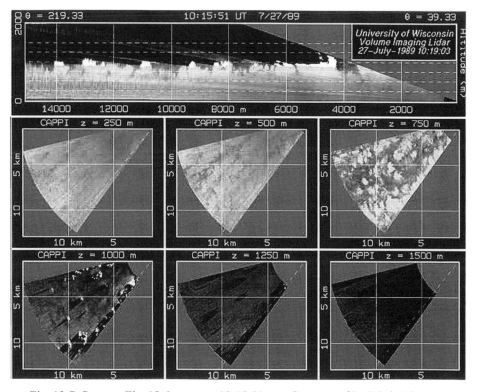

Fig. 12.7 Same as Fig. 12.6 except at 10:15:51 a.m. Courtesy of Dr Edwin Eloranta.

of a cumulus field in the morning. When some cloud cells grow taller, their more vigorous updrafts cause greater compensating downdrafts around them that may suppress the development of smaller clouds in that area. The net result is the decrease in number but an increase in size of the clouds. The cells at 500 m level remain more or less the same as in Fig. 12.6 but with somewhat larger size. Clouds are seen at the 1000 m level but not yet at the 1250 and 1500 m levels.

Fig. 12.9 shows the situation at 11:38:56 a.m., close to noon. The boundary layer thickness increases further and the top of the cloud now reaches 1500 m. Cloud widths also increase. The top panel shows that the clouds respond to the wind shear and tilt to the left. The aerosol field below 500 m appears to be well mixed, but the cellular structure can still be seen at the 1000 m level. Clouds can be seen in the 1250 and 1500 m scans.

12.9 Entrainment

The mixing between a rising air parcel and its environment is thought to be due to *entrainment*. Entrainment is especially important in the development of convective

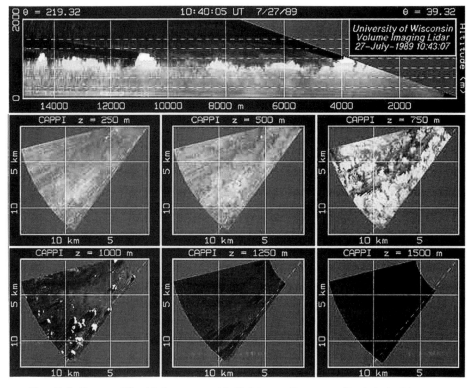

Fig. 12.8 Same as Fig. 12.6 except at 10:40:05 a.m. Courtesy of Dr Edwin Eloranta.

clouds because the larger updrafts of convective clouds tend to cause more vigorous mixing, and mixing changes the temperature and humidity profiles of the air parcel.

The fact that mixing does occur can be seen from the aircraft measurements of liquid water contents in cumulus clouds as depicted in Fig. 12.10. Here we see that the ratio of observed liquid water content (w_L) to that predicted by the pseudo-adiabatic theory ($w_{L,ad}$) is less than 1, indicating that cumulus cloud formation is never adiabatic. Furthermore, the ratio decreases with height, which implies that dilution due to entrainment becomes greater the higher in the cloud.

While the evidence for entrainment abounds, how entrainment occurs is much less clear and, in fact, is still debated to this date. In the following, we will examine some conceptual models of entrainment.

12.9.1 Lateral entrainment

The concept of entrainment was first introduced by Stommel (1947), and the idea was later adopted for the development of cumulus clouds by some cloud physicists (e.g. Scorer and Ludlam, 1953; Simpson, 1971). This idea is based on *lateral*

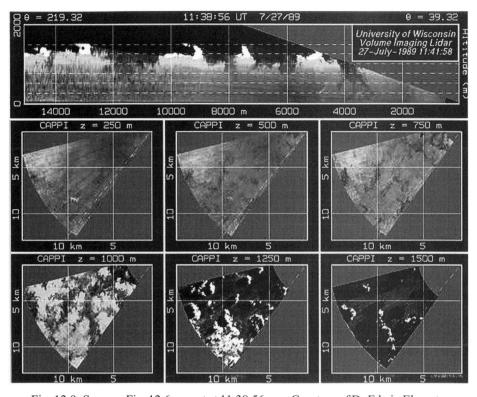

Fig. 12.9 Same as Fig. 12.6 except at 11:38:56 a.m. Courtesy of Dr Edwin Eloranta.

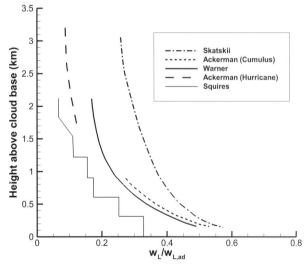

Fig. 12.10 The ratio of the observed mean liquid water content at a given height above cloud base to the adiabatic value. The values attributed to Skatskii (1965) were obtained from his published results assuming that the cloud base was at a height of 1 km and at a temperature of 8°C. Based on data from Warner (1970).

entrainment due to either plumes or bubbles (blobs). The lateral entrainment theory predicts that the fractional rate of entrained air into the cloud updraft varies with the cloud radius. This can be easily verified for the plume and thermal models. If we assume that a cloud system starts as a spherical bubble or thermal of radius r_B rising in a surrounding air environment, then the mass of the bubble is

$$m_B = 4\pi\rho_a r_B^3/3, \tag{12.22}$$

and hence the fractional entrainment rate is

$$\mu_B = \left(\frac{1}{m_B}\frac{dm_B}{dz}\right) = \frac{3(dr_B/dz)}{r_B} = \frac{C}{r_B}. \tag{12.23}$$

The term $dr_B/dz = C$ is constant because we assume that the bubble rises with a constant broadening angle.

Similarly, if the cloud starts as a continuous jet or plume with radius r_J, then the mass flux of the plume is given by

$$F_m = \pi r_J^2 \rho_a w, \tag{12.24}$$

where w is the updraft speed. Again, the fractional entrainment rate is

$$\mu_J = \left(\frac{1}{F_J}\frac{dF_J}{dz}\right) = \frac{2(dr_J/dz)}{r_J}, \tag{12.25}$$

where dr_J/dz is again constant for a fixed plume broadening angle.

If we take the bubble model result (12.23) and integrate, we obtain

$$\frac{dm_B}{dt} = \frac{dm_B}{dz}\frac{dz}{dt} = w\frac{dm_B}{dz} = \frac{Cwm_B}{r_B} = C'\rho_a r_B^2 w, \tag{12.26}$$

where C' is just another constant. Eq. (12.26) predicts that the mass of entrained air is proportional to the cloud area and its updraft velocity. The properties of such entrainment processes have been measured in water tank experiments in the laboratory and then applied to the modeling of cloud development.

12.9.2 Cloud top entrainment

While the lateral entrainment model has the benefit of simplicity and is intuitively appealing, it also has many problems. Various observations have shown that pure lateral entrainment does not produce the right profile of liquid water content. For example, Warner (1970) found that the standard lateral entrainment model cannot correctly predict both the height of the ascending air parcel and the liquid water content at the same time when applied to a one-dimensional cloud model. In addition, it cannot explain the decreasing $w_L/w_{L,ad}$ with height, as shown in Fig. 12.10. There are also other discrepancies between the observations and the predictions by the model. For example, Heymsfield *et al.* (1978) found undiluted air that originated from the cloud base at all levels in the cumuli. Paluch (1979) analyzed field observation data of cumuli in Colorado and showed that the entrained air at any observational level typically comes from a few kilometers above, and the mixed regions are typically found in downdrafts and weaker updraft regions. If the entrainment is lateral, the air should mostly come from below, instead of above, the observational level.

Another entrainment process was proposed by Squires (1958b) that is able to address some of these discrepancies. This is based on the concept that the entrainment mainly occurs at the cloud top by mixing. The action of this mechanism is the mixing of dry environmental air from above the ascending cloud top. Once the dry air is entrained into the cloud, it decreases the humidity and causes evaporative cooling. The result is that the air becomes colder than the surrounding cloudy air and sinks to lower level, causing the mixing effect. This is sometimes called the *penetrative downdraft model*.

Currently, both lateral and top mixings are considered in the more sophisticated numerical cloud models via parameterized forms.

Blyth *et al.* (1988) analyzed the data gathered by aircraft observations of small cumulus clouds in Montana. They concluded that cumulus clouds consist of thermal-like elements, from which the least buoyant mixed parcels are shed off and the most buoyant mixed parcels may continue with the general ascent. A more recent observation by Damiani *et al.* (2006) used airborne dual Doppler radar to study the velocity fields in vertical planes across cumulus turrets. They found clear evidence that the clouds evolved through a sequence of thermals with well-defined toroidal circulations (vortex rings). The largest updraft speeds were observed in the ring centers, but regions of turbulent ascending air extended behind the thermals to distances comparable with the toroid sizes. Patterns in the reflectivity and velocity fields indicated regions of major intrusions into the thermals, accompanied by entrainment of ambient air, or recycling of larger hydrometeors, depending on their location. In addition, at the upper cloud–environment interface, instability nodes contributed to further entrapment of cloud-free air. The findings of Damiani *et al.* (2006) largely confirmed the thermal shedding model proposed by Blyth *et al.* (1988).

12.9.3 *Effect of entrainment on convective clouds*

The net effect of entraining the colder and drier environmental air into the parcel is to reduce the buoyancy of the parcel so that the vertical velocity of the parcel is slower than that predicted by the adiabatic model. At the same time, the volume of the parcel becomes bigger due to the addition of the entrained outside air.

We can make a rough estimate of the influence on the parcel due to the entrainment. When the environmental air is entrained into the cloudy air parcel, it will expend an amount of heat $c_p\,(T-T')\,dm$ to warm up the colder new air, expend an amount $L_e\,(q_{v,\mathrm{sat}}-q'_v)\,dm$ to evaporate some drops to saturate the drier new air, and, at the same time, gain an amount $L_e m\,dq_{v,\mathrm{sat}}$ due to the latent heat released from condensation due to ascending cooling (which was already considered in the moist adiabatic process). The whole process is no longer adiabatic and therefore the conservation of energy now requires that

$$m\left(c_{p,a}dT + L_e dq_{v,\mathrm{sat}} - R_a T\frac{dp}{p}\right) = c_{p,a}(T-T')dm + L_e(q_{v,\mathrm{sat}} - q'_v)dm,$$

(12.27)

from which the lapse rate in a cloud with entrainment can be derived:

$$\Gamma_c = -\left(\frac{dT}{dz}\right)_c = \frac{g + \left(\frac{1}{m}\frac{dm}{dz}\right)\left[L_e(q_{v,\mathrm{sat}} - q'_v) + c_{p,a}(T-T')\right]}{c_{p,a} + L_e(dq_{v,\mathrm{sat}}/dT)}.$$

(12.28)

For an entraining cloud, the entrainment rate $\mu = (1/m)(dm/dz)$ is positive as well as the quantities in the square brackets. Thus, by comparing (12.28) with (12.5), we see that $\Gamma_c > \Gamma_s$, i.e. the lapse rate in an entraining cloud is greater than that of a pseudo-adiabatic cloud, as it should be because heat needs to be spent to warm up the colder air and evaporate some liquid drops.

12.10 Summary

What we have presented above are basic concepts of the dynamical processes for the formation of small cumulus clouds. The formulations are highly simplified so as to make them easy to understand and convenient for discussion. While useful, it is known that real cloud processes are much more complicated, and numerical cloud models are necessary to obtain more realistic and quantitative understanding of them. This is the subject of Chapter 13.

Problems

12.1. Prove that θ is conserved during a dry adiabatic process. Is θ conserved during a moist adiabatic process?

12.2. Prove that θ_e is conserved during a pseudo-adiabatic process. Is θ_e conserved during a dry adiabatic process?

12.3. Derive a set of stability criteria similar to (12.11) but using the wet-bulb potential temperature θ_w as the variable.

13

Numerical cloud models

13.1 Introduction

In Chapter 12, we examined the fundamental thermodynamic and dynamic processes leading to the formation of clouds, focusing on small convective clouds. We gained some insights into cloud formation and development using a few simple analytical expressions that describe the physical (mainly thermodynamic and dynamic) state of an air parcel in a prescribed environment. Our discussions focused on the processes occurring in the parcel as we followed the ascent or descent of the parcel. This type of discussion (i.e. following a particle or a parcel) is called the *Lagrangian description* of a dynamic system. Except in the slice model, where the compensating downdraft was treated in a crude manner, we did not consider what happened to the cloud environment in that description.

However, when an air parcel rises and forms a cloud, its dynamics and microphysics will influence the physical states of its environment, not only via forcing compensating downdrafts, but also via the changes in the latent heat that will eventually be transferred to the environment. These changes in the environment, in turn, will cause changes in the cloud by essentially setting different boundary conditions, and so on. In other words, there are feedbacks that the cloud and its environment impart upon each other. The parcel model as described in the last chapter cannot address these feedback effects realistically, especially in the time evolution. In addition, in a real cloud, there are interactions not only between the cloud and its environment, but also between different parts of the cloud as well. Such interactions are especially complicated in a mixed-phase system such as a cumulonimbus, in which both updrafts and downdrafts are present, and different kinds of hydrometeors (including water drops and ice particles, either growing or evaporating) are carried by the cloud circulation into different regions and interact with each other. One cannot hope to describe such interactions by the analytic equations derived in the previous chapter.

In fact, the division between "cloud" and "environment" is somewhat artificial. Before the formation of the cloud via condensation, the so-called cloud air parcel is also just part of the general atmosphere, and we have no *a priori* knowledge which part of this atmosphere will become a cloud. But in the Lagrangian description it is necessary to decide on a parcel that will become a cloud. This poses a limitation on the utility of the Lagrangian model. What would be much more satisfactory is to have a set of equations that describe the formation of a cloud from the general atmosphere without the prior knowledge of where and when the cloud will form. We could then solve these equations subject to some initial and boundary conditions, and their solutions will tell us the time and location of cloud formation and dissipation.

To achieve this goal, we need a system of more complicated differential equations that are coupled to each other in space and time. High-speed digital computers are the only means we know by which we can obtain any meaningful solutions for such a system. This requires the numerical cloud models that we will discuss in this chapter.

We will focus on the basic schematics of cloud models rather than on the details of the numerical model and dynamics involved. We will then use specific cloud models to illustrate how such models can be applied to study the basic cloud microphysical and dynamical processes of two cloud types – one convective (exemplified by a cumulonimbus) and one stratiform (exemplified by a cirrostratus cloud). Readers interested in more general cloud modeling techniques may want to consult specific papers, such as Srivastava (1967), Takeda (1971), Klemp and Wilhelmson (1978), and Schlesinger (1973), or books such as Cotton and Anthes (1989) and Houze (1993). A more recent book by Straka (2010) provides in-depth discussions of the numerical principles and parameterization of microphysics.

So far, the more developed cloud models have been convective cloud models, which is the main focus of the following discussions. However, such models can also be applied to studying stratiform clouds with suitable modifications, as we will see later. A numerical cloud model consists of at least the following components:

(1) *Dynamics* – sets of equations describing the motion of the air. These are essentially the equations of motion, but most models also include the treatment of turbulent mixing.
(2) *Thermodynamics* – sets of equations that describe the thermodynamic states and energy conservation.
(3) *Microphysics* – sets of equations describing the cloud microphysical processes, such as the initiation, growth, and dissipation of hydrometeors. These are essentially the equations of the conservation of various water substances.

The degree of sophistication of a model depends on the intended specific application, and that is usually reflected in the different treatments of microphysics. For example, some models include the effect of solar and infrared radiation on the cloud because the cloud–radiation interaction is of concern, while others include cloud electrification processes because the thunderstorm charging mechanisms are being studied, and so on.

A cloud model can be a steady-state model or a time-dependent model. The former is mostly for diagnostic purposes, where the relations between different cloud variables at a given time are to be studied. The latter is prognostic, and can be used to simulate the time evolution of a cloud system. Most of the so-called *cloud-resolving models* (CRMs) in use presently are prognostic models, and this is the type that we will discuss below.

13.2 Types of cloud models

Cloud models differ in many different aspects, such as dimensionality, numerical scheme, formulation of dynamics, treatment of microphysics, etc., that sometimes cause confusion. In the following we will briefly describe what these different models refer to.

13.2.1 Dimensions of cloud models

Cloud models can be one-, two- or three-dimensional depending on the applications. Historically, the 1D model is the first one to be developed, followed by 2D and then 3D models. This obviously has to do with both modeling experience and the advancement of computing technology. But this does not mean that 1D or 2D models become obsolete, as they can still serve useful purposes where higher dimensions are unnecessary.

One-dimensional models

One-dimensional cloud models, by definition, are concerned with the cloud properties only in one direction, which is almost always the vertical direction. A 1D prognostic model can produce, for example, the vertical distribution of cloud and rain water contents, updraft/downdraft, momentum and energy flux, etc., as a function of time. We can regard the value of a cloud variable at a vertical point z at time t as the horizontal average of the values of this variable at this time. A typical example of a 1D cloud model is the one by Srivastava (1967). It is also possible to produce size distributions of hydrometeors from a 1D model if a specific way of distributing the cloud water content, for example, the Khrgian–Mazin distribution, is assumed.

On the other hand, owing to the restriction to just one dimension, there are many processes that such a model cannot study. One notable example of such limitations is the missing wind shear effect, as there is no way to treat horizontal winds in such a model. Without wind shear, the effect of vorticity on the cloud development cannot be studied either. Observationally, it is known that wind shear is very important to convective cloud formation, especially the deep convective clouds. Even though they have such limitations, 1D models can still be a useful tool in treating clouds in a simplified setting, for example, for studying possible cloud response to a mesoscale or synoptic-scale disturbance or the effect on a cloud if the atmospheric radiative field is changed. One-dimensional models require small computing resources and are especially suitable for studies that involve a long time span but require only knowledge of highly averaged cloud properties, for example, cloud response in long-term climate changes.

Two-dimensional models

This type of model includes the vertical dimension and one of the horizontal dimensions. With the inclusion of a horizontal dimension, it is now possible to represent horizontal winds and hence wind shears. The model can calculate cloud variables in two dimensions as a function of time. On the other hand, cloud variables are assumed to be uniform in the third dimension.

Two-dimensional models can be either *slab symmetric* or *axially symmetric*. Slab symmetric models are those formulated in Cartesian coordinates (x, y, z). Cloud properties are allowed to vary in, say, the x and z directions, while they are assumed to be constant in the y direction. The simulated cloud would look like a slice of bread from a very long loaf. Slab models are suitable for studying stratiform clouds or squall lines whose properties are quasi-uniform along one direction. Axially symmetric models are those formulated in cylindrical coordinates (r, ϕ, z). Cloud properties vary in the z (vertical) and r (radial) directions but are constant in the ϕ (azimuthal) direction. The simulated cloud would look like a long cylindrical column. Axial symmetric models can be useful for studying clouds such as isolated small cumulus where vertical wind shear is not important.

Two-dimensional model results can also be obtained by using a 3D model but taking the derivatives of cloud variables in one of the horizontal directions to be zero.

Two-dimensional models certainly also have their limitations. The lack of a third dimension means that any cloud property cannot be changed in the third dimension. The simulated cumulus often looks like the speech balloon drawn by cartoonists due to the lack of interaction with the third dimension. The lack of this interaction is especially restrictive to the realistic simulation of thunderstorms, in particular, supercell storm development, where rotation is a crucial factor.

Three-dimensional models

Naturally, three-dimensional models are the most satisfactory, because they consider all 3D motions and physics. Many 3D simulations of clouds look realistic and can indeed illustrate the evolution of clouds, including deep convective storms, fairly well. The main drawback of 3D models is that the required computing resource to run such a model is much larger than for 1D and 2D models for the same sets of physics and dynamics.

13.2.2 Formulation of dynamics

The dynamics of cloud models is formulated based on the equations of motion of air, and there are various approaches. The equations can be written in terms of air velocity directly (the primitive equations) or in terms of the streamfunction, which is advantageous in 2D models (see Sec. 8.3 for a discussion of the streamfunction). Since the streamfunction formulation does not have any advantage in 3D models, virtually all 3D models are formulated in the primitive equation form.

   Another major difference is how the pressure perturbation is treated in the model. One of the important concepts of numerical cloud modeling is the view that the cloud represents a perturbed state of the atmosphere, which is otherwise in a basic state that is uniform horizontally and in hydrostatic balance vertically. When an air parcel ascends due to buoyancy, the initial pressure field will be perturbed, hence we need to consider the pressure perturbation. The perturbation is especially significant in the case of deep convective clouds, as the atmosphere during a strong convection deviates substantially from hydrostaticity. But sound waves propagating in air are also a form of pressure perturbation, because they are caused by the compression and rarefaction of air, which changes the density and hence the pressure of air. Consequently, when the equations of motion are solved numerically, sound waves also appear in the solutions. While sound waves do occur in the atmosphere, they are not relevant to cloud formation. Furthermore, they may appear as large-amplitude rapidly propagating waves superimposed on, and perhaps completely masking, the solutions of interest to cloud formation. Considerable efforts have been devoted to dealing with the problems of sound waves in numerical cloud models (as well as in other branches of fluid dynamics). There are two types of cloud models, each using different techniques to tackle this problem.

Anelastic models

The flux form of the continuity equation is

$$\frac{\partial \rho_0}{\partial t} + \nabla \bullet (\rho_0 \vec{u}) = 0, \qquad (13.1)$$

where ρ_0 is the basic state air density and $\vec{u}$ is the air velocity. If we set the local change of density $\partial\rho_0/\partial t = 0$, then we are assuming that no local air density at a fixed point is allowed to change with time, and the continuity equation becomes

$$\nabla \bullet (\rho_0\vec{u}) = 0. \qquad (13.2)$$

This prohibits the propagation of sound and eliminates the influence of sound waves on the numerical solutions. This is called the *anelastic assumption*, and the cloud models that make use of such an assumption are called *anelastic models*. Anelastic models were first introduced by Ogura and Phillips (1962), and many varieties have been formulated since then. The earliest 3D cloud models were mostly anelastic models (e.g. Wilhelmson, 1974; Schlesinger, 1973).

The anelastic assumption is related to a more restrictive assumption called the Boussinesq approximation, which assumes that ρ_0 is everywhere constant except when the buoyancy is calculated; hence the continuity equation reduces to $\nabla \bullet \vec{u} = 0$.

Even though the anelastic assumption works well for its purpose, it takes more computer resources to perform calculations because of the form of the pressure perturbation equation (elliptic differential equation) that results. Newer cloud models tend to be compressible models.

Compressible models

These newer models called *compressible models* use the full compressible form of the prognostic continuity equation, which allows more flexible and more efficient computations. Since the full compressible continuity equation includes sound waves, steps must be taken to prevent the sound wave solutions from masking the solutions of interest. Sound waves are the fastest travelling waves in the solutions and require very small time steps if we want to resolve them correctly, which requires large computing resources. There are two techniques that have been used to tackle the sound wave problem in compressible models.

- *Time splitting*. Klemp and Wilhelmson (1978) designed a time-splitting technique to deal with this problem. They split the solutions of the dynamic equations into a part related to sound wave modes and another part that is unrelated to sound. They then used a small time step for the sound wave modes and a larger time step for other modes. In this way they achieved computational efficiency better than the anelastic models and with results that are just as good.

- *Quasi-compressible models*. This type of model uses a technique that artificially slows down the propagation of sound waves in the model by assigning a slower pseudo-sound speed to the pressure perturbation equation. This was proposed by Anderson *et al.* (1985), who showed that the results using this technique are as good as those of the anelastic models as long as the pseudo-sound speed is greater than twice the maximum wind speed in the model domain.

13.2.3 Formulation of cloud microphysics

The dynamical and thermodynamic conditions determined by the model set the environmental conditions under which cloud microphysical processes would occur. Earlier cloud models considered only liquid drops (cloud drops and raindrops) and totally ignored the ice phase. This may be acceptable for small cumulus, but obviously is unsatisfactory for larger convective clouds, which usually contain ice as well as liquid drops. Formation of the ice phase appears to impact not only the microphysical characteristics of deep convective clouds but also their dynamic properties. Of course, it is necessary to have an ice phase if we wish to simulate cirrus clouds.

Ideally, a cloud model should contain all the detailed microphysical processes as represented by the most accurate differential equations and should couple them with the dynamic and thermodynamic equations. The model then solves the complete set of coupled equations at all grid points at each time step. Unfortunately, the computing resources required to achieve this are usually too prohibitive. The highly nonlinear processes of cloud microphysics make the inclusion of precise differential equations for cloud microphysics in cloud models impracticable. Except for some small-scale models designed to study specific aspects of the microphysics (e.g. cloud drop size distribution), most cloud models use a parameterization method to handle microphysical processes. Kessler (1969) was the first to propose the use of such a technique to include microphysical processes in cloud models. Straka (2010) gave a detailed discussion of microphysics parameterizations in cloud models.

Let us use the rain process as an example to illustrate the parameterization of cloud microphysical processes. Other hydrometeor species can be treated in a similar way. Following Orville and Kopp (1977), the rain process can be represented by the conservation of rain water mixing ratio q_r:

$$\frac{\partial q_r}{\partial t} = -\vec{u} \bullet \nabla q_r + \nabla \bullet K_m \nabla q_r + \sum_k R_k + \frac{1}{\rho_a} \frac{\partial}{\partial z} (V_t q_r \rho_a). \tag{13.3}$$

This equation considers what will happen to q_r within a "grid box". The first term on the right-hand side is the advection of rain water, the second term is the change due to turbulent mixing (or diffusion), with K_m the turbulent diffusion coefficient, the third term represents the sum of the production or depletion rates R_k of q_r due to a specific process k, and the last term is the fallout of rain water at terminal velocity required V_t. Thus the third term contains the microphysics that needs to be parameterized. For a cloud model that contains both liquid and ice microphysics, these processes can include the following.

- *Rain production* – collision and coalescence of cloud drops, collision and coalescence of cloud drop and raindrop, melting of snowflakes, melting of graupel,
  melting of hail, and shedding of rain water from melting hail (which is considered
  as different from melting).
- *Rain depletion* – evaporation, collection of rain by snow, collection of rain by
  graupel, and collection of rain by hail.

The number of microphysical processes to include in this term depends on the
sophistication of specific models.

It is impossible at the present stage to implement the differential equations
describing these processes directly into the model and solve them along with the
dynamical and thermodynamic equations at each grid point at each time step.
Instead, simple algebraic equations are implemented to substitute for these differential equations. This is the parameterization technique.

For example, instead of calculating the collision growth of cloud drops to form
raindrops individually using Eq. (11.6), Orville and Kopp (1977) used the following
parameterization of cloud drop to raindrop "autoconversion" based on the calculation results of Berry (1968):

$$R_{auto} = \rho_a (q_c - q_{c0})^2 \left[1.2 \times 10^{-4} + \frac{1.596 \times 10^{-12} N_1}{D \rho_a (q_c - q_{c0})} \right]^{-1}, \qquad (13.4)$$

where q_c is the cloud water mixing ratio, q_{c0} is the conversion threshold (set at $2\,g\,kg^{-1}$,
i.e. the part of q_c greater than $2\,g\,kg^{-1}$ will turn into rain water), N_1 is the number
concentration of cloud drops, and D is the dispersion of the drop size. The numerical
coefficients in the square brackets are obtained from fitting the numerical calculation
results. Other microphysical processes can be similarly parameterized.

Bulk parameterized models vs. bin models

In calculating the changes of hydrometeor mass, there are two types of models –
bulk parameterization models, and bin (or size category) models. Bulk parameterization models calculate the growth and depletion of a hydrometeor category as a
whole: for example, just calculate the mixing ratio at a grid point but do not keep
track of the growth of each size category. In contrast, a bin model divides a
hydrometeor category into several size bins and keeps track of the change of
each size bin and hence predicts the size distributions in a prognostic manner.
Obviously, bin models require more computer resources than bulk parameterization models.

It is also possible to determine size distributions from a bulk parameterization
model, but it can only be done in a diagnostic manner. For example, after the model
has computed the rain water mixing ratio q_r, we distribute it according to the

Marshall–Palmer distribution (2.10). Obviously, this is not as natural as the bin model because true raindrop size distributions do not always conform to the MP distribution.

One- and multiple-moment schemes

If a microphysical parameterization scheme predicts only one property of a hydrometeor, for example, mixing ratio, then it is called a one-moment scheme. If, instead, the scheme can predict two properties, say, the concentration and mean size of a hydrometeor category independently, then it is a two-moment scheme. In general, two-moment schemes produce better prediction of hydrometeor distributions than one-moment schemes.

There are also three-moment schemes, for example, the one designed by Milbrandt and Yau (2005) that contains shape parameterization of ice particles in addition to size and concentration.

13.3 Numerics

13.3.1 Subgrid turbulence

Since the cloud models are numerical, they can only resolve down to the grid size employed in the model. Quantities with scales smaller than the grid size are considered as subgrid turbulence and cannot be resolved by numerical methods. However, they can be parameterized and computed together with the solutions of the equations of motion. There are various parameterization methods for treating subgrid turbulence. The simplest one is to assign a constant turbulent mixing coefficient to the whole domain, but this is clearly unsatisfactory. One of the more popular methods was proposed by Klemp and Wilhelmson (1978), who calculated the turbulent mixing coefficient based on the turbulent kinetic energy (TKE). This is sometimes called the one-and-a-half-order turbulent closure scheme. There are also other higher-order turbulence closure schemes, but the benefits of such schemes are still being debated.

13.3.2 Initialization of the model

With all the equations specified, it is still necessary to provide an initial perturbation somewhere in the domain to start the convective motion. This is called the *initialization* of the model. There are various ways to design this initial perturbation. Some models use actual observed mesoscale surface conditions as the initialization method, while others may use upper level pumping to trigger the convection. But one of the most popular techniques for initializing a convection is to specify a temperature and moisture perturbation near the surface in the shape of a spherical bubble, called the "warm bubble", that is warmer and more moist than the rest of the domain. This provides initial buoyancy so that the bubble will begin to ascend when the model run starts. The ascent of the bubble then causes other dynamical, thermodynamic, and microphysical processes to occur subsequently. If the magnitude of the perturbation and the size of the bubble are chosen suitably, the simulated storm can resemble an actual storm and behave similarly. This is the technique used in the convective cloud example to be illustrated below.

Simulation of stratiform clouds can be initialized by prescribing a random velocity field in the domain instead of a warm bubble. We will demonstrate this in the cirrostratus cloud example later in this chapter.

In general, however, we usually do not know what really triggered a specific cloud, so we cannot always expect that the simulated cloud will behave exactly like the actual one. This is especially so when the actual time evolution of the storm is concerned, because the actual cloud might have been triggered by a totally different mechanism than the simulated one. On the other hand, a simulated cloud often exhibits many features similar to the actual cloud, and in that case the model physics can usually be utilized to explain the physical mechanisms responsible for the formation of these features. This is because some features can only be produced by specific physical mechanisms, so that the appearance of those features can be attributed to these mechanisms if the model is equipped with such physics.

13.3.3 Boundary conditions

We also need to specify proper boundary conditions for the cloud model in order to have particular solutions that we can calculate to obtain specific numbers in the domain. There are also various ways of specifying boundary conditions. The upper and lower boundary conditions are often assumed to be rigid. However, rigid boundaries will reflect gravity waves generated by the convection and produce standing wave patterns that may mask the solutions of interest. A usual way to minimize this is to design an absorption layer at the top to absorb the gravity wave energies. Similar considerations should be done for the lateral boundary conditions. For example, the so-called "radiation boundary condition" is often used for simulating convective clouds because it allows the gravity wave energies to be transmitted across the lateral boundaries without reflecting back into the computational domain. For simulating stratiform clouds, the periodic boundary condition can be applied.

13.3.4 An example of a three-dimensional cloud model formulation

As an example, we list the set of equations used by Schlesinger (1973) to form a 3D prognostic anelastic cloud model with only liquid-phase cloud microphysics:

$$\frac{\partial u}{\partial t} = -\vec{V} \cdot \nabla u - R_{\mathrm{d}}\theta_0 \frac{\partial P}{\partial x}, \tag{13.5}$$

$$\frac{\partial v}{\partial t} = -\vec{V} \cdot \nabla v - R_{\mathrm{d}}\theta_0 \frac{\partial P}{\partial y}, \tag{13.6}$$

$$\frac{\partial w}{\partial t} = -\vec{V} \cdot \nabla w - R_{\mathrm{d}}\theta_0 \frac{\partial P}{\partial z} + g\left(\frac{\theta'}{\theta_0} + \varepsilon q_{\mathrm{v}}' - q_\ell\right), \tag{13.7}$$

$$\nabla \cdot \rho_0 \vec{V} = 0, \tag{13.8}$$

$$\frac{\partial \theta'}{\partial t} = -\frac{1}{\rho_0}\nabla \cdot \rho_0 \theta' \vec{V} - w\frac{\partial \theta_0}{\partial z} + \frac{\theta}{c_p T}Q, \tag{13.9}$$

$$\frac{\partial q_{\mathrm{v}}}{\partial t} = -\frac{1}{\rho_0}\nabla \cdot \rho_0 q_{\mathrm{v}} \vec{V} - \frac{Q}{L_{\mathrm{e}}}, \tag{13.10}$$

$$\frac{\partial q_\ell}{\partial t} = -\frac{1}{\rho_0}\nabla \cdot \rho_0 q_\ell \vec{V} + \frac{Q}{L_{\mathrm{e}}} \tag{13.11}$$

Eqs. (13.5)–(13.7) are the equations of motion that describe how air moves in three dimensions. Note that the Coriolis force is not considered here. The non-dimensional pressure variable P in these equations is related to the perturbation pressure p' by

$$P = \frac{T_0}{\theta_0 p_0}p'. \tag{13.12}$$

The primed variables represent the perturbed states, while those with subscript "0" represent the basic state (assumed to be hydrostatic). The last term on the right-hand side of (13.7) includes the thermal buoyancy (first two terms in parentheses) and the liquid water drag, and $\varepsilon = (R_{\mathrm{d}} / R_{\mathrm{v}}) - 1 \approx 0.61$. Eq. (13.8) is the continuity equation for the air, and (13.9) is the thermodynamic equation. Eqs. (13.10) and (13.11) are the conservation equations for vapor and liquid water, respectively. In the last three equations, Q is the heating rate due to latent heat release.

The seven equations (13.5)–(13.11) form a self-contained cloud model and can be solved numerically with proper initial and boundary conditions. Even without the inclusion of the ice phase, this model has produced many features that are consistent with observations. This is especially so in the major dynamical processes.

However, deep convective storms ultimately involve ice processes, as the updrafts carry moisture into the higher troposphere. Hence, to be realistic, modeling ice processes is necessary. The ice processes are of course necessary for simulating cirrus clouds.

13.4 Cloud model simulation examples

In the following, we will show the simulation results of a convective cloud and a stratiform cloud to illustrate the simulation of cloud processes by cloud models. Both simulations use a single sounding as the initial condition.

The convective cloud simulation uses the cloud model WISCDYMM (Wisconsin Dynamical/Microphysical Model) developed in the author's research group. This is a 3D prognostic, primitive equation, non-hydrostatic model cast in quasi-compressible dynamic framework. It contains 38 cloud microphysical interactions among six categories of water substance – water vapor, cloud water, rain water, cloud ice, snow aggregates, and graupel/hail (treated as one category) – and can be in either bulk parameterization or bin mode. Detailed descriptions of the parameterizations of the cloud microphysics are given by Straka (1989), but a brief summary is also given by Lin *et al.* (2005).

The model used for simulating the stratiform cloud will be described in a later section.

Fig. 13.1 shows a schematic chart of the microphysical processes included in WISCDYMM. The boxes indicate the category of water substance, and the arrows represent the processes that convert one category into others. Table 13.1 explains the acronyms shown in Fig. 13.1. These are the microphysical processes that we have discussed in the preceding chapters, but they are represented by simplified parameterized forms rather than the precise detailed equations.

13.4.1 Simulation of the CCOPE supercell of 2 August 1981

In this section, we show the WISCDYMM simulation of a deep convective storm that passed through the Cooperative Convective Precipitation Experiment (CCOPE) network on 2 August 1981 in the US upper High Plains (Colorado–Wyoming–Montana–South Dakota) region. This storm turned out to be a special kind of long-lived storm called a "supercell". Aside from its longer life span and usually more intense updraft, a supercell storm is distinguished from regular isolated or multicellular storms by its obvious rotation (signified by the presence of a *mesocyclone*). Readers are referred to Doswell (2001) for more details about this type of storm. The simulated storm goes through different stages of convective cloud development and serves to illustrate the interaction between cloud microphysics and dynamics for convective clouds.

WISCDYMM used the sounding in Fig. 13.2 as the initial conditions and a warm bubble of 20 km (horizontal) by 4 km (vertical) and a maximum temperature perturbation $\Delta T = 3.5$ K to initialize the simulation of the supercell. The relative humidity of the bubble is assumed to be the same as that of the environmental unperturbed air. This implies, however, that the mixing ratio of water vapor in the bubble is greater than in

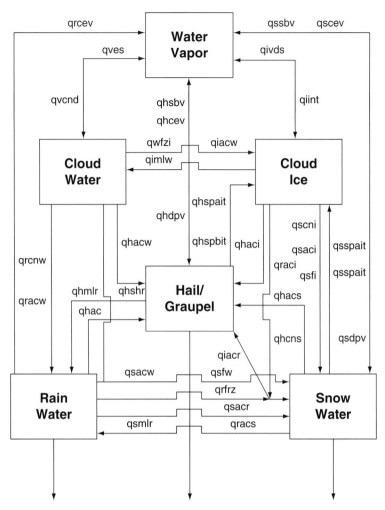

Fig. 13.1 Block diagram of the microphysical interactions in the WISCDYMM cloud model. The acronyms are explained in Table 13.1.

the environment because of its higher temperature. The following discussions are based on the simulations with grid resolution 1 km × 1 km × 0.2 km. Most of the results shown here are from a run with computational domain of 55 km × 55 km × 20 km, but a few are from a run with 120 km × 100 km × 20 km for the purpose of illustrating features further downwind. The simulated storm develops into a supercell storm, just like the observed actual storm, and its general behavior is also quite similar to the real one. More discussions of the comparison between the simulated and observed storms are given in Johnson *et al.* (1995).

A warm bubble of this magnitude is usually necessary for initializing thunderstorms. For smaller cumulus clouds, smaller bubble sizes and weaker temperature

Table 13.1 *The acronyms used in the WISCDYMM cloud model in Fig. 13.1.*

Acronym	Process
qhaci	Accretion of cloud ice by hail
qhacr	Accretion of rain by hail
qhacs	Accretion of snow aggregates by hail
qhacw	Accretion of cloud water by hail
qhcev	Condensation/evaporation of vapor to/from wet hail
qhcns	Autoconversion of snow to hail
qhdpv	Vapor deposition to hail
qhmlr	Melting of hail to rain
qhsbv	Vapor sublimation from hail
qhshr	Rain water shed from hail
qhspait	Secondary production I of cloud ice from hail
qhspbit	Secondary production II of cloud ice from hail
qiacr	Accretion of rain by ice to form snow or hail
qiacw	Accretion of cloud water by cloud ice
qiint	Nucleation of pristine cloud ice
qimlw	Melting of cloud ice to cloud water
qivds	Vapor deposition/sublimation to/from cloud ice
qraci	Accretion of cloud ice by rain to form snow or hail
qracs	Accretion of snow by rain to form snow or hail
qracw	Accretion of cloud water by rain
qrcev	Evaporation of rain
qrcnw	Autoconversion of cloud water to rain
qrfrz	Probabilistic freezing of rain to form snow or hail
qsaci	Accretion of cloud ice by snow
qsacr	Accretion of rain by snow to form snow or hail
qsacw	Accretion of cloud water by snow
qscev	Condensation/evaporation of vapor to/from wet snow
qscni	Autoconversion of cloud ice to snow
qsdpv	Vapor deposition to snow
qsfi	Bergeron process, transfer of cloud ice to snow
qsfw	Bergeron process, transfer of cloud water to snow
qsmlr	Melting of snow to rain
qssbv	Vapor sublimation from snow
qsspait	Secondary production I of cloud ice from snow
qsspbit	Secondary production II of cloud ice from snow
qvcnd	Condensation/deposition to/on cloud water/cloud ice
qves	Evaporation/sublimation of cloud water/cloud ice
qwfzi	Homogeneous freezing of cloud water to cloud ice

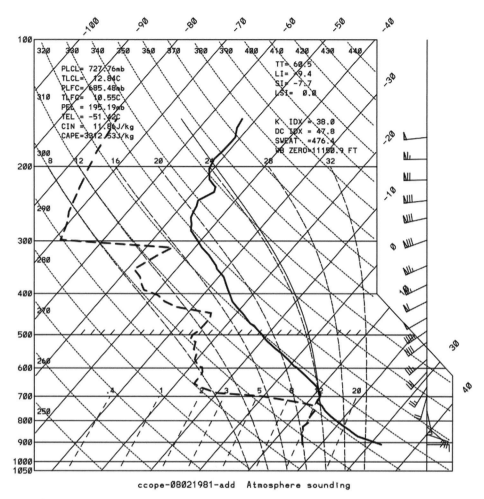

Fig. 13.2 The Knowlton, Montana, sounding on 2 August 1981 (taken at 17:46 Mountain Daylight Time), used as the initial condition for the CCOPE storm simulations illustrated in this chapter. The solid curve is for temperature and the dashed curve is for dewpoint. The portion of the dewpoint curve above 300 hPa, which was missing in the original sounding, is constructed using an average August 1999 water vapor profile over 40–60°N obtained from the Halogen Occultation Experiment (HALOE) onboard the UARS satellite.

perturbations can be used. Different cloud models respond differently to a perturbation of a certain magnitude, and many tests need to be performed to determine the sensitivity of a model to these perturbations.

Note that, since the storm is propagating downwind, the simulated storm will also move downwind as time goes on and will soon be out of the computational domain. Thus an algorithm is implemented to keep the storm within the domain. Hence the horizontal coordinates in the figures to be discussed below refer to a coordinate box

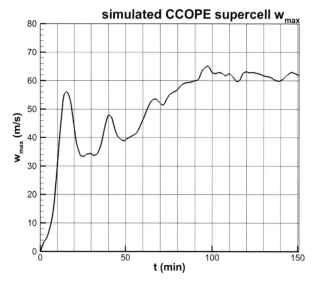

Fig. 13.3 Time evolution of the simulated CCOPE supercell, showing the quasi-steady state after $t \sim 90$ min.

that has been shifted with time to follow the storm. The history of the maximum updraft of the simulated storm is shown in Fig. 13.3. We see that w_{max} reaches a maximum of more than 50 m s^{-1} at ~13 min. This large updraft is induced by the initial warm bubble perturbation that provides fairly large buoyancy and also the release of latent heat due to condensation. At this time, the amount of hydrometeors is relatively small and so the downward drag they exert on the ascending air is small, hence the very rapid updraft. This is often artificial due to the initial bubble setting and may or may not reflect the real cloud behavior. This is the spin-up stage of the simulated storm, and many features present at this stage are not necessarily characteristic of many observed storms and hence are often discounted. However, this does not mean that the convection during this stage is unphysical, just that the initialization by a warm bubble does not necessarily correspond to the actual situation.

If the simulation is successful, however, the storm features after the spin-up stage are often consistent with observations. Once the deep convective system is established, its characteristics are mainly determined by the environmental conditions as specified by the sounding and have little to do with the initialization technique. In the present case, the w_{max} decreases to 30–40 m s^{-1} at ~25 min. The decrease of w_{max} is due to the loading of precipitation particles that form within this period. After about 60 min into the simulation, w_{max} becomes quasi-steady at around 60 m s^{-1} throughout the simulation period of 2 h.

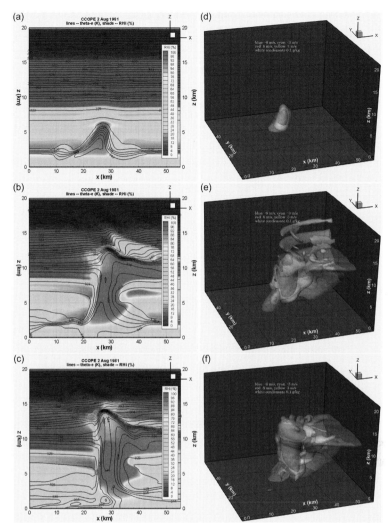

Fig. 13.4 (left) The θ_e field (black contours) in the central east–west cross-section of the storm at (a) $t = 10$ min, (b) $t = 30$ min, and (c) $t = 60$ min. The color shades represent the relative humidity with respect to ice (RH_i) field. (right) The 3D structure of the vertical velocity field at (d) $t = 10$ min, (e) $t = 30$ min, and (f) $t = 60$ min. Isosurface colors: blue, -8 m s^{-1}; cyan, -3 m s^{-1}; yellow, 3 m s^{-1}; red, 8 m s^{-1}. Faint white isosurface: total condensate mixing ratio 0.1 g kg^{-1}, which is used to approximately represent the visual cloud boundary. See also color plates section.

Vertical structure of the simulated thunderstorm

Fig. 13.4 shows three frames ($t = 10$, 30, and 60 min) each of the θ_e field (left panels) in the central east–west (x–z) vertical cross-section ($y = 27$ km) and the 3D w field (right panels) of the simulated storm. The faint white isosurface of 0.1 g kg^{-1} total condensate mixing ratio in the right panels roughly represents the visual cloud boundary, which corresponds approximately to the high RH_i (relative humidity with respect to ice) area represented by the red shades.

Fig. 13.4(a) shows that, at $t = 10$ min, the instability carried by the warm bubble is ascending. The lower part of the closed θ_e curve represents the region where $\partial\theta_e / \partial z < 0$, i.e. unstable air. The perturbation is transported not only upward but also outward, as can be seen by the wavy θ_e curves in the lowest layer extending to the left and right. The top of the cloud has reached about 6 km at this time and the body of the storm is tilted to the east in response to the wind shear. The cloud is still in the cumulus stage of the storm (Byers and Braham, 1949), although it is already larger than typical fair-weather cumulus.

The corresponding w structure in Fig. 13.4(d) indicates that it is all updraft inside the cloud, which is the definition of the cumulus stage as defined by Byers and Braham (1949). Downdrafts around the cloud are still very weak at this time.

At $t = 30$ min, the cloud has developed all the way to the tropopause and beyond, forming a cumulonimbus, as seen in Fig. 13.4(b). The energy of cloud growth comes not only from the original perturbation but mainly from the continuous inflow of unstable air at low levels as the storm advances to the east (right). An overshooting top and an anvil system have been developed; both are visibly identifiable thundercloud characteristics. The boundary layer air of $\theta_e \sim 345$ K is carried upward by the strong updraft in a 4 km wide core and reaches the tropopause without much dilution. The θ_e gradient is nearly zero in the updraft core, indicating little mixing. On the other hand, θ_e gradients are very large at the cloud top. Dramatic gravity wave breaking is occurring at the top of the storm, producing a phenomenon called "jumping cirrus", which will be discussed more in Chapter 15.

Fig. 13.4(e) shows the corresponding w field structure at this time. The main body of the storm coincides with the strong updraft core as represented by the yellow and red regions, which also tilts to the east in response to the ambient wind shear. Many downdraft regions (cyan and blue shades) surround the updraft core. Those in the middle and upper levels are the compensation downdrafts caused by the updrafts, but the one at low level (below the updraft core) is due to the precipitation. The w structure above the anvil, the blue and yellow arcs, reflects the motions due to the gravity waves. Thus the cloud now has both updrafts and downdrafts, which fits the definition of the mature stage of a thunderstorm of Byers and Braham (1949). It has also developed an overshooting top and a rather extensive anvil.

At $t = 60$ min, the storm has already become a quasi-steady-state supercell, as was observed for the actual CCOPE supercell. A regular storm would be in the dissipating stage with nearly all downdrafts in the cloud at this time, but a supercell like the one simulated here develops a circulation that tends to sustain the continuous development, and the storm remains quasi-steady at even 4 h into the simulation.

This indicates that a well-formulated model can indeed simulate the physics and dynamics of the cloud system very well. The maximum updraft remains at around 50–60 m s^{-1} after that. Fig. 13.4(c) shows that the storm is less tilted than before and the anvil becomes larger (far beyond the domain). Otherwise the θ_e field looks similar to that in Fig. 13.4(b). The w field as shown in Fig. 13.4(f) is a very typical one for a mature thunderstorm. The nearly vertical strong updraft core reaches slightly beyond the tropopause. The upper part of this core is covered by a dome of strong induced downdrafts, like a downdraft cap. There are also downdraft regions downshear of the main updraft core. On the other hand, a precipitation-caused downdraft region is present at the low-level upshear of the updraft core.

Remember that the simulated storm (as well as the real one) is traveling to the east (right) all these times such that the unstable moist air keeps entering the storm system following the updraft path from the low-level downstream region, whereas the main precipitation particles are shed from the storm in the upstream downdraft region. During the run it is necessary to subtract the mean motion of the storm from the output so as to keep the simulated storm in the center of the computational domain.

Fig. 13.5 shows the snapshots of the temperature (left panels) and pressure perturbation (right panels) in the central east–west cross-section at $t = 10$, 30, and 60 min. Fig. 13.5(a) shows that, at $t = 10$ min, the warm bubble carries the warmer and moister air upward to higher than 6 km. While the ascending motion causes the cloud air to be cooled, the condensation releases latent heat to warm the air parcel. There is an overall warming effect in the cloud. The purple contours show the region that is supersaturated with respect to ice. We see that the top part of the cloud has the highest *RH* and is supersaturated with respect to ice. Very few downdrafts are present at this time.

Fig. 13.5(e) shows the perturbation pressure p' field at $t = 10$ min. The upshear side of the cloud is characterized by positive p' and the downshear side shows negative p'. This pattern is reminiscent of the pressure distribution for flow past an obstacle, e.g. a drop or an ice particle, that we described in Chapter 8, namely, high pressure (stagnation) upstream and low pressure in the wake (see Fig. 8.22). This type of blocking pattern has been described by Newton and Newton (1959), who indicated that, when vertical shear is present, a hydrodynamic (non-hydrostatic) pressure field is induced by relative motions near the boundaries of a large convective system that does not move with the ambient winds, which is the case in the simulated storm. Note, however, that the cloud is not a solid object, and the effect cannot be totally the same as for a solid obstacle. Note that the pressure perturbation pattern is not just induced by the dynamic effect of the updraft. Two other factors, the thermal buoyancy and the hydrometeor loading, also contributed to the pressure perturbation. The relative contributions of these factors vary in time and space depending on specific situations.

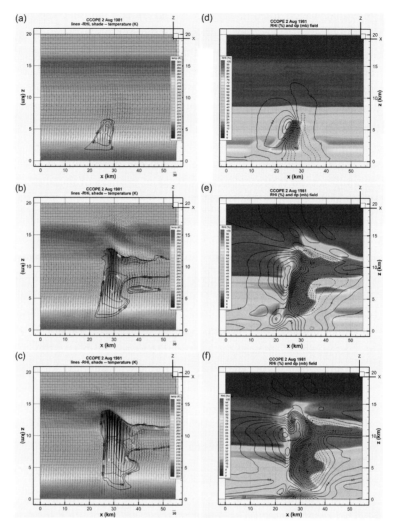

Fig. 13.5 (left) The temperature field (K) in the central east–west cross-section of the storm at (a) $t = 10$ min, (b) $t = 30$ min, and (c) $t = 60$ min. Black contours are for $RH_i = 70$–90% while purple contours are for $RH_i = 100\%$ or greater. Black arrows show the wind vectors projected onto this x–z plane. A reference vector of 30 m s^{-1} is shown in the lower right corner. (right) The perturbation pressure (p') in the central east–west cross-section of the storm at (d) $t = 10$ min, (e) $t = 30$ min, and (f) $t = 60$ min. Solid (dashed) curves represent positive (negative) pressure perturbations. The color shades represent the RH_i field. See also color plates section.

At $t = 30$ min, as shown in Fig. 13.5(b), the storm has developed an overshooting dome or top (OT). Inside the dome the temperatures are very low. It is also very dry, as most of the water vapor is used up to form condensate. Immediately downwind of the OT at the anvil level, however, is an ice-supersaturated region. Most of the moisture, however, remains below ~9 km, and the part of the storm above this level is generally dry. Very large updrafts as well as very large downdrafts are present in the storm and around it. Very turbulent air flow and vigorous gravity wave activity can be seen above the storm top, especially the area around the OT. The wave activity causes drastic changes in the initial horizontally uniform temperature field. The region above the anvil and immediately downwind of the OT is a warm region related to this wave activity.

The corresponding p' field is shown in Fig. 13.5(e), which is more pronounced than at $t = 10$ min. The main high-perturbation center is at $z \sim 10$ km and a smaller high center is at $z \sim 2$–3 km associated with the precipitation. The main low center is located at $z \sim 5$ km, but there is another low center at $z \sim 12$ km at the cloud top immediately downwind of the OT; both are located close to the high RH_i centers. The quasi-blocking pressure pattern at the tropopause level causes many storm top features such as the above-anvil plumes (AAP), jumping cirrus (JC), and ship waves that have been observed by meteorological satellites and research aircraft. These phenomena are also associated with the pressure perturbation pattern in the strato-sphere above the storm that is the signature of internal gravity waves induced by the strong convection. Further discussions of the AAP and JC will be given in Chapter 15. The ship wave feature is shown in Fig. 13.6, which plots the w field at the storm top (using $RH_i = 50\%$ to represent the cloud top surface). Actual satellite

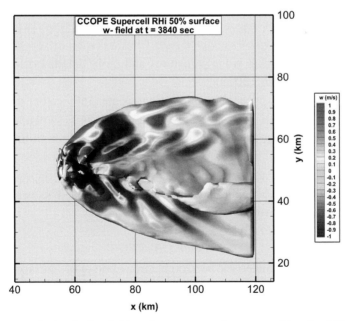

Fig. 13.6 Vertical velocity field at the storm top (represented by the $RH_i = 50\%$ isosurface) showing the ship wave signature. See also color plates section.

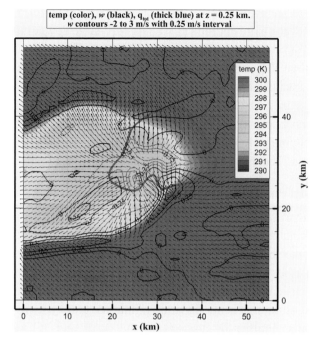

Fig. 13.7 Temperature field (color) and vertical velocity field (black contours: solid = positive, dashed = negative) at the surface of the simulated CCOPE supercell. The thick blue curve is the cloud boundary represented by the 0.1 g kg^{-1} total condensate mixing ratio (sum of all condensed forms of water). See also color plates section.

storm images showing ship wave signatures and more detailed discussions have been given by Wang *et al.* (2010).

At $t = 60$ min, the OT pushes even higher and is generally colder than at 30 min. The storm enters the quasi-steady state of a supercell storm as shown in Fig. 13.5(c). The perturbation pressure field shown in Fig. 13.5(f) shows a high-pressure perturbation right in the core of the OT and another even more intense high at the surface. The latter is mainly due to the evaporation of precipitation particles (mainly raindrops) in highly subsaturated air and is called the *mesohigh*. The two low-pressure centers remain approximately at the same position.

Horizontal circulation

Fig. 13.7 shows the characteristic horizontal circulation at the surface at $t = 60$ min when the storm has reached quasi-steady state. The white contour is the 0.1 g kg^{-1} total condensate mixing ratio. The horizontal circulation enclosed by the white contour is characterized by strong divergence, which coincides with the heavy downdraft and heavy precipitation. This divergent area is cold relative to other

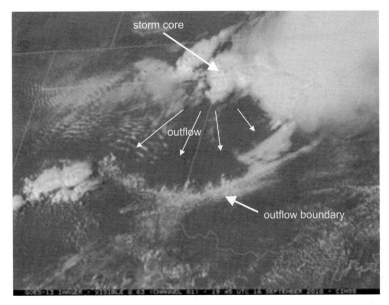

Fig. 13.8 A satellite visible image of a storm system showing the associated surface outflow and the arc-shaped outflow boundary. The updrafts at the outflow boundary cause the formation of a line of new convective clouds. Image courtesy of NOAA.

areas because of the evaporative cooling of the precipitating particles, and is called the *cold pool*. The belt-shaped region to the right (east) of the advancing edge of the cold pool is the gust front, which is an updraft region. This is the inflow region where surface unstable air is fed into the storm.

There is also a weaker updraft region in the trailing (northwest) edge of the cold pool caused by the compensating motion of the downdraft. This compensating updraft may trigger the formation of clouds or even new storms if the magnitude is large enough. When this happens, the updraft region forms a visible outflow boundary of the storm, often in the shape of a circular arc of clouds. Between the arc and the core of the storm is a clear area that is almost devoid of clouds. This phenomenon is frequently observed in radar storm echo and also meteorological satellite storm images, for which Fig. 13.8 is an example.

Fig. 13.9 shows the horizontal section at $z = 6$ km. The mid-level storm circulation is characterized by the intense updraft core, which coincides with a warm core that is mainly caused by latent heat release due to condensation. Fig. 13.10 shows the horizontal circulation at $z = 10$ km. Here the updraft reaches a maximum and the core is also warm as in the case of the mid-level section. The updraft core behaves like an obstacle to the ambient flow so that the latter appears to be diverted to the sides, forming a V-shaped flow pan. In addition, the effect of gravity waves becomes more pronounced at this level as it is closer to the tropopause. The stable

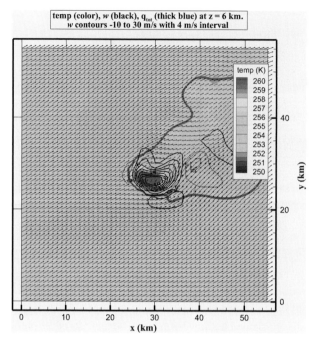

Fig. 13.9 Same as Fig. 13.7, except for *z* = 6 km. See also color plates section.

stratification of the stratosphere prevents the vertical motions of the storm from going too far into the stratosphere, but in doing so causes gravity wave motions in the upper troposphere and lower stratosphere (UTLS) region. The upstream edge of the cloud is characterized by a downdraft belt, which is relatively warm mainly due to the compression warming of the sinking air. This belt is a signature of the gravity waves in the presence of an ambient flow.

Hydrometeor distributions

In the following, a few snapshots of the hydrometeor fields of the simulated storm will be shown. These figures were rendered from a run with a horizontal domain of 120 km × 100 km so that a larger downstream portion of the anvil can be shown. The grid resolution, however, is still 1 km × 1 km × 0.2 km.

Fig. 13.11 shows the comparison between the outward appearances of the simulated storm as if seen by naked-eye observation (Fig. 13.11a) and the corresponding distribution of the hydrometeors (Fig. 13.11b). As expected, the upper level of the cloud consists of mostly cloud ice and snow aggregates, but the anvil, which extends far downstream, consists of mainly snow aggregates. On the other hand, the overshooting top consists of mainly cloud ice. Rain water exists at lower level, as expected. Graupel/hail is widely distributed in the middle to high levels.

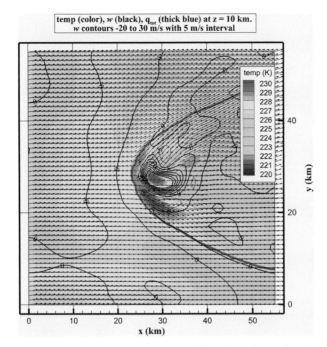

Fig. 13.10 Same as Fig. 13.7, except for $z = 10$ km. See also color plates section.

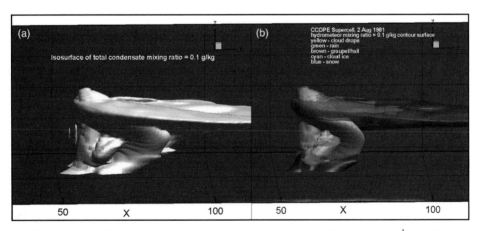

Fig. 13.11 (a) The isosurface of total condensate mixing ratio $= 0.1 \, \mathrm{g \, kg^{-1}}$. (b) The corresponding distribution of hydrometeors represented by the $0.1 \, \mathrm{g \, kg^{-1}}$ isosurface of the respective hydrometeor. Isosurface colors: yellow, cloud water; green, rain; red, graupel/hail; cyan, cloud ice; blue, snow. See also color plates section.

Fig. 13.12 shows a few snapshots of the hydrometeor distribution in the central cross-section of the storm. We see that at $t = 10$ min, most of the cloud volume consists of cloud drops. Rain water has formed in the central cloud region but has not yet reached the ground. Ice particles have also appeared at higher levels,

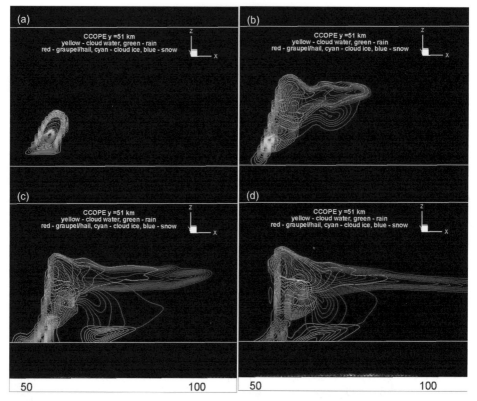

Fig. 13.12 Snapshots of hydrometeor distribution in the central east–west cross-section ($y = 51$ km) at (a) 10 min, (b) 20 min, (c) 30 min, and (d) 60 min. Color and range of mixing ratio contours (g kg^{-1}): yellow, cloud water (0.1–2.8); green, rain (0.1–4.2); red, graupel/hail (0.1–8); cyan, cloud ice (0.1–2); blue, snow (0.1–1.3). See also color plates section.

though not developed extensively. The maximum graupel/hail mixing ratio occurs at a level higher than rain, as would be expected. At $t = 20$ min, extensive ice particle formation has occurred as the cloud pushes into the higher troposphere, as can be seen from the cloud ice and snow aggregates at the upper level and the wide distribution of graupel/hail in the mid-level. Rain has reached the ground at this time and is apparently contiguous to the graupel/hail region, making it possible that the rain production is closely related to the graupel/hail process. This is further demonstrated by the cross-section at $t = 30$ min, where both rain regions on the surface are spatially associated closely with the graupel/hail regions above them. One of the rain regions is immediately upwind of the main updraft. This rain region coincides with the main downdraft region on the surface where the storm cold pool occurs. Between the cold pool and the surface updraft is the gust front.

Hook echo and bounded weak echo region

Aside from the above basic dynamical and microphysical features, other features that appear under more specific conditions have also been simulated successfully. For example, it is frequently observed that tornado-spawning thunderstorms in the US Great Plains often exhibit "hook echo" and an associated vault called the bounded weak echo region (BWER) in the low-level scan of radar plane-position indicator (PPI) images. Both the hook echo and BWER have been simulated by WISCDYMM and other similar models.

13.4.2 *Cirrostratus model and simulated cirrus clouds*

In this section, we will demonstrate the simulation of a stratiform cloud using a cloud model. The equations governing the cloud microphysical and dynamical processes are really the same, and hence the same model framework can be used. Only the initial and boundary conditions and the techniques of initializing the model need to be modified to suit the stratiform cloud. Stratiform clouds are those with more or less uniform horizontal structures and hence it is often adequate to simulate them in two dimensions. Here we will show the simulation of an optically thin cirrostratus cloud that consists completely of ice particles – pristine ice crystals and snow aggregates.

One of the main purposes of studying thin cirrus clouds is to understand their potential impact on the radiative budget of the Earth–atmosphere system because of the strong interaction between ice crystals and visible–infrared radiation. Although these clouds are optically thin, they can be rather thick vertically and very expansive horizontally, and hence their total radiative impact can be considerable. Thus, cirrus models often come with specially designed radiation modules to examine the ice–radiation interaction in greater detail.

The example to be discussed below was taken from Liu *et al.* (2003a,b). They adopted the dynamics module of WISCDYMM but modified the physics to suit better the purpose of cirrus simulations. The modifications of the physics include (a) changes in treating ice particle microphysics and (b) the implementation of a radiation module, as qualitatively described below. The afore-mentioned papers have the detailed formulations.

Modifications in ice microphysics

Since a cirrus cloud consists of all ice particles, the model considers only the microphysics of ice particles, and hence all liquid-phase processes are excluded. Moreover, it is better to have a two-moment scheme for ice particles because both the particle size and the concentration impact strongly upon the radiative properties of the cirrus. In addition, the nucleation process has to be examined more carefully here than for a deep convective storm case because the simulated cirrus behavior is very sensitive to the nucleation parameterization. All these were considered by Liu *et al.* (2003a,b).

The radiative properties of cirrus also depend strongly on the habit of the ice particles. So ideally a three-moment scheme that considers the size, concentration, and shape of ice particles simultaneously should be implemented. However, the habit issue is still not yet quite well resolved, and it is difficult to realistically simulate the evolution of all three ice properties. Consequently, Liu *et al.* (2003a,b) considered the ice habit effect by assuming that the whole cloud consists of just one kind of ice particle, for example, hexagonal plates or ice columns, etc., instead of predicting the habit explicitly. On the other hand, by considering the habit separately, it is easier to understand how the habit impacts on the cloud radiative properties because that factor is clearly isolated. The two moments they considered are the mixing ratio and concentration. The size can be deduced from these two parameters.

Radiation module

The radiation module includes two main submodules. The first is how individual ice particles of various habits interact with the incoming radiation (i.e. the absorption, reflection, and refraction of radiation by ice particles). For this, Liu *et al.* (2003a,b) adopted a scheme called modified anomalous diffusion theory (MADT) developed by Mitchell (2000) that treats the interaction explicitly.

The second submodule is a radiative transfer module that determines how the radiation goes through the cloud layer. For this, Liu *et al.* (2003a,b) adopted the two-stream formulation developed by Ackerman and Stephens (1987) to determine the upwelling and downwelling of the radiation flux, and, by doing so, the net radiation energy gained (or lost) at specific points in the cloud can be determined. Note that, since the radiative flux is calculated only in the vertical direction, this is equivalent to saying that vertical atmospheric columns are independent of each other. Consequently, interactions of radiative flux among adjacent columns are ignored.

The interaction between the microphysics and radiation module is illustrated in Fig. 13.13.

13.4.3 Initial and boundary conditions and initialization

A cirrostratus is a layer cloud that is more or less uniform horizontally, and therefore it may be adequate to consider a 2D model instead of a 3D model. The boundary condition is assumed to be periodic. Also, unlike a strong convective cloud, where there is often a dominating strong updraft core, a stratiform cloud is characterized by much weaker and randomly distributed updraft regions in the cloud. Thus, instead of using a single warm bubble to initiate cloud formation, the initialization is achieved by using a randomly

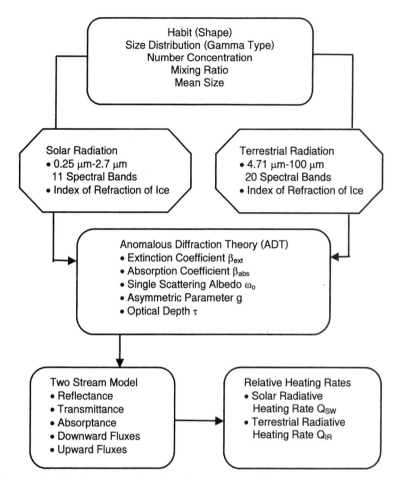

Fig. 13.13 Links between the microphysics and radiation modules of the cirrus model given by Liu *et al.* (2003a). Reproduced by permission of the American Meteorological Society.

distributed temperature perturbation from -0.02 to $+0.02\,^{\circ}$C to initialize the convective disturbance. This random perturbation field is imposed on a background with uniform vertical velocity of 3 cm s^{-1}, which is typical for large-scale lifting. Fig. 13.14 shows the schematic of the model domain configuration and grid resolution.

Cirrus clouds can occur in a wide variety of atmospheric environments. The two examples illustrated here use soundings representing the warm unstable and cold unstable environmental conditions, respectively, made available by a cirrus observation group at the National Aeronautics and Space Administration (NASA) Global Energy and Water Cycle Experiment (GEWEX) Cloud System Study (GCSS) Working Group 2 (WG2), which is considered to be representative of cirrus formation in spring. The warm unstable condition represents the mid-latitude cirrus. It is

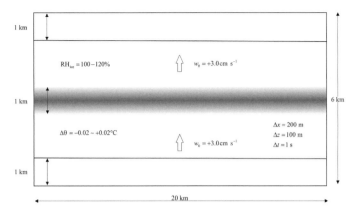

Fig. 13.14 Schematic of the model domain configuration, grid resolution, and initial perturbations of temperature, updraft, and RH_i. From Liu *et al.* (2003a). Reproduced by permission of the American Meteorological Society.

called "warm" because the cirrus clouds there often occur at lower altitudes and hence are warmer than their tropical counterparts. The cold unstable condition represents, of course, the tropical situation. Fig. 13.15 shows these two soundings. The potential temperature profile shows that $(\partial\theta/\partial z) > 0$ at all levels except the unstable layer, where $(\partial\theta/\partial z) < 0$. The wind shear effect is not considered in this simulation.

This unstable layer is located in the supersaturated layer where the cirrus cloud is to develop. For both cases, the temperature lapse rate is ice pseudo-adiabatic for the lower 0.5 km of the supersaturated layer, while the lapse rate is $1°C\,km^{-1}$ greater than the ice pseudo-adiabatic lapse rate for the upper 0.5 km. As the model run starts, the randomly imposed temperature perturbations would allow water vapor in the most supersaturated layer to nucleate into ice, and then the cloud starts to evolve according to the physics specified by the model. Ice crystals in this simulation are assumed to be hexagonal ice columns. Simulation results by Liu *et al.* (2003b) show that cirrus development under unstable conditions is more vigorous than under stable conditions, as would be expected.

Fig. 13.16 shows the ice water content contour field of the warm cirrus at $t = 60$ min. It reveals that the cloud has developed the cellular structure and ice fall streaks that are characteristic for typical cirrus as observed by lidar in Fig. 13.17, demonstrating that the model can produce realistic cirrus structure.

Fig. 13.17 shows an example of cirrus observation made by radar and lidar. Note that this is not a snapshot of the cirrus cloud but rather the time-series plot of the vertical profiles of the cirrus at a fixed location. Both the radar and lidar were pointing vertically and the cloud drifted over the observation site. However, if we can regard the layer-like cirrus clouds as in quasi-steady-state condition, as they

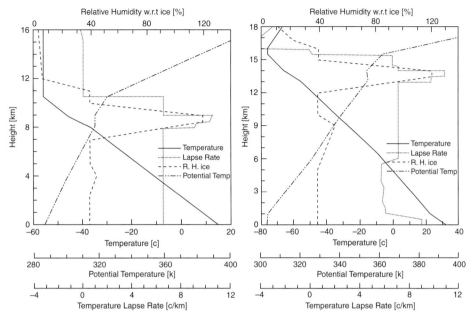

Fig. 13.15 The warm unstable (left) and cold unstable (right) soundings used to simulate the formation of tropical cirrus. From Liu *et al.* (2003b). Reproduced by permission of the American Meteorological Society.

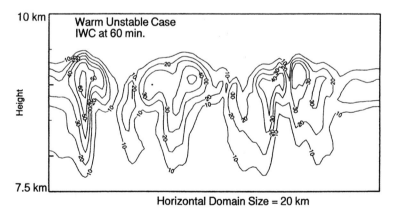

Fig. 13.16 The ice water content field of the simulated cirrus at *t* = 60 min formed in a warm unstable environment. From Liu *et al.* (2003b). Reproduced by permission of the American Meteorological Society.

often are, then the time-series plot is also approximately equivalent to a snapshot. Thus the cellular structure in the time-series plot should also be present in the cirrus at a fixed time. The features in Fig. 13.16 compare well with those in Fig. 13.17 (except for the wind shear, which is not considered by the model).

Numerical cloud models

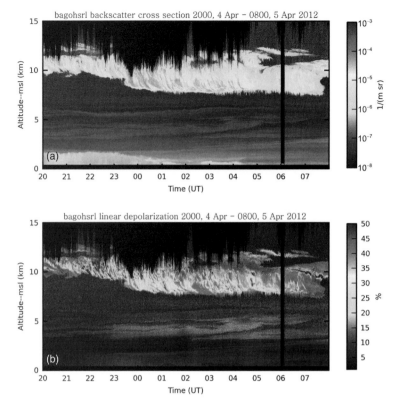

Fig. 13.17 Lidar image of cirrus development from 20:00 UTC 4 April to 08:00 UTC 5 April 2012 at Madison, Wisconsin (UTC = Coordinated Universal Time). (a) Aerosol backscattering cross-section σ. The color scale is $\log_{10}\sigma$ in unit of m^{-1} sr^{-1} (sr = steradian). (b) Aerosol linear depolarization (%). Courtesy of Dr Edwin Eloranta. See also color plates section.

Fig.13.18 shows the time evolution of the horizontally averaged ice water content (IWC) and ice number concentration vertical profiles in the 3 h of simulated warm cirrus development. Fig. 13.18(a) shows that the peak IWC evolves from ~ 7 mg kg^{-1} at $t = 20$ min to a maximum value of ~ 35 mg kg^{-1} at 40 min. After that, the value decreases and also peaks at a lower level. At $t = 90$ min, the IWC peaks at two height levels, indicating that part of the cloud is extending downward due to the fall of larger ice crystals that grow larger through either diffusion growth or aggregation. At $t = 180$ min, the vertical extent of IWC is ~ 3 km, i.e. three times as thick as the initial supersaturated layer. Figs. 13.18(b,c) show the vertical profiles of pristine ice and aggregate IWC separately. Comparing these two panels to Fig. 13.18(a), we see that the lowest part of the cloud consists of mainly ice aggregates, which seems to be reasonable. Figs. 13.18(d–f) show the corresponding evolution of ice concentration vertical profiles, whose behaviors are quite similar to that of IWC.

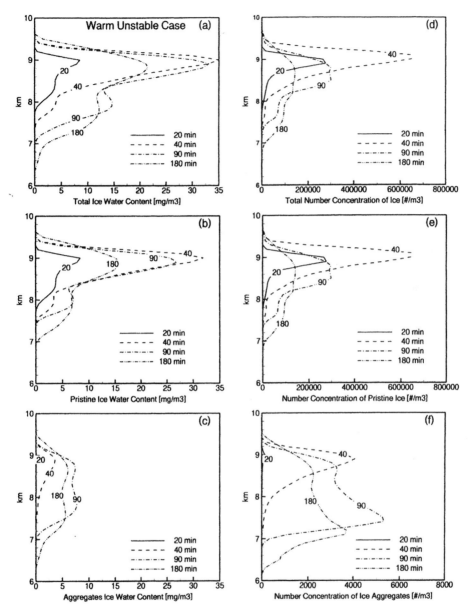

Fig. 13.18 Profiles of horizontally averaged ice water content and number concentration of ice, for the warm unstable case: (a) total ice water content, (b) pristine ice water content, and (c) aggregates ice water content; (d) total number concentration of ice, (e) number concentration of pristine ice, and (f) number concentration of ice aggregates. From Liu *et al.* (2003b). Reproduced by permission of the American Meteorological Society.

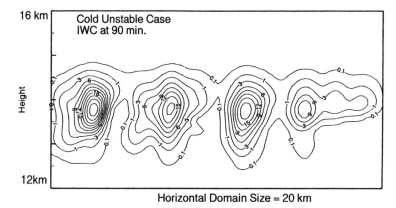

Fig. 13.19 The ice water content field of the simulated cirrus at $t = 90$ min formed in a cold unstable environment. From Liu *et al.* (2003b). Reproduced by permission of the American Meteorological Society.

Fig. 13.19 shows the IWC contour field of the cold cirrus at $t = 90$ min. The IWCs of this cloud are much less than those of the warm cirrus in Fig. 3.17.

Fig. 13.20 shows the time evolution of the horizontally averaged ice water content (IWC) and ice number concentration vertical profiles in the 3 h of simulated cold cirrus development. It is clear that the cold cirrus has much less IWC and number concentration of ice crystals than the warm cirrus shown in Fig. 13.18. Thus the cold cirrus is said to be optically thin as compared to the warm cirrus. The optical properties of cirrus clouds obviously influence how the cloud interacts with radiation. The issue of radiative heating due to cirrus–radiation interaction and its implications on the global climate will be discussed in Chapter 15.

Problems

Numerical cloud model results are sometimes difficult to interpret because there are many nonlinear processes interacting simultaneously. Nevertheless, it is useful to use basic physics and dynamics to understand why the cloud behaves the way it does. Because the results are numerical rather than analytical, the conclusions made are often just "very likely" instead of absolutely certain, but it is a very useful exercise trying to interpret numerical results. The following provides some specific topics for students to perform such exercises.

13.1. Look at Fig. 13.4. Discuss what causes the variations in the spacing between the θ_e isotherms in panels (a), (b), and (c).

13.2. Where is the region of instability in Fig. 13.4(a)?

13.3. The gradient of θ_e is very small in the updraft core of Fig. 13.4(b) and (c). What does that imply?

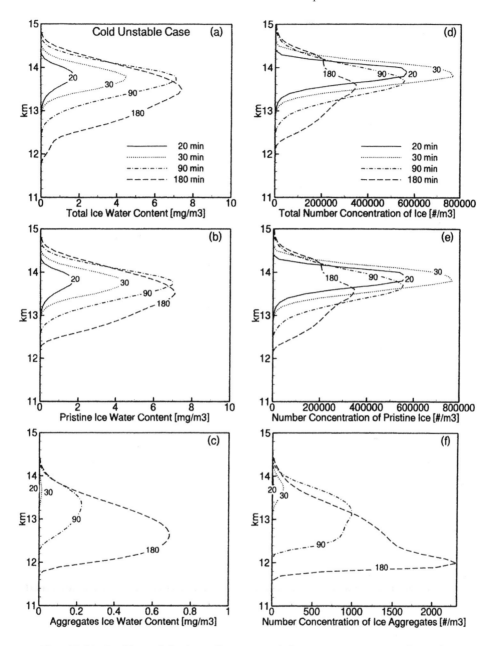

Fig. 13.20 Profiles of horizontally averaged ice water content and number concentration of ice, for cold unstable case: (a) total ice water content, (b) pristine ice water content, and (c) aggregates ice water content; (d) total number concentration of ice, (e) number concentration of pristine ice, and (f) number concentration of ice aggregates. From Liu *et al.* (2003b). Reproduced by permission of the American Meteorological Society.

13.4. Can you identify the strong diabatic regions in Fig. 13.4(b) and (c)? What are the possible processes causing such phenomena?

13.5. There is a small region of closed θ_e contours near the center bottom of Fig. 13.4(c) ($z \sim 1$ km, $x \sim 28$ km). What does that imply and why?

13.6. In both Figs. 13.5(b) and (c), the coldest temperature occurs near the top of the overshooting dome, which is much colder than its surroundings. Why?

13.7. In the downstream region of the overshooting dome, but at about the same level, the temperature is relatively warm. What is the possible cause of this?

13.8. Figs. 13.5(d)–(f) show positive pressure perturbations in the upstream and negative pressure perturbations in the downstream side of the storm. This is the situation in the presence of wind shear. What would you expect the pressure perturbation to be if there were no wind shear?

13.9. What are the processes that may be associated with the temperature patterns shown in Figs. 13.9 and 13.10?

14

Cloud electricity

Thunderstorms are among the most impressive weather phenomena with their strong winds and sometimes heavy rain and hail. But what distinguishes them from other storms is the lightning and thunder that are their trademark.

It is well known that it is lightning that causes thunder. But the lightning itself is caused by charge separation in a cloud, a process called *cloud electrification* that is still not totally understood at present. In this chapter, we will first examine the electricity in air during fair weather, and then we will discuss the electrical phenomena occurring during a thunderstorm. Finally, we will discuss the mechanisms that may be responsible for the electrification of thunderclouds.

14.1 Fair-weather electricity

Electricity in the atmosphere is not limited to stormy weather. During a clear day, a downward-pointing static electric field of about $130 \, \mathrm{V \, m^{-1}}$ (volts per meter) near the Earth's surface can be measured and is called the *fair-weather electric field* (see Fig. 14.1). This is just a highly averaged condition, and large variations may occur for specific locations and specific times – the field may vary from less than $50 \, \mathrm{V \, m^{-1}}$ to more than $300 \, \mathrm{V \, m^{-1}}$. In general, this fair-weather field is usually larger over land than over the ocean surface.

The electric field $\vec{E}$ is a vector field and by convention its direction points from positive to negative. The observation that the fair-weather electric field points downward toward the Earth's surface indicates that the surface is overall negatively charged. Measurements indicate that the Earth's surface contains an average charge density of about $-1.1 \times 10^{-9} \, \mathrm{C \, m^{-2}}$ (coulombs per square meter). Since the total surface area of the Earth is about $5 \times 10^{14} \, \mathrm{m^2}$, the total global fair-weather charge on the Earth's surface is about $-5.1 \times 10^5 \, \mathrm{C}$.

That amount of surface charge would cause the Earth's surface to possess a certain electric potential ϕ_0. We can determine the electric potential ϕ distribution

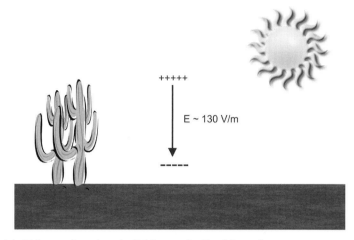

+++++

E ~ 130 V/m

- - - - -

Fig. 14.1 Fair-weather electric field near the Earth's surface.

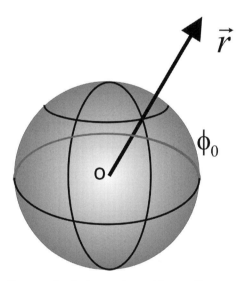

$\vec{r}$

ϕ_0

O

Fig. 14.2 The Earth as a spherical capacitor with a surface potential ϕ_0.

around the Earth by approximating the Earth as a spherical capacitor with a charge of -5.1×10^5 C uniformly distributed over its surface (Fig. 14.2). If there are no space charges (i.e. no other electric charges in the air), then the electric potential ϕ around such a capacitor should satisfy the Laplace equation:

$$\nabla^2 \phi = 0. \tag{14.1}$$

This is the same equation as for the diffusion growth problems that we dealt with in Chapter 9. The zero on the right-hand side of (14.1) indicates that there is no space charge, i.e. no electric charges in the atmosphere. The solution of (14.1) in spherical coordinates subject to the boundary conditions

$$\begin{aligned} \phi &= \phi_0 \quad \text{at } r = R, \\ \phi &= 0 \quad \text{at } r \to \infty, \end{aligned} \tag{14.2}$$

is

$$\phi = \frac{\phi_0 R}{r}, \tag{14.3}$$

where R is the radius of the Earth and r is the radial distance from the center of the Earth. The electric field is defined as the negative gradient of the electric potential,

$$\vec{E} = -\nabla\phi = \frac{\phi_0 R}{r^2}\hat{e}_r, \tag{14.4}$$

i.e. the electric field magnitude will decrease with height at a rate proportional to $1/r^2$.

Actual measurements of the vertical variation of $\vec{E}$ do not agree with (14.4), but rather show a quasi-exponential decrease with height, although there are substantial fluctuations at specific times and places. Gish (1944) suggested the following empirical fit for the electric field E as a function of height z:

$$E(z) = -(81.8e^{-4.52z} + 38.6e^{-0.375z} + 10.27e^{-0.121z}), \tag{14.5}$$

where the unit of E is $V\,m^{-1}$ and z is in km.

A vertical profile of E given by Eq. (14.5) implies that the electric potential cannot be a solution of the Laplace equation (14.1). On the other hand, such an exponential-type function can be a solution of the Poisson equation (in SI units):

$$\nabla^2\phi = \rho_e(z)/\varepsilon_0, \tag{14.6}$$

where $\rho_e(z)$ is the space charge density and ε_0 is the permittivity of vacuum. Substituting (14.4) and (14.5) into (14.6), we can obtain $\rho_e(z)$ as

$$\rho_e(z) = 20.4e^{-4.52z} + 0.8e^{-0.375z} + 0.069e^{-0.121z}, \tag{14.7}$$

which is the space charge density (in unit of $e\,cm^{-3}$) responsible for generating the electric field (14.5). Here e is the elementary charge: $1\,e = 1.602 \times 10^{-19}\,C$.

On the surface ($z=0$), the value of ρ_e would be about 21 e cm^{-3} or $3.36 \times 10^{-12}\,C\,m^{-3}$. Note that this is the space charge density in the air, and it should not be confused with the surface charge density mentioned at the beginning of this section.

14.2 Electric charges in the atmosphere

Measurements show that there are both positive and negative charges in the atmosphere. There are two main mechanisms by which these charges are produced, and both are related to the ionization of neutral molecules.

- *Radioactive emanation from the Earth's surface*. The Earth's crust contains many radioactive isotopes that emit high-energy charged subatomic particles (alpha particles = helium nuclei; and beta particles = high-energy free electrons) and ionizing radiation (gamma rays = high-energy electromagnetic radiation). The typical energy of these particles is about 1 MeV (10^6 eV), where 1 eV (electron-volt) = 1.602×10^{-19} J (joule). Both alpha and beta particles have short mean free paths (a few centimeters and a few meters, respectively) in the lower atmosphere due to the relatively high pressure and will not contribute much to the charges in the atmosphere. It is mainly the gamma rays that ionize neutral air molecules into positively charged (positive ions) and negatively charged (free electrons) pairs. Owing to their high mobility, the electron quickly attaches to a neutral molecule to become a negative ion. The positive ions are large and have smaller mobility.
- *Cosmic rays*. These are actually charged high-energy subatomic particles entering the Earth's atmosphere from either the Sun (solar cosmic rays, SCR, also called solar energetic particles, SEP) or further out in space (galactic cosmic rays, GCR), so the name "rays" is actually a misnomer. About 90% of them are protons (the bare nuclei of hydrogen atoms), 9% are helium nuclei, and the remaining 1% are other nuclei. When these energetic particles enter the atmosphere, they collide with air molecules and ionize them, again creating ion pairs. Cosmic ray particles have energies exceeding 10^{20} eV, whereas the most powerful man-made particle accelerators can produce only 10^{12}–10^{13} eV particles at present. Because of their extremely high energies (many orders of magnitude higher than those particles emitted by radioactive isotopes), they produce huge numbers of secondary particles such as pions, kaons, mesons, and muons. This phenomenon is called *air showers*.

Ionization caused by radioactive emanation only contributes about half of the charges near the surface. On the other hand, ionization due to cosmic rays contributes the other half of the charges near the surface and nearly all the charges in both the lower and upper atmosphere (ionosphere). At altitudes higher than 1 km or so, virtually all ions are produced by cosmic rays.

When ion pairs are produced by ionizing radiation, both positive and negative ions are produced. The charges on some of the ions can be neutralized by recombination of oppositely charged particles, and no doubt some charges are removed in this way soon after they are created. However, two types of charged ion still exist in the atmosphere, because other ions attach to molecules or aerosol particles and produce small and large ions:

- *Small ions* – these are formed by the ions attaching themselves to neutral molecules. They are essentially charged molecules. Since they are smaller, they have higher mobility.
- *Large ions* – also called Langevin ions. These are either simple ions or small ions attached to aerosol particles. They have lower mobility because of their size.

The ratio of the positive ion concentration to the negative ion concentration in the air near the surface is ~1.2. This indicates that there are about 20% more positive charges than negative charges in the air.

14.2.1 Leakage current

Because of the presence of the fair-weather electric field, these ions in the air will be subject to the electric force and will accelerate in a direction according to the sign of their charges. For a configuration as in Fig. 14.1, positive ions will drift downward and negative ions upward. The flow of charges constitutes an electric current, called the *conduction current* in this case. This conduction current is mainly due to the movement of small ions. Large ions move relatively slowly because of their much larger masses and hence contribute little to this current. Naturally, the magnitude of the current varies in time and space, but usually the current density is between 2×10^{-12} and $4 \times 10^{-12}\,\mathrm{A\,m^{-2}}$ (amps per square meter), so a global average value of $3 \times 10^{-12}\,\mathrm{A\,m^{-2}}$ can be assumed to represent the current density. Multiplying this value by the Earth's surface area and we get a global current of about 1500 A. The net result of this current is to bring positive charges to the surface to neutralize the negative charges there. This current is called the *leakage current*. This is in the same sense as the current that leaks through the protective ground conductor to the ground in electrical installations. It is also called the leakage current in that case.

We can make an estimate of the time it requires to totally discharge the global charge of $5.1 \times 10^5\,\mathrm{C}$ by solving the equation governing the discharging process, and the value turns out to be $\sim 385\,\mathrm{s}$ or $\sim 6.5\,\mathrm{min}$. This means that, if there were no other charging mechanism to supply the negative charges to the Earth's surface, the Earth's surface would become neutral and the fair-weather field would disappear in about 6.5 min. But the fact that the fair-weather electric field is present constantly indicates that there must be some mechanisms that supply negative charges to the surface to maintain this field. The mechanism turns out to be the thunderstorm activity around the world.

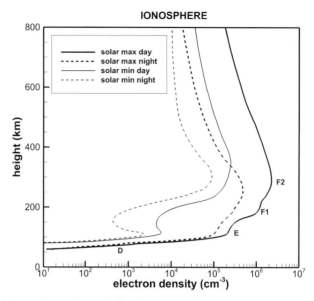

Fig. 14.3 Free electron density in the atmosphere as a function of height. The changes in the density profile allow the division of the ionosphere into D, E, and F layers, each with different electric and chemical characteristics. During the daytime, the F layer can be subdivided into F_1 and F_2 layers. The D layer disappears at night because of the very low ionization.

14.2.2 Electrosphere (ionosphere)

As we mentioned above, ions in the higher atmosphere are almost exclusively produced by cosmic rays. As the air density decreases, the mean free path of air also increases and the ions' lifetime becomes longer because they will not recombine with other ions or collide with air molecules as frequently as in the lower atmosphere, and hence the concentration of free ions increases. This is usually measured by the electron density, as shown in Fig. 14.3.

The ions in the higher atmosphere consist of both positive and negative charges of about equal concentrations. While some recombination does occur, enough free ions are present at any time that the ensemble of ions behaves as a fluid called a *plasma*. So a plasma is an electrified fluid that is overall neutral (because of equal concentrations of positive and negative charges) but will respond to external electric and magnetic fields.

One consequence of the higher density of free ions in the air is the increasing electrical conductivity. This means that, if you impose an external electric field on the plasma, the ions of opposite charges will move in a manner so as to generate an induced electric field that will tend to cancel the field imposed on the plasma. This is precisely what happens when you impose an electric field on a conductor such as a copper rod. The conduction electrons in the copper will rearrange themselves so as to cancel the imposed electric field inside the copper. Consequently, inside the copper rod, the electric field is zero. Because the electric field is the negative gradient of the electric potential, the zero electric field implies that there is no potential gradient, and the whole copper rod is an equipotential body.

Since the plasma in the higher atmosphere behaves in the same way, we understand that it is a conducting layer and hence also an equipotential layer. This layer is usually taken to begin at $z \sim 60$ km and is called the *electrosphere*, which also overlaps with the ionosphere. The two terms are often used interchangeably. We usually consider that the atmosphere above this layer is conducting and hence is equipotential throughout the layer.

The Earth's surface, including both land and water surface, is also conducting (though not perfect) and hence can also be approximated as an equipotential surface. On the other hand, the atmosphere below the ionosphere, including both the clear and cloudy regions, is usually not conducting and has significant electrical resistance.

14.2.3 Global electric circuit

In view of the electrical nature of the different atmospheric layers as described above, it is useful to summarize the global electrical equilibrium in terms of an electric circuit. This is called the *global electric circuit* or simply the *global circuit* (Fig. 14.4), first proposed by Wilson (1903, 1920).

Here the atmosphere is sandwiched between two spherical conducting (equipotential) surfaces: the Earth's surface and the ionosphere. The potential difference between these two surfaces is about 200–300 kV. The atmosphere located between these two surfaces serves as an insulating layer. But the atmosphere is not a perfect insulator, and hence there is a leakage current of 1500 A (or leakage current density ~ 3 pA m^{-2}) in the clear regions. The resistance in the atmosphere is not uniform but depends on local conditions; thus clear air, cloud, and polluted regions all have different resistances. Thunderstorms represent charge generators (or capacitors) that serve to replenish the charges on the Earth's surface that are lost due to the leakage current.

Fig. 14.4 also shows that the atmosphere behaves as a cavity between two conducting surfaces. Electromagnetic signals can propagate in this cavity but will be reflected by the conducting surfaces. Lightning during a thunderstorm generates many electromagnetic signals that will propagate in the atmospheric cavity, causing a phenomenon called the *Schumann resonance* (Schumann, 1952) because signals bounce back and forth between the Earth's surface and the ionosphere. The

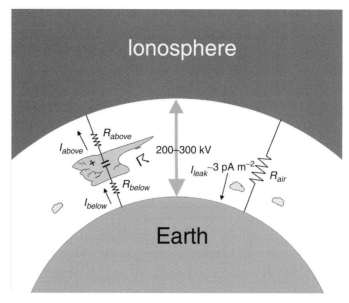

Fig. 14.4 A schematic of the global electric circuit. The lower atmosphere is depicted as a cavity between the Earth's surface and the ionosphere, with a leakage current I_{leak} of about $3\,pA\,m^{-2}$. The potential difference between the surface and ionosphere is between 200 and 300 kV, with an average of 250 kV, depending on location and time. Thunderstorms behave as electric capacitors that replenish the charges lost due to leakage.

Schumann resonance refers to the extremely low-frequency (ELF) band (3–30 Hz) of this oscillation of the electromagnetic field. ELF waves can go around the Earth several times before being damped out and can be detected by one single detector anywhere on Earth. The largest peaks are at 8, 15, and 20 Hz (MacGorman and Rust, 1998). The variations of the Schumann resonance have been used as an indicator of the Earth's electrical climate (see e.g. Tinsley *et al.*, 2007; Williams, 2005; Markson, 2007).

14.3 Thunderstorm electricity

Even smaller clouds can be electrified and show certain electrical structure, but the electric fields in them are usually small. Thunderstorms generate a large quantity of electric charges and transport them to the ground to maintain the global electrical equilibrium by means of lightning discharge, and they are the focus of our study here. Recent studies of thunderstorm electricity have been greatly facilitated by the accumulation of observation data via the global lightning networks. We will summarize some observational results that are useful for our purpose here.

14.3.1 Electric charge distribution in thunderclouds

Lightning is an extreme form of electric discharge that generates intense light and sound known to us as thunder. For lightning to occur, the electric field should be large enough to cause an electrical breakdown, which, in turn, requires that there is enough buildup of charge in the cloud. A thundercloud is a highly complex entity and the actual charge distribution can vary greatly from time to time, and from cloud to cloud, so the following just represents the most common situation when the condition is favorable for lightning to occur.

Many studies show that most thunderstorms have a *tripole* structure of charge distribution in their updraft core, as depicted in Fig. 14.5. In the upper layer of the cloud, where the temperatures are typically lower than −20°C, the particles (which are mainly cloud ice or snow aggregates) mostly carry positive charges. In the middle level, on the other hand, particles (which may contain both raindrops and graupel/hail) carry mostly negative charges. There is a region of concentrated negative charges in the level between about −10 and −20°C. In the lower level near the cloud base, there is often a smaller positive charge region (see e.g. Williams *et al.*, 1989). Some researchers ignore this smaller lower positive center and consider only the upper positive and lower negative charge centers. Then the thunderstorm can be thought of as having a *dipole* structure.

There are also observations showing that some thunderstorms have inverse electrical structure, namely, a main negative charge center is located above a main

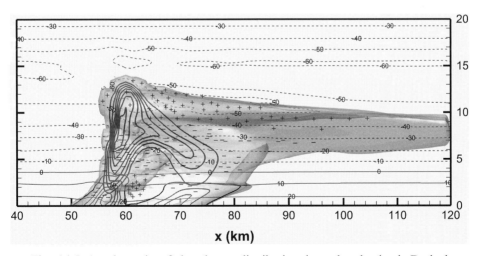

Fig. 14.5 A schematic of the charge distribution in a thundercloud. Dashed contours are isotherms in °C. Thick solid contours are the graupel/hail mixing ratio, whereas thin solid contours are rain mixing ratio.

positive charge center located in the mid-level ($\sim -10°C$), just the opposite to the structure depicted in Fig. 14.5.

These charge centers are responsible for the buildup of high electric field for breakdown. But the exact magnitude of the critical electric field necessary for the breakdown to occur depends on other factors in addition to the amount of charge accumulated. An electrical breakdown indicates that the resistance of air has been drastically reduced so that sudden increase of charge transfer occurs, a phenomenon sometimes called *electron avalanche*. The reduction of the resistance depends on many factors, such as air density, temperature, humidity, trace chemicals, ion concentration, etc. The typical environmental field during the breakdown is estimated at about $100–300\,V\,m^{-1}$, but this is not the critical field at the point of breakdown, which is probably much higher. Once the breakdown and electron avalanche occur, the lightning process will follow.

The electric charges in thunderclouds are carried largely by cloud and precipitation particles. Takahashi (1973) found that the mean maximum charge carried by a water drop in a thundercloud is about $\pm 2a_0^2$ with the charge in esu and the radius a_0 in cm.

14.3.2 *Lightning and thunder process*

As required by global electrical equilibrium, lightning should transport negative charges to the surface to replenish those lost as a result of the fair-weather leakage current. The recent study by Liu *et al.* (2010) demonstrates that this does indeed occur globally on average. However, this does not mean that each individual lightning strike behaves like others. Indeed, there are various kinds of lightning, some transporting negative charges but others transporting positive charges to the surface. In addition, some lightning does not transport charges to the surface as it occurs only in clouds. In the following, we will briefly summarize the main features of the lightning process. A standard reference for readers interested in this topic is Uman (1987). For more recent treatments, see MacGorman and Rust (1998) or Rakov and Uman (2003).

We can divide lightning discharges into four broad categories: (1) cloud-to-ground flash, (2) intra-cloud flash, (3) inter-cloud flash, and (4) cloud-to-space flash. We will describe the cloud-to-ground flash in some detail, but only look very briefly at the other three categories.

Cloud-to-ground flash

The cloud-to-ground flash is probably the most stereotypical "lightning" in most people's minds – a blinding flash of light descending from a dark cloud above to hit the ground. A few seconds later, the sound of the rumbling thunder arrives. But the real sequence is more complicated, and usually goes as follows.

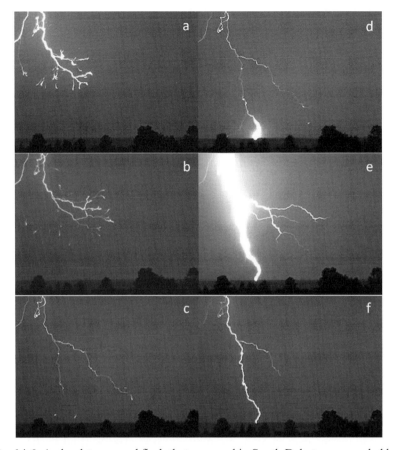

Fig. 14.6 A cloud-to-ground flash that occurred in South Dakota as recorded by a high-speed video camera running at 7200 frames per second. Panels (a), (b) and (c) show the stepped leaders; panel (d) is the streamer; and panels (e) and (f) show the return stroke. Stills from video taken by Tom C. Warner. Reproduced by permission of www.WeatherVideoHD.tv.

(a) *Stepped leader.* These are luminous (but not very bright) channels lowering from the cloud base toward the ground in a zigzag (stepped) and forked manner. At the tip of each fork is a luminous blob of freshly ionized plasma that will cause more ionization in front of its path. This stage is called the "stepped leader" stage and is shown in Fig. 14.6(a)–(c). While the general motion is downward, some individual leaders move upward. As can be seen from the images from the high-speed video camera, the motion of the stepped leaders resembles that of tadpoles swimming in the sky. Wurden and Whiteson (1996) estimated that the velocity of the stepped leader shown here was about $10^5\,\mathrm{m\,s^{-1}}$ or more.

(b)   *Streamer*. When the stepped leader approaches to within a few tens of meters above the ground (Fig. 14.6c), there is often a "streamer" going up from the ground to meet the head of the leader, which produces a continuous channel of ionized gas between the cloud base and the ground. Fig. 14.6(d) shows this connection stage. Streamers sometimes can occur independently without the leader. Up to this stage, the luminosity of the channels is usually low and not easily noticed by the naked eye.

(c)   *Return stroke*. This is the main event – what most people call "lightning". It is characterized by a very bright flash going *upward* from the ground into the cloud, hence the name "return stroke". Figs. 14.6(e) and (f) show two frames of the return stroke. In most cases, while the electric current (and the light) is going upward, negative charges are transported downward to the ground. This is called the *negative flash*. Occasionally, though, lightning can occur between a positively charged region in a cloud and the ground, and positive charges are transferred to the ground after the lightning. This is called the *positive flash*. Positive flashes are much rarer than negative flashes. For example, Orville (1994) found that fewer than 10% of US cloud-to-ground flashes are positive. However, certain thunderstorms in specific regions during specific seasons – for example, winter storms over the coastal Sea of Japan, or the intense storms over the US High Plains that are associated with the dryline condition (see Doswell (2001) for discussions of dryline thunderstorms) – tend to produce predominantly positive cloud-to-ground (PPCG) flashes (see e.g. Takeuti *et al.*, 1978; Lang *et al.*, 2004). The positive flashes in these storms exceed 50% and may approach 100% of all flashes produced by them.

The speed of the return stroke is much faster than the speed of the stepped leaders, being about $1.1 \times 10^8 \, \mathrm{m\,s^{-1}}$ on average, which is "lightning fast" indeed. The typical current of a negative return stroke is about 30 000 A or 30 kA, which carries 5 C of charge to the ground. But there are also reports of currents greater than 100 kA and charge greater than 350 C. The temperature in the lightning channel can reach 20 000 K or higher. The rapid heating causes rapid expansion of the air, and the resulting compression waves propagate initially at supersonic speed in the form of shock waves. They quickly decay into normal acoustic waves and form the sound of thunder. The rumbling of thunder comes from the reflection off clouds and ground objects such as terrain and buildings.

(d)   *Dart leader and second return stroke*. The return stroke can occur more than once. For multiple-return-stroke lightning, the ionization of the channel after the first return stroke is accomplished by the "dart leader" instead of the stepped leader. The luminous channel goes straight from the cloud to the ground without a tortuous path. The speed is about $10^7\,\mathrm{m\,s^{-1}}$. After the dart leader, the "second return stroke" follows. Some lightning may produce more than two return strokes, each with a dart leader in between. The multiple return strokes give us a sensation that the lightning is "blinking".

Intra-cloud flashes

This refers to lightning flashes that occur in the same cloud, and in principle an intra-cloud usually occurs between the upper positive and lower negative charge centers in the cloud. Intra-cloud flashes generally outnumber cloud-to-ground flashes in summer thunderstorms and in semi-arid regions where the cloud bases are high. Because an intra-cloud flash is entirely surrounded by cloud, its channel structure is not as clear as for the cloud-to-ground flashes, but rather appears diffuse and hence is sometimes called *sheet lightning*.

Inter-cloud flashes

Lightning flashes can also occur between different clouds as long as the electric field between the two is stronger than the breakdown requirement. But inter-cloud flashes are much fewer than intra-cloud flashes.

Upper atmospheric lightning – sprites, jets, elves, etc.

Aside from the above flashes, recent studies have also verified some long suspected (and occasionally observed by aircraft pilots), but for a while unsubstantiated, reports of upward lightning to the sky from large thunderstorm systems. These are now called either *upper atmospheric lightning* or *transient luminous events* (TLE), and they consist of phenomena that are called *red sprites*, *blue jets*, *sprite halos*, *gigantic jets*, and *elves*. They do not resemble lightning in the troposphere but are more akin to light glow in plasma. A discussion of this subject is beyond the scope of this book, but the reader is referred elsewhere (e.g. Rakov and Uman, 2003) for a general understanding of this subject.

Volcanic lightning, dust storm lightning, and lightning in other planetary atmospheres

Lightning can also occur in an environment other than thunderclouds. Volcanic eruptions as well as some dust storms can also trigger lightning. This happens as long as there are effective charge separation mechanisms that cause the buildup of a large electric field. Similarly, lightning has been observed in the atmospheres of Jupiter and Saturn, whose atmospheric chemical compositions are very different from ours.

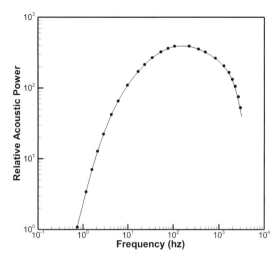

Fig. 14.7 Composite acoustic power spectrum of thunder (an average of 23 thunder power spectra). Adapted from Few *et al.* (1967).

Thunder

Thunder is a result of the intense heating of about 20 000°C in a few to a few tens of microseconds due to a lightning discharge. Theoretical studies indicate that such rapid heating causes the air plasma in the lightning channel to expand rapidly at supersonic speeds, thus forming a shock wave that pushes into the surrounding air. This shock wave causes a very loud sound similar to an explosion or a sonic boom caused by the flight of a supersonic jet. This is the thunder we hear during thunderstorms. Many observational studies have shown that the peak acoustic energy of thunder is centered around 200 Hz, although the spectrum is rather broad, as shown in Fig. 14.7 (Few *et al.*, 1967). Uman (1969) provided a summary for earlier studies of thunder. The rumbling sounds are caused by the reflection and scattering of the original boom by clouds and other environmental objects (topography, buildings, etc.).

Even though thunder is a secondary phenomenon caused by lightning, it is curious that thunder seems to impress the people of many cultures more than lightning. The name "thundercloud" rather than "lightning cloud" is perhaps a good indication of this observation. In many ancient Eastern and Western cultures, the gods associated with thunderstorms are identified more with thunder than with lightning. In East Asian mythology, the thunder god is called Lei Gong (Chinese) or Raijin (Japanese), both of which can be translated directly as "God of Thunder". In Greek mythology, one of the symbols of Zeus is a thunderbolt (not called a lightning bolt!). The same goes for Jupiter, the King of the Gods of the ancient Romans, and for Thor in the Norse mythology.

14.4 Cloud electrification

We have said above that a large electric field is necessary for lightning to occur. So our next question is this: How does such a large electric field occur? The large electric field in a thundercloud is produced by the separation of a large amount of electric charges of opposite polarity. The process of charge separation is called *cloud electrification*.

Before we delve into the discussions of cloud electrification mechanisms, it is useful to examine some observational facts about the relation between lightning and clouds, so as to set some proper constraints on the possible separation (or charging) mechanisms.

Fig. 14.8 shows the global distribution of lightning flashes. Here we use the lightning density to represent thunderstorm density. This chart shows that the lightning-producing storms are concentrated mostly over land, and the three main centers are Central Africa, South America, and Southeast Asia. Storms over the ocean produce relatively few lightning discharges. On the other hand, heavy precipitating storms are distributed over both continents and oceans more or less evenly in the tropics in the upwelling regions of the Hadley circulation (Williams, 2005).

So, why the difference between land and ocean? The answer lies in the updraft speed. Storms over land generally have higher updraft speeds than storms over the ocean, and this is generally related to the higher convective available potential energy (CAPE) of storms over land, too. Baker *et al.* (1995) used a simple one-dimensional model to show that the lightning frequency in cloud can be written as

$$f \approx R w^6 \overline{V}_i, \tag{14.8}$$

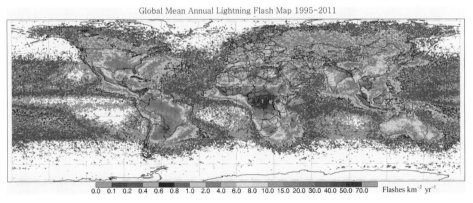

Global Mean Annual Lightning Flash Map 1995–2011

0.0 0.1 0.2 0.4 0.6 0.8 1.0 2.0 4.0 6.0 8.0 10.0 15.0 20.0 30.0 40.0 50.0 70.0   Flashes km^{-2} yr^{-1}

Fig. 14.8 The mean annual flash rate from combined space-borne Optical Transient Detector (OTD) and Lightning Imaging Sensor (LIS) data at 0.5° grid resolution from May 1995 to December 2011. Data from NASA, courtesy of Ms Sherry Harrison. See also color plates section.

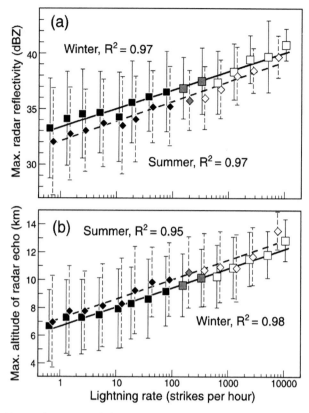

Fig. 14.9 (a) Maximum precipitation radar (PR) reflectivity and (b) maximum height of radar echo. Squares and solid lines are for winter data, and diamonds and dashed lines for summer data. Filled black symbols are PacNet data, open symbols are LIS data, and gray symbols are combined PacNet and LIS data. Abscissa shows the number of lightning flashes per hour normalized over $10^4 \, \mathrm{km}^2$. The error bars are $\pm 1\sigma$. From Pessi and Businger (2009). Reproduced by permission of the American Meteorological Society.

where f is the lightning frequency, R is the cloud radius, w is the updraft speed, and $\overline{V}_i$ is the volume of ice in the cloud. While the exact functional dependence may be different from this idealization, the equation provides a useful qualitative understanding of the relation between f and other cloud parameters. Thus f is a very strong function of w, which can interpret the observation in Fig. 14.8 to a good extent.

Thus to produce frequent lightning, it is necessary to have high updraft speed, large cloud radius, and large ice volume. In contrast, a storm cloud does not need to have high updraft to produce large rainfall, and this is the case for oceanic storms. Large updrafts produce taller and larger clouds whose upper part would extend much higher than the freezing level and hence produce larger ice volume and more

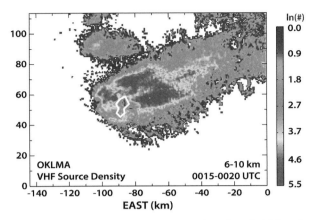

Fig. 14.10 Density of mapped VHF sources produced at 6–10 km from 00:15 to 00:20 UTC on 30 May 2004, during the Thunderstorm Electrification and Lightning Experiment (TELEX). The density color scale is a $\log_e = \ln$ scale. The white outline delineates the BWER. Note that the minimum lightning density in the hole is within the BWER and the core of maximum updraft speed. Coordinates are relative to the location of the KOUN radar. From MacGorman *et al.* (2008). Reproduced by permission of the American Meteorological Society. See also color plates section.

frequent lightning. This seems to corroborate the recent observation that the lightning frequency strongly correlates positively with the maximum radar reflectivity and maximum radar echo altitude, as shown in Fig. 14.9. Thus, aside from high updraft speed, these observations also suggest that the occurrence of lightning, and hence the electrification of cumulonimbus cloud, is closely related to the formation of ice.

Fig. 14.10 shows another important observational feature called *lightning holes* first noted by Krehbiel *et al.* (2000) in intense thunderstorms (MacGorman *et al.*, 2008). This is the region bounded by the white outline, where lightning is non-existent, and many observations have essentially shown the same. This lightning hole practically coincides with the bounded weak echo region (BWER). The BWER is a result of the rapid ascent of boundary air within which there is a lack of large precipitation particles to cause strong radar echoes. The same reasoning can explain the lightning hole phenomenon. On the other hand, we see that the areas surrounding the BWER and the lightning hole are characterized by dense lightning (very high-frequency, VHF, sources). These areas are also the regions with high concentrations of graupel and hail. This coincidence corroborates the importance of large ice particles on the electrification.

In the following paragraphs, we will summarize some mechanisms that have been suggested to lead to charge separation in a thundercloud. With the exception of convective charging (see later), which involves cloud dynamics directly, all the other mechanisms essentially operate on the microscale – either molecular or cloud particle scale. Their validity is best judged by testing them in realistic cloud models with adequate microphysics to see if they can indeed produce enough separated charges in the right places in a reasonable amount of time. The recent advance of sophisticated cloud models with more realistic dynamics and microphysics provides a reasonable platform to perform such tests.

Aside from the convective charging theory, which is in a category by itself, the rest of the charging mechanisms can be divided into two broad categories: *inductive* and *non-inductive*. The inductive mechanisms are those operating under the influence of an externally imposed electric field (such as the fair-weather field). The non-inductive mechanisms, on the other hand, can operate without an external electric field. We will describe them separately below.

14.4.1 Convective charging theory

This mechanism operates as follows. During the development of a convective cloud in fair-weather conditions, positive charges are carried up into the cloud by updraft, forming a positive charge center in the cloud interior. This charge center soon causes a negative charge screening layer to form at the top and edge of the cloud because the cloud drops there would capture downward-drifting negative ions. During the convective growth, the cloud top and edge would be downdraft regions, and hence negatively charged drops would be carried to the lower level of the cloud whereas the positive charge center continues to rise to the upper level. In this way, an upper positive and lower negative charge separation is established in the cloud. Fig. 14.11

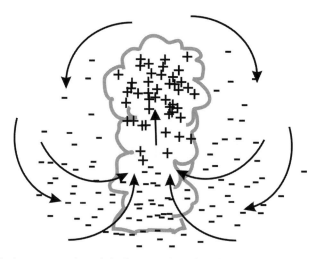

Fig. 14.11 A conceptual model of convective charging.

shows a schematic of this process. For more details, see Vonnegut and Moore (1958).

There are many qualitative and quantitative discrepancies between the observations and theoretical predictions, as pointed out by many investigators (see e.g. Chiu and Klett, 1976; Latham, 1981; Williams *et al.*, 1989).

14.4.2 Inductive charging mechanisms

An object placed in an external electric field will have charges induced in it. Thus cloud or precipitation particles, whether water drops or ice particles, will have charges induced in them in a pre-existing electric field such as the fair-weather electric field. Charge separation mechanisms due to the action of the induced charges are called *inductive charging mechanisms*.

Selective ion capture

In the presence of a fair-weather electric field, a cloud droplet, being conducting, will be polarized in the manner shown in Fig. 14.12. Now imagine that this droplet is falling in air filled with ions of both signs. Since the underside of the droplet is positively polarized, it will attract negative ions and repel positive ions. This will cause the underside to acquire a net negative charge. But what about the topside, which is negatively polarized and supposed to attract positive ions from above? The theory assumes that the droplet is falling and the positive ions have to catch up with the droplet, so the charging rate cannot be as fast as that of the negative ions. Consequently, after the droplet has fallen a certain distance, it will acquire a net negative charge. Eventually, negative charges are concentrated in the lower levels,

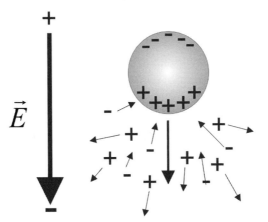

Fig. 14.12 Selective ion capture. The external electric field causes the drop to be polarized. The drop selectively captures negative ions as it falls.

which results in large-scale charge separation, with positive charges above and negative charges below. This enhances the original fair-weather field and eventually leads to the dipole structure in thunderclouds. This mechanism was first proposed by Wilson (1929).

While the theory produces the correct polarity, it was found that it is only effective under a weak electric field ($E < 570\,\mathrm{V\,cm^{-1}}$), which is inadequate in producing lightning. The theory is thus quantitatively unrealistic.

Charging by drop breakup

Under certain suitable conditions, large raindrops may break up, as we have described in Chapter 10. If the drop is polarized in an existing electric field, as mentioned above, then it is possible that, upon breakup, the upper portion of the drop may carry negative charges away while the lower portion carries positive charges away. During the "bag breakup" of a large drop, the top part would break up into many fine droplets that would carry negative charges, whereas the lower part would form a water ring and break up into fewer but larger drops carrying positive charges. Upon gravitational separation of the two groups of drops, we would have positive charges settled in the lower level and negative charges in the upper level. This seems to produce a configuration with the wrong polarity according to the bipolar model, so it is not considered to be effective, at least for building up the main charge centers. But some researchers think that it may contribute to the small positive charge center near the cloud base according to the tripolar model.

However, most large drop breakup in clouds is due to drop–drop collisions and not to bag breakup, and the charge separation in that process is unknown. Consequently, the effectiveness of charging by drop breakup is still questionable.

Charging due to rebound of particles during collision

As described in Chapter 10, colliding cloud particles (either water or ice) may end up rebounding from each other instead of joining together. If those particles are also polarized in a pre-existing electric field, would the rebound result in charge separation? When a large particle collides with a smaller particle from above, and the two are in contact, as the traditional collision process is usually portrayed, some positive charges on the lower half of the large particle are neutralized by the negative charges on the upper half of the smaller one (Fig. 14.13). After rebound, the large particle will carry a net negative charge and the smaller particle will acquire a net positive charge. Upon gravitational separation of the two, an upper positive and lower negative charge structure is established. Some early estimates claimed that this charging mechanism can be efficient enough to produce lightning.

But several problems must be overcome before this mechanism can be accepted as effective. Among them is the rebound problem itself. The collision does not just

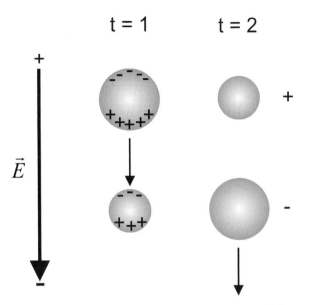

Fig. 14.13 Charging due to rebound of two particles during collision. At $t = 1$, the collision of two drops causes the larger drop to gain net negative charge while the smaller drop gains net positive charge. At $t = 2$, the gravitational separation results in upper positive and lower negative configuration of electric charges.

occur in the simple configuration as shown in Fig. 14.13, where the line connecting the collision pair is parallel to the fair-weather electric field vector, but can be any angle. The induced charge distribution on the polarized particle is also a function of angle. Thus the collision and rebound can produce not only charge separation but also discharging. Furthermore, if the collision pair consists of ice and a supercooled droplet, then rebound is very unlikely. If the pair consists of ice and ice, then the long relaxation time for charge conduction through ice may prevent the occurrence of charge separation. A more detailed quantitative discussion can be found in Pruppacher and Klett (1997, chapter 18, pp. 818–821).

14.4.3 Non-inductive charging mechanisms

There are mechanisms that can cause charge separation without the need for an external electric field. These are called *non-inductive charging mechanisms*. Some of them are currently considered by many researchers as the major charging mechanisms responsible for thundercloud electrification.

Thermoelectric effect

It was observed in laboratory experiments that, when a hail pellet collides with an ice crystal of different temperature, the pellet becomes charged. This was first reported by Reynolds *et al.* (1957). Brook (1958) suggested that this is due to the effect of the temperature gradient over the point of contact. At that point, the H^+ ions (i.e. protons) in the warmer particle have a higher mobility in the ice lattice than OH^- ions and hence there will be a net transfer of positive charges moving over to the colder body. Consequently, the warmer particle will gain a net negative charge whereas the colder particle will carry a net positive charge. Gravitational separation (assuming that warmer particles fall to lower levels) then results in cloud electrification. This charging mechanism is called the *thermoelectric effect*.

Later studies, however, point out that it requires very large impact velocities, temperature difference, and contact areas to produce enough charges on the particles, which are unrealistic in clouds.

Charging due to collision with rimed particles (riming electrification)

This mechanism appears to be one that can explain many observed electrical features in thunderstorms, and hence is currently considered by many as the major charging mechanism that leads to lightning. This is not to say that it precludes the operation of other mechanisms, but merely indicates that it is likely the most important. This mechanism is often called *riming electrification*.

The initial observation upon which this mechanism is based is the same as that of the thermoelectric effect, i.e. Reynolds *et al.* (1957). Many researchers felt that the charge separation observed by Reynolds *et al.* (1957) was not really due to the thermoelectric effect but rather due to other reasons. Among the reasons, the concept of contact potential is often mentioned. This has something to do with charging due to the contact of two bodies with different electric potential, but the details are still unknown. Among the most systematic studies to elucidate this mechanism are those by Takahashi (1978), Jayaratne *et al.* (1983), and Takahashi and Miyawaki (2002), who performed laboratory experiments to investigate the charging due to collision between rimed particles and ice crystals.

Fig. 14.14, taken from Takahashi (1978), summarizes the experimental results. Thus it was found that substantial electric charges can deposit on a riming graupel when it is colliding with ice crystals in an environment where supercooled droplets are present. The magnitude and sign of the electrification are highly dependent on the temperature and cloud water content (CWC, includes both ice and water). Small CWC results in small amount of ice and produces little electricity. But when CWC is large, graupel will not be highly charged either. The rarity of lightning in storms over tropical oceans seems to corroborate this observation. The optimal CWC is $1-2\,\mathrm{g\,m^{-3}}$, when the graupel will be highly charged. Most importantly, there exists a *charge reversal temperature* $T_{rev} \sim -10°C$. Graupel will be charged positively if $T > T_{rev}$ and

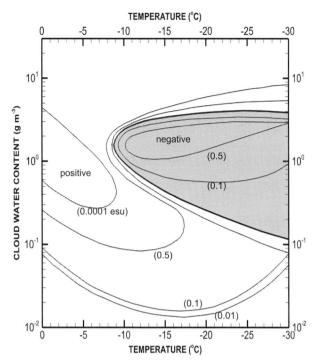

Fig. 14.14 Electrification of rime in the presence of supercooled water drops. Dark shaded region represents the conditions when negative charges are produced, whereas the unshaded region is when positive charges are produced. There are a few uncharged cases located in the unshaded region also. The electric charge of rime per ice crystal collision is shown in units of 10^{-4} esu. Adapted from Takahashi (1978).

negatively if $T < T_{\mathrm{rev}}$ (dark shaded region in Fig. 14.14). Ice crystals, on the other hand, will mostly carry positive charges, and they will be most concentrated higher up (i.e. the colder part) in the thundercloud. Graupel will fall at the periphery of the updraft column, where the updraft speed is weak. If the graupel falls to a lower level where $T > T_{\mathrm{rev}}$, it will be positively charged, and may contribute to the small positive charge center near the cloud base. This scenario is completely consistent with the observed tripole structure in Fig. 14.5.

Thus riming electrification is capable of producing electric fields large enough to cause lightning. This scenario seems to be consistent with recent observations of lightning as exemplified in Fig. 14.10.

While this mechanism seems to be the prime candidate that can explain the observed cloud electrification both qualitatively and quantitatively, it is unclear *why* the collision between ice crystals and graupel should lead to the charging behavior outlined above.

Charging due to freezing of aqueous solution drop
(Workman–Reynolds effect)

This effect is caused by the large potential difference occurring at the ice–solution boundary during the freezing of dilute aqueous solutions. The electric potential drives the ions, which are selectively taken by the ice to incorporate into its crystal lattice, and as a result charge separation occurs – the ice part may get charged one sign and the liquid part the other. The sign and magnitude of the interface potential depends on several variables, such as the ion concentration, the freezing rate, and the type of ions. If, at a later time, the unfrozen liquid parts shed away from the ice part, then gravitational separation leads to charge separation. However, the charges so separated appear to be quite small and unlikely to lead to lightning.

Charging due to drop breakup

This should not be confused with the inductive drop breakup charging in the presence of an external electric field as discussed earlier. Even without an external electric field, the breakup of a neutral drop can lead to charging. Usually the main bodies of the water become positively charged whereas the surrounding air becomes negatively charged. Judging from the tripolar model, this mechanism at most contributes only to the lower positive charge center, but even for that its charging rate is deemed too slow to be of any significance.

Charging due to fracture of freezing drops

When a drop is freezing, its outer ice shell may fracture or produce spikes that splinter into the air. Charges of either sign have been observed, but the small fragments or splinters are usually positively charged, while the main ice particles are often negatively charged. This seems to produce the right polarity configuration according to the dipole model. However, the problem lies in the fracture itself, which does not occur very often and, if it does, the amount of fragments is probably highly variable. This mechanism is presently not considered as an important charging mechanism for lightning.

Charging due to splintering during riming

Laboratory studies simulating the formation of graupel show that, when small droplets of 20–90 μm collide with a 5 mm diameter ice sphere, positively charged ice splinters are ejected and the ice sphere becomes negatively charged (Latham and Mason, 1961). Thus it is capable of producing the correct polarity, but the charging rate is deemed too small to play a major role.

Charging due to air bubbles escaping from melting graupel
(Dinger–Gunn effect)

When a graupel is melting, the air bubbles trapped in it will be released into the environment in the form of bubble burst. The bursting creates tiny negatively charged droplets and the remaining ice acquires a net positive charge. This was first reported by Dinger and Gunn (1946) and is sometimes called the *Dinger–Gunn effect*. This effect can produce enough charge to contribute to the lower positive charge center.

14.5 Modeling electrification using cloud models

In judging the effectiveness of the charging mechanisms, the conventional way of thinking – hydrometeors become charged and then separate gravitationally – is somewhat limited because the circulation inside a thunderstorm is very complicated. Large hydrometeors such as raindrops, graupel, and hail are not always located in the lower levels. At certain stages of the storm evolution, they can be present in the middle or higher level of the cloud, and rapid charging may occur there as well. The best way is to incorporate the charging mechanisms with a storm model with explicit cloud microphysics such as the one introduced in Chapter 12. The most ideal way would be to write down the precise differential equations for each charging mechanism and evaluate their charging rates at each grid point at each time step. This is not yet possible, so at present the charging mechanisms are parameterized before they are implemented in the model. Like the parameterization of other cloud microphysical processes, some schemes offer more sophistication than others. Straka (2010) has provided some information about such parameterizations. As an example, Mansell *et al.* (2010) simulated a small thunderstorm using a two-moment cloud microphysical scheme and studied its electrification. Figs. 14.15 and 14.16 illustrate what such a model can simulate about the electrification of thunderclouds. Fig. 14.15 shows that the simulated storm exhibits a three-pole electric structure consistent with general observations. Fig. 14.16 shows the modeled charges on different hydrometeor categories before a cloud-to-ground flash occurred.

14.6 Impact of cloud electricity on microphysical processes

14.6.1 Electric effect on the collision efficiency of cloud droplets

Many cloud microphysical processes will be influenced by the presence of enhanced electricity – electric charges and external electric fields – in thunderstorms. An obvious and important example is the electric influence on cloud growth. Schlamp

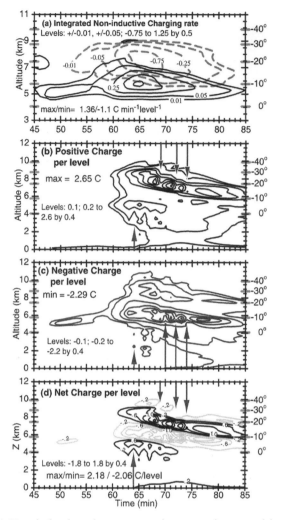

Fig. 14.15 (a) Non-inductive charge separation rates, integrated by model level (125 m thickness), between graupel and ice crystals–snow. Polarity indicates the sign of charge gained by graupel. (b) Layer total positive charge. (c) Layer total negative charge. (d) Layer net charge. Arrows in (b)–(d) indicate the first lightning flash just before 64 min and lightning events with obvious effects on the layer charge. From Mansell *et al.* (2010). Reproduced by permission of the American Meteorological Society.

et al. (1976, 1979) performed numerical studies on the effect of both electric charges and external electric fields on the collision efficiency of cloud drops. It is easy to understand the effect of electric charges. If the two particles are charged with opposite signs, then they will attract each other, which results in the increase of the collision efficiency, and vice versa. This effect disappears if both drops are

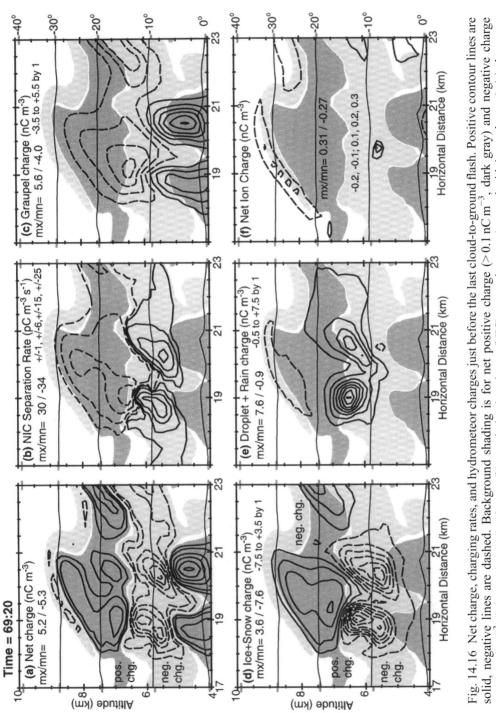

Fig. 14.16 Net charge, charging rates, and hydrometeor charges just before the last cloud-to-ground flash. Positive contour lines are solid, negative lines are dashed. Background shading is for net positive charge (>0.1 nC m^{-3}, dark gray) and negative charge (<-0.1 nC m^{-3}, light gray). (a) Net charge, (b) non-inductive charge (NIC) separation rate (graupel with ice and snow), (c) charge on graupel, (d) charge on ice crystals and snow, (e) charge on cloud droplets and rain, and (f) net free ion charge (note different scale on contours). Maximum and minimum values (mx/mn) are indicated on each panel. From Mansell *et al.* (2010). Reproduced by permission of the American Meteorological Society.

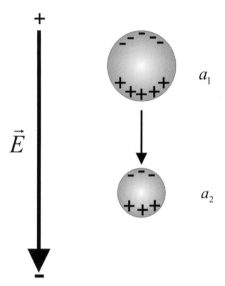

Fig. 14.17 Collision of two drops in the presence of an external electric field. Both drops are polarized.

electrically neutral. But in the presence of an external electric field, even neutral drops will be affected. This is because of the polarization of the drop due to the external $\vec{E}$ that causes charges of opposite signs to be induced on the drops, as shown in Fig. 14.17. The lower positive charges of the a_1 drop are closer to the negative charges of the a_2 drop, and an attractive Coulomb force will result, enhancing the collision efficiency.

Schlamp *et al.* (1976) studied the case where a positively charged larger drop is falling to collect a negatively smaller drop in the presence of an external electric field. The drop size range is 11.4–74.3 μm and the electric field ranges from 0 to 3429 V m^{-1}. The electric charges on the droplet range from 0 to 1.1×10^{-4} esu or 3.7×10^{-14} C. They found the following.

(1) External electric fields in the presence of electrically neutral drops invariably enhance the collision efficiency of cloud drops, with the effect being most pronounced for the smallest collector drops. This implies that the critical electric field strength necessary to affect the collision efficiency increases with increasing collector drop size. Thus an external field of 500 V cm^{-1} significantly raises the collision efficiency of an 11.4 μm collector drop, while it has a negligible effect if the radius is ~ 30 μm. On the other hand, an external electric field of 3000 V cm^{-1} significantly increases E for all collector drops of $a_1 < 50$ μm, while for $a_1 > 50$ μm the effect of these high-intensity fields is negligible.

(2) Electric charges residing on drops in the absence of an external electric field invariably raise the collision efficiency of cloud drops if the two interacting drops are oppositely charged, the effect being most pronounced for the smallest collector drops. This implies that the critical electric charge necessary to affect the collision efficiency increases with increasing collector drop size. Thus, oppositely charged drops that carry a charge of magnitude equal to one-tenth the mean thunderstorm charge, as defined above, significantly raises the collision efficiency for collector drops of $a_1 < 40\,\mu m$ while having negligible effect on drops of $a_1 > 40\,\mu m$.

(3) In contrast to the results when *either* electric fields *or* charges are present, their results indicate a reduction in the collision efficiency to near-geometric values when *both* electric fields *and* charges are present. This reduction is due to the substantially decreased interaction time during which mutually attractive charges on the two drops can act to force a collision. This decreased interaction time is the result of the vertical component of the relative velocity of the two drops, which is caused by the external vertical field acting on the charges carried by the drops.

(4) For $a_1 > 70\,\mu m$ the effect of electric fields and charges on collision efficiency is negligible even though the field strengths and charges may be as large as those found in thunderclouds close to electrical breakdown.

Later, Schlamp *et al.* (1979) extended their earlier study by assuming the a_1 drop to be negatively charged and the a_2 drop to be positively charged. They also considered two configurations: (i) the a_1 drop is initially above the a_2 drop; and (ii) the a_1 drop is initially below the a_2 drop. The sizes of the drops considered range from 1 to 118 μm in radius. The magnitude of the electric charges on the drops ranges from 0 to 2.8×10^{-4} esu, and the electric fields range in strength from 0 to 3429 V cm^{-1}. The geometrical configuration of the collision actually matters, and different impacts are seen. This time they found that for a negatively charged a_1 drop and a positively charged a_2 drop interacting in the presence of a downward-pointing electric field, the electrostatic forces are responsible for determining the shape of the collision efficiency curves. The hydrodynamic flow interactions, in this case, are of secondary importance. In contrast, the collision efficiencies derived for the electrostatic configurations considered by Schlamp *et al.* (1976) are controlled by the hydrodynamic interaction of the flow fields around each drop. The electrostatic forces, on the other hand, play an important, but secondary, role, in that they enhance the value of the collision efficiency but appear not to change the basic shapes of the collision efficiency curves.

Apparently, the presence of electricity will change the collision behavior of cloud drops in a somewhat complicated manner, and more studies are needed to fully understand the impact. In general, it appears that electric charges and fields work to enhance the collision of cloud drops.

Czys and Ochs (1988) studied the coalescence of an electrically charged drop pair of 340 μm and 190 μm, respectively, and concluded that, in a highly electrified thundercloud, such drop pairs would always coalesce, but rebound may occur in less electrified clouds.

Similar studies for the electric charge effect on the collision between planar ice crystals and supercooled droplets were reported by Martin *et al.* (1981), who showed that collision efficiencies are enhanced substantially if the charges on both the ice and the drop are greater than $0.8a^2$, where a is the radius (in cm) of either the ice plate or the drop.

14.6.2 Effect of electric field on the shape of large raindrops

The high electric fields associated with thunderstorms also influence large rain-drops. In this section, we will discuss the influence on the shape of raindrops. We have mentioned previously that drop shapes are important for radar detection of precipitation. Changing drop shapes will also change the flow fields around them and hence impact on their collision behavior.

We have seen in Chapter 8 that a water drop falling at terminal velocity assumes an equilibrium shape due to the balance of the surface tension and hydrodynamic force. In an external electric field, the drop will be polarized and the charges induced in it will be subject to the external electric force. Unlike a solid, whose shape is fixed, a liquid drop will respond to the electric stress and change its shape. If the electric field exceeds a critical value E_{0c}, the electric stress becomes so large that it will cause the drop to become mechanically unstable and break up.

Let us consider first the force balance between the critical electric stress and surface tension for an initially spherical drop, and ignore the hydrodynamic force for the time being. This balance is

$$\frac{E_{0c}^2}{8\pi} = \frac{2\sigma_w}{a},\tag{14.9}$$

which implies that

$$E_{0c}(a/\sigma_w)^{1/2} = X_c = \text{const.},\tag{14.10}$$

where X_c is a dimensionless constant characterizing the balance. Eq. (14.9) only applies to a sphere. The equilibrium shape of a large raindrop falling in air will deviate from a sphere, and hence the value of X_c will also be different. Fig. 14.18 shows the plot of the calculated axis ratio (b/a) from the drop shape that should satisfy (14.10) as a function of X_c if an initially spherical drop is subject to a vertical

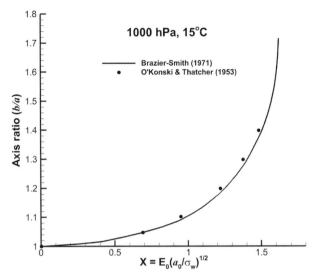

Fig. 14.18 Variation of theoretically predicted shape of an originally spherical water drop in air assuming that the drop is at rest, with $X = E_0(a/\sigma_w)^{1/2}$. Based on results presented in the literature, after Rasmussen *et al.* (1985).

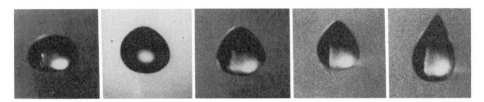

Fig. 14.19 Photographs from a 16 mm motion picture of water drops ($a_0 = 0.23$ cm) in an airstream being exposed to various external electric fields. From left to right: $b/a = 0.725$, 0.925, 0.970, 1.12, and 1.52, for $E_0 = 0$, 6.75, 8.25, 9.0, and 10.0 kV cm^{-1}, respectively. From Rasmussen *et al.* (1985). Reproduced by permission of the American Meteorological Society.

electric field. We see that the drop will be elongated in the z direction in response to the electric stress, and the elongation is the more pronounced the higher X_c. This implies that, for a fixed E_{0c}, a larger drop will elongate more than a smaller drop, which is intuitively plausible.

The problem of Eq. (14.9) or (14.10) and therefore Fig. 14.18 is that they totally ignore the hydrodynamic force around the water drop. Several wind tunnel experiments have been performed (e.g. Richards and Dawson, 1971; Rasmussen *et al.*, 1985) to determine the shape of water drops suspended in a vertical electric field. Fig. 14.19 shows some example photographs taken by Rasmussen *et al.* (1985) for neutral drops suspended in wind tunnel.

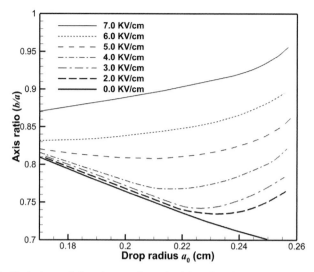

Fig. 14.20 Variation of the drop axis ratio b/a with drop radius a_0 for various external electric field strengths. Adapted from Rasmussen *et al.* (1985).

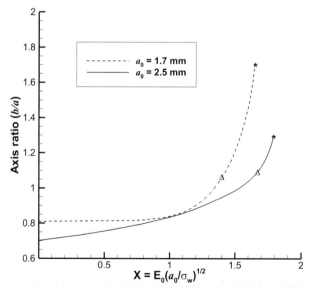

Fig. 14.21 Variation of the drop shape (axis ratio) b/a with X for drops of $a_0 = 2.5$ and 1.7 mm. The stars mark the critical breakup values obtained by Richards and Dawson (1971). The section of the curve between the triangle and the star is extrapolated from the experimental data. Adapted from Rasmussen *et al.* (1985).

The experiments show that the drop becomes more and more distorted as the field strength increases, and the shape changes from a quasi-spheroid to conical. Breakup of drops occurred from the top side of the drop rather than on the bottom side, as in the case of regular breakup in the absence of an electric field. Sometimes the drops landed on the lower charging grid that was used to generate the external electric field and became charged, and the high electric field would draw a water string out of the drop.

Fig. 14.20 shows the plot of the measured drop axis ratio as a function of drop radius under the influence of various external field strengths. We see that the situation is more complicated and the axis ratio does not increase monotonically with drop size as implied by Fig. 14.18. Naturally this is due to the effect of the hydrodynamic force that complicates the situation. Fig. 14.21 is a plot of b/a vs. X_c for drop radius 1.7 and 2.5 mm. Compared with Fig. 14.18, we see that, while the shape of the curves looks similar, the b/a vs. X_c relationship is not unique as implied by Fig. 14.19 but depends on drop size. The extrapolation (the section between the triangle and the asterisk) of the curves would smoothly reach the critical points (marked by the asterisk) where drop breakup occurs as determined by Richards and Dawson (1971).

Such elongation of raindrops should occur dramatically right before the lightning occurs. If a dual polarization radar is scanning through the right cross-section at the right moment, it would observe the sudden change in Z_{DR} due to the elongation. After the lightning, the electric field would diminish and the Z_{DR} would go back to regular values.

Problems

14.1. (a) Calculate the electric field given by Eq. (14.4) assuming that at the surface of the Earth the field is $-130.67\,\mathrm{V\,m^{-1}}$.
 (b) Plot the results you obtain in (a) as a curve in a chart with E as the horizontal axis and z as the vertical axis.
 (c) Plot a curve on the same chart using Eq. (14.5).
 (d) Compare the two curves and explain their difference in terms of the physics of electromagnetism.
14.2. Prove that the vertical profile of the space charge density that causes this electric field in Eq. (14.5) is indeed given by Eq. (14.7).

15

Clouds–environment interaction

Up to now we have been studying the internal physics and chemistry of cloud systems – some large and some small – without thinking too much about how clouds influence their environments. In other words, we consider a cloud as if it is an isolated system. But we know that clouds cannot be really separated from the rest of the atmosphere, and the formation or dissipation of a cloud will ultimately change its environment. How clouds influence the atmosphere, or, more fundamentally, how clouds and their atmospheric environments interact with each other, is a complicated question that requires more research to clarify. In this chapter, we will discuss some relevant issues in this area.

15.1 Clouds and atmospheric chemistry

The formation of clouds has much to do with chemistry, as we have discussed previously, especially in the chapters on aerosols and nucleation. Conversely, the formation of clouds also greatly influences the chemistry of the atmosphere. First of all, once an aerosol particle nucleates a cloud drop or an ice crystal, it begins to be involved in the cloud and precipitation process, and ultimately it may be removed from the atmosphere if it falls with the precipitation. Even those aerosol particles that are not involved in the nucleation process (called the interstitial aerosol) may be washed out by the cloud and precipitation particles. Similarly, trace gases such as SO_2, NH_3, NO_2, and CO_2, etc., can be absorbed by cloud and raindrops and turn into other chemicals, which may then be involved in further chemical reactions (such as the production of acid precipitation). In the following sections, we will discuss some aspects of the cloud impacts on atmospheric chemistry.

15.1.1 Wet removal of aerosol particles

We have discussed the production and removal of aerosol particles in the atmosphere briefly in Chapter 6. Among all the removal processes, the most efficient is removal by cloud and precipitation processes, and this is called the wet removal process. It is also called precipitation scavenging or wet deposition.

The study of precipitation scavenging received initial attention after World War II because of concerns over radioactive fallout from the debris of nuclear weapons tests in the atmosphere, as such material may be washed out by precipitation. Studies concerned with the meteorological processes involving radioactive debris were categorized as *nuclear meteorology*, a discipline that has since become dormant even before the last nuclear test in the atmosphere conducted in 1980. However, the nuclear power plant accidents in Chernobyl, Ukraine, in 1986, and Fukushima, Japan, in 2011, serve as a sober reminder that man-made radioactivity in the atmosphere can still be a problem. Later, it became clear that modern industrial activities inject huge amounts of aerosol particles into the Earth's atmosphere that may cause inadvertent climate change, and the main thrust of precipitation scavenging is now focused on whether or not the cloud and precipitation process can efficiently cleanse the atmosphere.

In the following, we will examine the basic physics of the removal of aerosol particles by cloud and precipitation particles. There are two main categories of precipitation scavenging mechanisms:

(1) *Nucleation scavenging* – refers to the mechanism that aerosol particles serve as either CCN or IN to nucleate a water drop or an ice crystal, and hence may be removed from the atmosphere if the drop or ice crystal precipitates. We have already discussed the nucleation phenomenon in Chapter 7 and hence there is no need to repeat the discussion here.
(2) *Impaction scavenging* – refers to those aerosol removal mechanisms that occur during the cloud and precipitation process that are not related to nucleation. These mechanisms are all mechanical in nature because they are all related to the motions of the drop/ice and aerosol particles. It is this category that we will discuss below.

15.1.2 Physical mechanisms of impaction scavenging

There are several mechanisms that can cause aerosol particles to impact and become collected by a surface. Some of these mechanisms are similar to those involved in the dry removal processes explained in Chapter 6.

Brownian diffusion of aerosol particles

As we have mentioned in Chapter 6, the vast majority of aerosol particles that are suspended for a long period in the atmosphere are typically smaller than 1 μm in size. Such small particles are easily perturbed by the bombardment of the air molecules that surround them and perform motions due to momentum transfer by the air. In the absence of wind (i.e. mean motion of the air), the bombardment is in principle random and isotropic, and consequently the aerosol particles also perform random zigzag motions, known as *Brownian motion*. If there is a cloud drop nearby, the Brownian motion of the aerosol particle may cause the particle to collide with the droplet and become attached to it, and is thus removed from the atmosphere.

Inertial impaction

When a water drop (or ice particle) is falling in air, it creates a flow field around itself and hence exerts hydrodynamic drag forces on aerosol particles in the vicinity of its fall path. If an aerosol particle has no mass or size, it will never collide with the drop because it will just follow the streamlines like air molecules. However, since it does have a mass and hence inertia, its trajectory will deviate from the streamlines and may end up colliding with the drop and hence become removed from the atmosphere. This mechanism is called *inertial impaction*.

Interception

An aerosol particle also has finite size, and hence the collision occurs when the center of the aerosol particle is a radius away from the drop surface. This is, of course, purely due to the size factor and it exists even in the case of no flow field.

Phoretic effects

The phoretic effects work when the environmental condition is not saturated. Let us take a water drop as an example. When a drop is evaporating in a subsaturated environment, its surface is cooler than the air around it. Hence a temperature gradient exists between the air and the drop surface. This thermal gradient will produce a "force" that pushes small aerosol particles in the vicinity to move toward the anti-gradient direction, which, in this case, is toward the drop surface. These particles thus become collected by the drop. Such a phenomenon is called *thermophoresis* and the force is called the *thermophoretic force*. Clearly, in a supersaturated environment, when the drop is growing, the thermophoretic force works in the opposite direction.

But when a drop is evaporating, there also exists a vapor density gradient around the drop, and this gradient also produces a force, called the *diffusiophoretic force*, which also pushes an aerosol particle in the anti-gradient direction. This phenomenon is called *diffusiophoresis*. But this time, the force is directed away from the drop surface, and thus it tends to prevent the aerosol particle from colliding with the drop. In other words, diffusiophoresis counteracts thermophoresis, and whether or not the particle can be collected by the drop depends on the relative strengths of the two forces.

Both phoretic forces are microscopic and work only over a very small distance from the drop surface. Nonetheless, they are effective when the particles are close to the drop surface.

Electric forces

Many aerosol particles are electrically charged and hence their motions can be influenced by electric forces. We have seen in Chapter 14 that cloud and precipitation particles can be highly charged during thunderstorms, and therefore it is quite possible that electric forces may greatly enhance the precipitation scavenging of aerosol particles at this time.

Turbulence

Turbulence is present in the atmosphere constantly and certainly influences the motions of aerosol particles that are suspended in the air and hence the efficiency with which particles are scavenged by hydrometeors. Currently, we do not yet have a deep understanding of the impact of turbulence on the scavenging of aerosol particles.

15.1.3 Mathematical model of precipitation scavenging

Wang *et al.* (1978) formulated a simple mathematical model to determine the collection efficiencies with which aerosol particles are scavenged by water drops. This model consists of two complementary submodels. The first one is the *flux model*, which is based on the convective diffusion equation of aerosol concentration around a falling water drop and is applicable to small aerosol cases where the Brownian motion effect is important. The other is the *trajectory model*, which is based on the equation of motion of an aerosol particle moving in the flow field of a falling water drop and is applicable to larger aerosol cases where the Brownian motion is unimportant. They are described in the following.

Flux model

When the aerosol particles are smaller than about 1 μm, their inertia is insignificant in deciding how much they can be collected by hydrometeors. Instead, the effect of Brownian motion becomes more and more important as particles become smaller and smaller. Einstein (1905) showed that the Brownian motion of an ensemble of such small particles can be treated mathematically in a manner similar to the diffusion of gases, and this realization is sometimes called Brownian diffusion. Thus, we can borrow the diffusion theory of water vapor we developed in Chapter 9 to treat the Brownian diffusion of aerosol particles here.

Consider a cloud of small spherical aerosol particles of radius r_p and number concentration n around a falling drop of radius a in an environment of pressure p and certain humidity condition represented by water vapor density ρ_v. Since the particles are small, we can neglect their inertia and hence the aerosol concentration n can be described by the steady-state convective diffusion equation (cf. Eq. (9.34)):

$$D_p\nabla^2 n - \vec{u}_{\text{drift}} \cdot \nabla n = 0, \tag{15.1}$$

where D_p is the Brownian diffusivity of the aerosol and $\vec{u}_{\text{drift}}$ is the drift velocity of the aerosol cloud responding to a net external force $\vec{F}_{\text{ext}}$, the two being related by

$$\vec{u}_{\text{drift}} = B_p\vec{F}_{\text{ext}}, \tag{15.2}$$

where B_p is called the mobility of the aerosol particles.

The external forces that we will consider here are thermophoretic, diffusiophoretic, and electrostatic forces, and their mathematical forms are, respectively:

$$\vec{F}_{\text{th}} = -\frac{12\pi\eta_a r_p(k_a + 2.5k_p N_{\text{Kn}})k_a \nabla T}{5(1 + 3N_{\text{Kn}})(k_p + 2k_a + 5k_p N_{\text{Kn}})p}, \tag{15.3}$$

$$\vec{F}_{\text{df}} = -6\pi r_p \eta_a \frac{0.74 D_v M_a \nabla \rho_v}{(1 + \alpha N_{\text{Kn}})M_w \rho_a}, \tag{15.4}$$

$$\vec{F}_e = q_p\vec{E}_a = -q_p \nabla \varphi_a. \tag{15.5}$$

In these equations, k_a and k_p are the thermal conductivities of air and the aerosol particle, respectively, φ_a is the electric potential due to the drop's electric charge q_a, q_p is the electric charge of the aerosol particle, and N_{Kn} is the Knudsen number, defined as

$$N_{\text{Kn}} = \lambda_a/r_p, \tag{15.6}$$

where λ_a is the mean free path of air. The factor α is defined as

$$\alpha = 1.257 + 0.4 \exp(-1.10/N_{\text{Kn}}), \tag{15.7}$$

which serves to correct the drag force experienced by the aerosol particle in air when its size is on the order of or smaller than the mean free path of air.

Since the inertia is ignored, the drag force on the particle due to the flow field is also ignored. On the other hand, the flow field will result in enhancement of the temperature and vapor density gradients in (15.3) and (15.4) and has to be considered. That is to say, they should be corrected by the ventilation coefficients $\bar{f}_h$ and $\bar{f}_d$, respectively. Similarly, the Brownian diffusivity D_p is also enhanced by a factor $\bar{f}_p$ since the

Brownian motion is also coupled to the flow field. On the other hand, there is no ventilation enhancement to the electrostatic force since it is not coupled to the flow.

With all these considerations, Eq. (15.1) can be written as

$$\bar{f}_p D_p \nabla^2 n - B_p (\bar{f}_h \vec{F}_{th} + \bar{f}_v \vec{F}_{df} + \vec{F}_e) \bullet \nabla n = 0, \tag{15.8}$$

which is assumed to be a radially symmetric spherical system, since we have removed the asymmetry of the forces due to the flow fields by the empirical ventilation coefficients. The boundary conditions suitable for this system are:

$$\begin{aligned} n &= 0 &&\text{at } r = a, \\ n &= n_\infty &&\text{at } r \to \infty. \end{aligned} \tag{15.9}$$

The first boundary condition simply states that all the particles that arrive at the drop surface are completely removed from the atmosphere (i.e. the drop surface is a perfect sink). The second states that the particle concentration remains a constant far away from the drop.

The system (15.8) and (15.9) was solved by Wang *et al.* (1978); the solution is

$$n(r) = n_\infty \left\{ \frac{\exp[(B_p \Phi / D_p \bar{f}_p a)(1 - a/r)] - 1}{\exp(B_p \Phi / D_p \bar{f}_p a) - 1} \right\}, \tag{15.10}$$

where

$$\begin{aligned} \Phi = -\,&\frac{12\pi \eta_a r_p (k_a + 2.5 k_p N_{Kn}) k_a (T_\infty - T_a) \bar{f}_h}{5(1 + 3N_{Kn})(k_p + 2k_a + 5k_p N_{Kn}) p} \\ &- \left[6\pi r_p \eta_a \frac{0.74 D_v M_a a (\rho_{v,\infty} - \rho_{v,a}) \bar{f}_v}{(1 + \alpha N_{Kn}) M_w \rho_a} \right] + q_p q_d. \end{aligned} \tag{15.11}$$

Eq. (15.10) allows us to calculate the total flux of particles that arrive at the drop surface per unit time, so the collection kernel of the aerosol particles by the drop is

$$K = -\frac{1}{n_\infty} \frac{\partial n}{\partial t} = -\frac{1}{n_\infty} \oint_{\text{surface}} \left(n B_p \vec{F}_{ext} - \bar{f}_p D_p \bullet \nabla n \right) \bullet d\vec{s}, \tag{15.12}$$

where $d\vec{s}$ is an infinitesimal surface area increment. The collection efficiency is again obtained from $E = K/K^*$ (see Eq. (10.1)). For a more detailed derivation, see Wang *et al.* (1978). For a more general treatment, including scavenging by ice crystals, see Wang (2002).

Trajectory model

The flux model ignores the contribution by inertial impaction and hence will severely underestimate the collection efficiency for aerosol particles larger than 1 μm. For such cases, the trajectory model can be used to obtain the collection efficiency. The mathematics of the trajectory model is entirely analogous to the collision growth of droplets that we treated in Chapter 10. Namely, we solve the equation of motion of an aerosol particle in the vicinity of a falling drop:

$$m_p \frac{d\vec{v}_p}{dt} = m_p \vec{g} - \frac{6\pi \eta_a r_p}{1 + \alpha N_{Kn}} \times (\vec{v}_p - \vec{u}) + \vec{F}_{th} + \vec{F}_{df} + \vec{F}_e, \tag{15.13}$$

where quantities with subscript p represent that of the aerosol particle. The second term on the right-hand side of (15.13) is the hydrodynamic drag of the flow field on the aerosol particle. The solution of this equation determines the particle trajectory and eventually leads to the collision efficiency $E = K/K^*$. The trajectory model generally works for particle size larger than 1 μm, as it ignores the Brownian diffusion, which is important for submicrometer particles.

The two models are complementary to each other and the merging of their results gives the collection efficiency of the whole spectrum of aerosol sizes. The merging is not done arbitrarily, but rather the two curves overlap with each other in the submicrometer size range. Hence the merging occurs rather smoothly. Fig. 15.1 shows a typical example. In the follow discussion, we will assume that the collision efficiency is the same as the collection efficiency (this implies that, once collision occurs, the particle is collected by the drop).

We see that the collision efficiency can be generally divided into three regimes:

- *Brownian diffusion regime* – relevant to particles smaller than 0.1 μm, but the specific range varies with the environmental conditions. In this regime, the dominant scavenging mechanism is Brownian diffusion, which is more active the smaller the particle size. The collision efficiency decreases with increasing size in this regime.
- *Inertial impaction regime* – relevant to particles larger than ~ 1 μm. The dominant scavenging mechanism is inertial impaction, and hence the larger the aerosol, the greater the collision efficiency.
- *Greenfield gap regime* – for particle size between ~ 0.1 and ~ 1 μm. In this regime, the particles are too large for Brownian motion to be dominant yet too small for inertial impaction to be effective. Consequently, the scavenging becomes ineffi-cient here and the curves form a minimum, which has been called the *Greenfield gap* after Greenfield (1957), who predicted the existence of this phenomenon.

All the above arguments apply to the scavenging of aerosol particles by water droplets. However, as Prodi (1976) showed, ice crystals also scavenge aerosol particles (Fig. 15.2), and, as shown by Miller and Wang (1989), their collision efficiencies also exhibit the three-regime structure (Fig. 15.3).

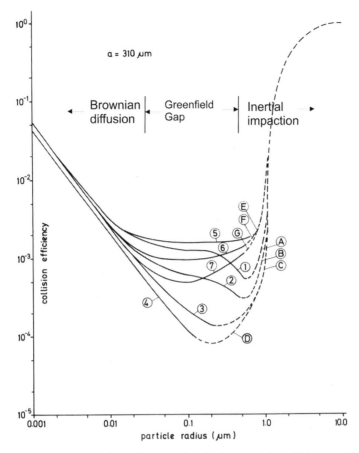

Fig. 15.1 Effect of relative humidity and electric charge on the efficiency with which aerosol particles of various sizes collide with water drops of $a = 310$ μm, in air at $10°C$ and 900 mb. Combination of the trajectory (dashed lines) and flux (solid lines) model results. Curves 1–4 and A–D are for no electric effect cases, and $RH = 50, 75, 95,$ and 100%, respectively. Curves 5–7 and E–G are for $q_a = \pm2.0a^2$, $q_p = \mp2.0r_p^2$, and $RH = 50, 75,$ and 95%, respectively. Note that the electric forces are assumed to be attractive in this study. From Wang *et al.* (1978). Reproduced by permission of the American Meteorological Society.

This three-regime feature is also present in the collision efficiency for other drop sizes. In the Greenfield gap regime, the particles are too large for Brownian diffusion to be effective and yet too small for inertial impaction to be significant. The only forces that are operating more or less effectively are the phoretic and electric forces, as we see from Fig. 15.1. This is the reason why the efficiency curves generated by the two models can merge so smoothly over this range. In the Greenfield gap regime, in general, the smaller the relative humidity, the higher the collision efficiency, because the thermophoretic force dominates in this size range and it

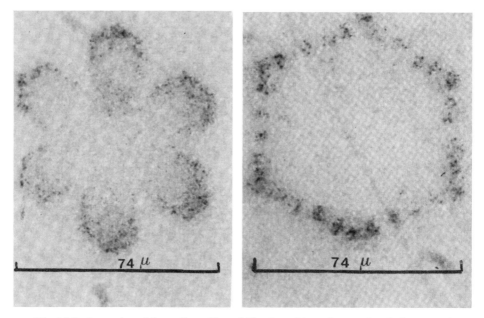

Fig. 15.2 Aerosol particles collected by a falling broad-branch crystal and a hexagonal plate. Photos from Prodi (1976), courtesy of Prof. Franco Prodi. Reproduced by permission of the American Meteorological Society.

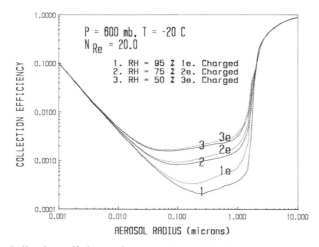

Fig. 15.3 Collection efficiency for aerosol particles by columnar ice crystals at 600 mb and $-20°C$ as a function of aerosol radius predicted by an ice scavenging model that combines the trajectory and flux models. Curves with *RH* of (1) 95%, (2) 75%, and (3) 50% without and with (e) electrostatic charges $q = 2$ esu cm^{-2} are presented for $N_{RE} = 20$. From Miller and Wang (1989). Reproduced by permission of the American Meteorological Society.

favors particle collection. But even though they can raise the efficiency by more than one order of magnitude, the gap is still prominent. The presence of the Greenfield gap implies that the mechanical scavenging process is unable to remove submicrometer particles efficiently from the atmosphere. In Chapter 6, we learned that the aging of aerosol particles leads to a concentration maximum of the submicrometer particles. We now know that, once the submicrometer particles are produced in the atmosphere, they are difficult to remove even by wet processes.

In a cloud, the most efficient removal mechanism is nucleation scavenging. This, of course, requires that the aerosol is either a good CCN or a good IN. Below the cloud, impaction scavenging dominates. To determine the actual amount of aerosol particles that can be removed by precipitating scavenging, it will be necessary to incorporate collection efficiency information such as that presented in Figs. 15.1 and 15.3 into the cloud model and perform simulation studies. Fig. 15.4 shows the cloud model estimate of various scavenging mechanisms given by Flossmann (1998).

More extensive studies have been performed in this direction using not only 2D but also 3D cloud models. In addition to precipitation scavenging, they also considered the more general aspects of the cloud–aerosol interaction as well. They found that nucleation incorporates about 90% of the initial aerosol particle

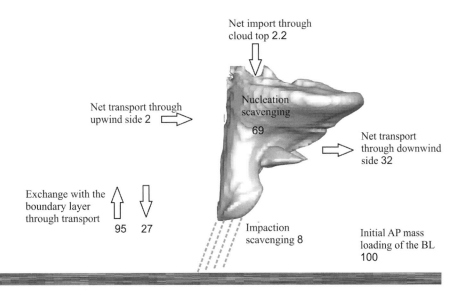

Fig. 15.4 Aerosol mass transfer and scavenging integrated over 1 h of simulation time for the GATE 261 cloud; values represent percentages and pertain to a 2D cloud unity slice. They are normalized with respect to the initial boundary layer loading set to 100. Adapted from Flossmann and Wobrock (2010).

mass inside the cloud drops. Impaction scavenging can probably be neglected inside clouds. Below clouds, impaction scavenging contributes around 30% to the particle mass reaching the ground by a rainfall event. The exact amount depends on the precise case studied. Nucleation and impaction scavenging directly by the ice phase in mixed-phase clouds seems to play a minor role with respect to the particle mass that enters the ice particles via freezing of the liquid phase.

Clouds are very efficient in pumping up the boundary layer aerosol, which essentially determines the cloud properties. For example, in a marine case studied by Flossmann (1998), the net pumping depleted about 70% of the aerosol from the section of the boundary layer considered. The larger particles (and thus 70% of the mass vented up) got activated inside the cloud. A recent review is given by Flossmann and Wobrock (2010).

Pollutant particle–cloud interaction

The above observations are based on the life cycle of a single cloud and focus on the aspect of aerosol scavenging. In reality, there are other complicated processes operating in clouds. To understand the total impacts of clouds on the atmosphere we would need to consider all these processes and over a larger space scale and a longer time period. For example, some of the aerosol particles scavenged may become re-suspended after the hydrometeors evaporate and thus may be recycled into another cloud. Clouds are also able to produce new particles and transport them out of the cloud (Hegg *et al.*, 1990) and may also play a role in the formation of new clouds.

One of the important issues that relate to the precipitation scavenging study is whether or not man-made pollutant particles may significantly impact upon natural precipitation. For example, based on a simple parcel model study, one would tend to conclude that increasing particulate pollution and decreasing solubility suppresses rain formation. But if we put this question in a broader context, the conclusion may be different. Flossmann and Wobrock (2010) performed mesoscale simulations that included the processes mentioned in the previous paragraph and integrated over a longer time period. They found that, by taking into account entire cloud fields over longer periods of time, the results show strong spatial and temporal variability, with isolated regions of inverse correlation of the effects. Even though, in general, initially the expected behavior was found, after several hours of simulation, the overall precipitation amounts of the more polluted cases caught up. This suggests that changing pollution will affect the spatial and temporal pattern of precipitation, but will probably not reduce the overall long-term precipitation amount, which might be entirely governed by the moisture state of the atmosphere.

This finding underlines the importance of a complete system approach when studying environmental problems – linear extrapolation of a single mechanism may lead to erroneous conclusions because the various components of an environmental system are not isolated but are related to each other in manners that are sometimes quite unexpected.

15.2 Acid rain – an example of wet removal of trace gases

Clouds and precipitation remove not only aerosol particles but also certain trace gases as well. The formation of acid rain is precisely such a process. Acid rain is actually just one component of acid precipitation, as not only water drops but also ice particles in clouds can become acidic via the same process. Here we will use rain as an example for the discussions.

"Natural" rain is acidic because it contains carbonic acid and has a pH value ~ 5.6. Carbonic acid (H_2CO_3) occurs when raindrops take up CO_2 in air via the reaction

$$CO_2(g) + H_2O(l) \rightleftharpoons H_2CO_3(l) \tag{15.14}$$

but it is considered harmless, as this is part of the natural equilibrium and the environment is supposed to have adjusted to that long before the appearance of humans. It is the additional acidity induced by human industrial activity that causes concern. Thus acid rain is defined as rain with pH < 5.6.

The two most common acid species in acid rain are sulfuric acid (H_2SO_4) and nitric acid (HNO_3). The partition of these two species in acid rain in the USA is about 60% and 35%, respectively, while other acids make up the remaining 5%. The precursor of H_2SO_4 is sulfur oxides (SO_x, mainly SO_2), while the precursor of HNO_3 is nitrogen oxides (NO_x, mainly NO and NO_2). When these precursors are released in the atmosphere, they may be taken up by cloud and precipitation particles and turn into acid precipitation via heterogeneous chemical reactions. Fig. 15.5 shows a schematic of this process.

In the following, we will look into more detail of the formation of a sulfuric acid drop. We will first look at the chemical reactions involved. When a water drop absorbs SO_2, the following set of reactions occur (Pruppacher and Klett, 1997; Seinfeld and Pandis, 2006):

$$SO_2(g) + H_2O(l) \rightleftharpoons SO_2 \cdot H_2O(l) \tag{15.15}$$

$$SO_2 \cdot H_2O \rightleftharpoons H^+ + HSO_3^- \tag{15.16}$$

$$HSO_3^- \rightleftharpoons H^+ + SO_3^{2-} \tag{15.17}$$

$$SO_3^{2-} \xrightarrow{\text{oxidation}} SO_4^{2-} \tag{15.18}$$

Fig. 15.5 A schematic illustrating the formation of acid rain.

The first reaction describes the absorption of gaseous SO_2 into the water drop and formation of $SO_2 \cdot H_2O(l)$ in the drop. The reaction (15.16) describes the first ionization, where the $SO_2 \cdot H_2O(l)$ ionizes into a hydrogen ion and a bisulfite ion HSO_3^-. Reaction (15.17) is the second ionization, where another H^+ is produced along with a sulfite ion SO_3^{2-}. In the presence of oxidants (such as the hydroxyl radical OH or H_2O_2 or O_3), SO_3^{2-} can be oxidized into sulfate ion SO_4^{2-}, which completes the formation of a sulfuric acid drop. The sulfur species $SO_2 \cdot H_2O$, HSO_3^-, and SO_3^{2-} are collectively called S(IV) – the oxidation state 4 – whereas SO_4^{2-} is often labeled as S(VI). Reaction (15.18) is sometimes written as

$$S(\text{IV}) \xrightarrow{\text{oxidation}} S(\text{VI}) \tag{15.19}$$

Each of the above reactions has a reaction rate constant and an equilibrium constant associated with it. The proportion of the equilibrium concentration S(IV) species depends on the pH value of the solution, as shown in Fig. 15.6. As the acidity increases to $pH \leq 5.5$, the proportion of the neutral species $SO_2 \cdot H_2O$ becomes higher and the electrical neutrality in the solution is maintained by $[H^+] = [HSO_3^-]$ only.

However, the chemical reactions alone cannot determine how fast acid raindrops are produced because they contain no information about transport. In order for chemical reactions to occur, the reactants must be transported to a location so that they are in contact with each other. It is well known that the chemical reaction rate depends on the concentrations of the reactants, which, in turn, depends on the transport.

To determine the transport, we will need a transport model similar to the flux model that we developed for aerosol scavenging. There is no need for a trajectory model

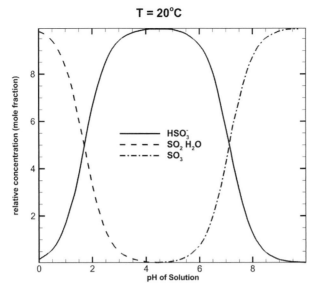

Fig. 15.6 Relative concentration of S(IV) species at equilibrium at 20°C as a function of the pH of the solution. Adapted from Baboolal *et al.* (1981).

since gases are assumed to have no inertia. However, unlike the aerosol particle flux model where we did not consider what happens inside the drop, it is now necessary to consider such processes because the formation of acid occurs inside the drop.

Suppose we have a spherical cloud drop of radius a absorbing SO_2. To understand the transport process, we will need to consider the transport outside and inside the drop separately.

Outside the drop

The situation inside the drop can be described by

$$\frac{\partial C_{SO_2}}{\partial t} = D_{SO_2} \nabla^2 C_{SO_2} - \vec{u} \cdot \nabla C_{SO_2}, \tag{15.20}$$

where C_{SO_2} represents the concentration and D_{SO_2} the diffusivity of SO_2, and $\vec{u}$ is the fall velocity of the drop. If, for the simplicity of discussion, we assume that the drop is stationary and the diffusion is steady state, then (15.20) becomes

$$\nabla^2 C_{SO_2} = 0. \tag{15.21}$$

If the SO_2 concentration is $C_{SO_2,a}$ at the drop surface and $C_{SO_2,\infty}$ far away from the drop, then we get the solution (cf. Eq. (9.9))

$$C_{SO_2} = C_{SO_2,\infty} + (C_{SO_2,a} - C_{SO_2,\infty})\frac{a}{r}. \tag{15.22}$$

The assumption of a constant finite concentration is called the *well-mixed model*. If we assume that the concentration is zero at the surface, then we would obtain the perfect-sink model solution:

$$C_{SO_2} = C_{SO_2,\infty}\left(1 - \frac{a}{r}\right). \tag{15.23}$$

Regardless of the model, it is clear that the concentration would decrease with $1/r$ toward the drop surface. Apparently, (15.22) is more realistic, as it is unlikely that the uptake of SO_2 can be infinite at the drop surface. The surface concentration, in principle, can be determined by Henry's law.

The non-steady convective diffusion equation (15.20) can be solved by the separation of variables method, but we will not go into details here. Interested readers should consult standard books on transport phenomena, such as Bird *et al.* (1960) or Jost (1960).

Interfacial transport

The SO_2 molecules arriving at the drop surface just touch the outer surface of the drop. They are adsorbed but do not automatically and instantaneously flip into the internal drop surface. To achieve that, they have to go through the interfacial transport process, whose details are currently not well understood, but Schwarz (1986) and Lamb and Verlinde (2011) have provided some in-depth discussions on this issue.

Inside the drop

The situation inside the drop is more complicated. Now there are not only transport processes but also chemical reactions occurring in the drop. The diffusion equation with first-order chemical reaction (the reaction rate is proportional to the first order of concentration) for any species of S(IV) is

$$\frac{\partial C_S}{\partial t} = D_S\nabla^2 C_S - kC_S, \tag{15.24}$$

where C_S represents the concentration of any S(IV) species. Remember, though, that this applies only to a stationary drop.

Effect of drop motion

All the above formulations are limited to a stationary drop. In reality, cloud and raindrops are all falling relative to the air, so that the effect of motion has to be taken into account. The ventilation effect in the gas-phase transport can be treated similarly to (15.8). Inside the drop, there will be some internal circulation (see Chapter 8), which will affect the convective transport and hence the reaction rates of S(IV) species.

Walcek and Pruppacher (1984) performed a study on the scavenging of SO_2 by cloud and raindrops and considered both external and internal flow when solving the convective diffusion equations. Fig. 15.7 shows an example of their results. Here a 300 μm radius drop is falling in air containing 10% SO_2. The right-hand side shows the SO_2 concentration distribution outside the drop. We see that the concentration gradients in the upstream side are enhanced by the ventilation effect. The left-hand side shows the concentration distribution inside the drop at different times. It is clear that the internal convective pattern significantly alters the way in which SO_2 is

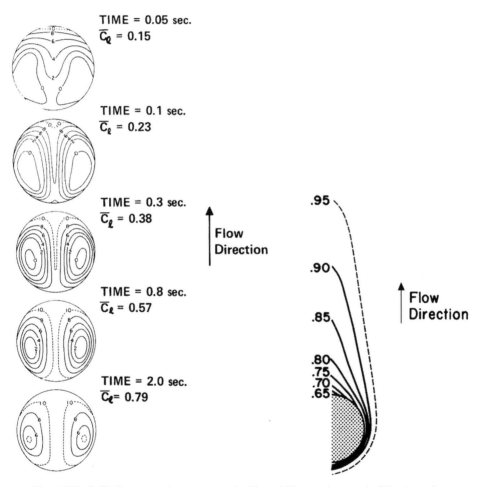

Fig. 15.7 (left) Concentration contours inside a 300 μm drop as it falls through a 10 vol.% mixture of SO_2 in air. Actual concentrations are normalized with respect to $c_{l,sat}$; and $\bar{c}_l$ is the volume-averaged concentration. (right) External SO_2 concentrations outside a 300 μm drop 0.003 s after its exposure to a uniform 10 vol.% mixture of SO_2 in air. Gas concentrations are normalized with respect to $c_{g,\infty}$. From Walcek and Pruppacher (1984). Reproduced by permission of Springer-Verlag.

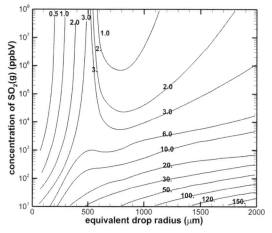

Fig. 15.8 The time (seconds) that it takes a drop falling through an environment of given SO_2 concentration to reach 63% of saturation, as a function of drop size and SO_2 concentration. Adapted from Walcek and Pruppacher (1984).

carried into the drop, if we compare the present results with those for a stagnant drop, which absorbs SO_2 in a radially symmetric pattern. Initially, SO_2 is carried by the internal circulation to the rear of the drop, where it is injected into the interior of the drop by the return circulation pattern. Within about 0.1 s, streamlines and lines of constant concentration become coincident. Afterward, molecular diffusion carries SO_2 perpendicular to the streamlines until the drop is in equilibrium with its environment.

Fig. 15.8 shows the time needed for a drop of certain size falling through an environment of a given SO_2 concentration to reach saturation. In general, small drops falling in an environment of high SO_2 concentration will reach saturation very rapidly. This would be the case for the formation of acid fog in highly polluted areas. On the other hand, large raindrops falling through the SO_2 concentrations typical of the atmosphere may take a distance as much as 1000 m to do so. Information generated by this kind of study can be incorporated into a cloud model to produce more accurate predictions of acid rain formation.

15.3 Radiative impact of cirrus cloud microphysics

One of the biggest current environmental concerns is global climate change, especially the global warming trend that has been observed since the beginning of the twentieth century. The current mainstream theory attributes this warming trend to the increasing amount of CO_2 in the atmosphere due to human industrialization, especially the large-scale burning of fossil fuel for power generation. The trend of increasing CO_2 level in the atmosphere has been observed since the 1950s. Since CO_2 is a well-known greenhouse gas, it seems to be logical to associate the CO_2 concentration in the atmosphere with the global warming because more CO_2 would trap more terrestrial infrared radiation in the atmosphere, causing the warming of the Earth's surface environment.

However, it is also clear that the global climate system is highly nonlinear and there is no guarantee that an initial perturbation of a certain magnitude will yield a response of the system proportional to this magnitude. Indeed, it is even possible to yield a negative response from an initial positive perturbation. Let us present a simple argument as an example. While an increase in the amount of CO_2 will trap more infrared radiation in the atmosphere, causing an initial warming, the warming may ultimately increase the general level of convection, which eventually leads to an increase of cloud cover globally, hence a reduction in the solar radiation that can reach the Earth's surface, and thus may lead to global cooling. This is to say that, in such a scenario, the convectively generated clouds have a negative feedback onto the climate system. The real climate system is highly complicated and consists of many components, and cloud is but one of them. Other components will have their feedbacks, and it will be necessary to know all the feedbacks to come up with an accurate quantitative assessment of the final climatic state.

About the only way that we can hope to comprehend to any degree of accuracy how the climate system responds to a perturbation, or forcing, is to use a realistic climate model and perform numerical simulations. This is the strategy adopted by the Intergovernmental Panel on Climate Change (IPCC) of the United Nations for studying the possible climate change induced by CO_2 increase. Needless to say, such models are highly complicated and necessarily numerical in nature. Many physical processes have to be implemented in the model in highly parameterized forms to render solutions possible. How realistic these parameterizations are will certainly impact on the accuracy of the simulation results. Highly simplified parameterizations are likely to lead to large errors that render the interpretation of the results uncertain. The IPCC relies not just on one model but on an ensemble of climate models, and all of them contain parameterizations of various kinds. This is to say that the simulation results of these climate models all contain uncertainties, and these uncertainties need to be understood clearly.

According to the IPCC's *Climate Change 2007: The Physical Science Basis, Contribution of Working Group I to the Fourth Assessment Report of the Intergovernmental Panel on Climate Change* (often called IPCC AR4 WG1; edited by Solomon *et al.*, 2007), the largest uncertainty in the radiative forcing terms of the simulation results lies in the "cloud albedo effect" (see Fig. 15.9). This is largely related to the so-called "indirect aerosol effect", namely, the nucleation capability of aerosol particles to form clouds.

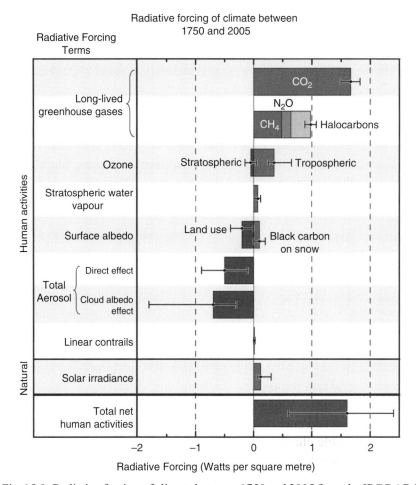

Fig. 15.9 Radiative forcing of climate between 1750 and 2005 from the IPCC AR4
WG1 report (Solomon *et al.*, 2007).

More recently, Lohmann *et al.* (2010) argued that a quantification of the aerosol
forcings in the traditional way is difficult to define properly. They suggested that
fast feedbacks should be included because they act quickly compared with the
time scale of global warming. They showed that for different forcing agents
(aerosols and greenhouse gases) the radiative forcings as traditionally defined
agree rather well with estimates from a method called the radiative flux perturbations (RFP), which takes these fast feedbacks and interactions into account.

It is not easy to resolve this uncertainty, and more research is still needed to
understand the impact of clouds on the radiative forcing. It is beyond the scope of
this book to discuss the radiative properties of clouds, but readers can consult
standard textbooks in atmospheric radiation for such information (e.g. Liou,
1992). It is known that the radiative forcing varies with different cloud types, height,

and particle composition. For example, it is generally thought that low clouds such as stratocumulus clouds, which often cover wide areas over the ocean, exert a cooling effect on the climate system because their main radiative effect is to reflect sunlight back into space. On the other hand, high cirrus clouds are generally considered to have a net warming effect because they can absorb long-wave radiation from the surface that is emitted at higher blackbody temperature but re-radiate some of the absorbed energy into space at a colder blackbody temperature.

But even among the same type of clouds, there can be large differences in their radiative properties. To illustrate this point, we will utilize some simulation results by Liu *et al.* (2003b), who used the cirrus model discussed in Chapter 13 to perform a series of sensitivity studies to understand the impact of cloud environment and particle composition on the cloud radiative properties.

Fig. 15.10 shows the time evolution of the heating profiles of cirrus cloud developed in the warm unstable environment as described in Chapter 13. The profiles of solar heating are plotted in Fig. 15.10(a). The solar heating rate is a sensitive function of ice water content and number concentration. In general, the solar heating becomes stronger as the ice water content and number concentration increase. The solar heating rate gradually increases upward from cloud base to a maximum slightly below the cloud top, then decreases sharply from there to the cloud top. Clearly, the solar heating is most pronounced high in the cloud layer.

The profiles of the IR heating, shown in Fig. 15.10(b), change dramatically with time. The IR heating rate is significant throughout the cloud layer, and it is very sensitive to the content of the cloud. In general, its profiles show warming in the lower part of the cloud and cooling in the upper part. Consequently, there is differential heating within the cloud deck, because the cirrus in this case is optically thick enough that most of the radiation from below is absorbed before it reaches the cloud top. Thus, the IR warming decreases with height and eventually becomes cooling. This differential IR heating is the strongest at around 40 min. As the profiles of ice water content and number concentration change with time, the main radiative cooling and warming layers also evolve accordingly. The magnitude of the maximum IR warming decreases as the IR warming maximum moves downward after 40 min. In contrast to the maximum warming in the lower cloud layer, the upper maximum IR cooling does not change its elevation significantly, yet undergoes obvious change in magnitude. This implies that the radiation-induced upward motion is reduced as ice crystals fall out of the initial supersaturated layer. During the later stages of cloud development, the peak IR warming in the lower part of the cloud wanes but the peak IR cooling in the upper part remains strong, as the solar heating systematically decreases with time.

Fig. 15.10(c) shows the evolution of latent heating profiles. The latent heat absorption due to sublimation is most pronounced just above the cloud base, and

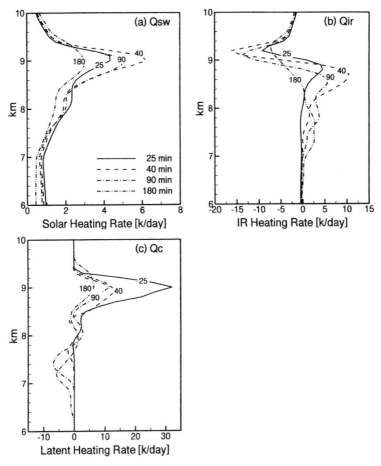

Fig. 15.10 Profiles of horizontally averaged (a) solar radiative heating rate Q_{sw}, (b) IR heating rate Q_{ir}, and (c) latent heating rate Q_c for the warm unstable case. From Liu *et al.* (2003b). Reproduced by permission of the American Meteorological Society.

its magnitude exceeds the radiative heating rate there, so that the diabatic warming near the cloud base is replaced by diabatic cooling during this stage. There is still diabatic warming in the middle of the cloud, but it becomes much weaker, while the diabatic cooling near the cloud top becomes smaller. The profile of latent heating indicates that heterogeneous nucleation and diffusional growth still take place near the cloud top, and that the upper part of the cloud layer remains quite moist by the end of the simulation.

To contrast the cirrus clouds formed in the warm unstable environment, Fig. 15.11 shows the corresponding heating profiles of cirrus developed in the cold unstable environment. In general, the solar heating rate (Fig. 15.11a) gradually increases upward from the cloud base to slightly below the cloud top, and then

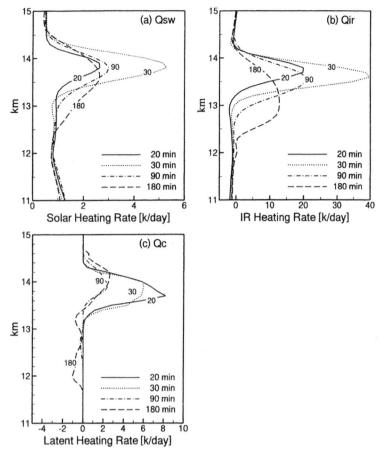

Fig. 15.11 Profiles of horizontally averaged (a) solar radiative heating rates Q_{sw}, (b) IR heating rates Q_{ir}, and (c) latent heating rates Q_c, for cold unstable case. From Liu *et al.* (2003b). Reproduced by permission of the American Meteorological Society.

decreases sharply up to the cloud top. It is apparently sensitive to the number concentration profile, reaches a maximum around 30 min, and is concentrated in the upper cloud layer. The IR heating profiles are shown in Fig. 15.11(b). The IR heating increases upward to a maximum value near 13.5 km and then decreases toward the cloud top. The IR cooling at the cloud top is negligible, whereas the IR warming is significant throughout the whole cloud layer. This is because the simulated cirrus is optically thin, so that the radiation from the lower and warmer layer can penetrate to the cloud top. Thus, the cloud layer is radiatively warmed during the simulation. The IR warming rates in the cloud layer are much larger than the solar warming rates, reaching as much as $40°C$ day^{-1}.

The profiles of latent heating rates, shown in Fig. 15.11(c), have much smaller amplitudes than the IR heating rates. This is caused by the small particle growth rates at cold temperatures. Subcloud sublimation is insignificant until 90 min into the simulation. The latent and solar heating rates are also much smaller than the IR heating rate throughout the simulation. The solar heating is smaller than the latent heating rate until 20 min, but by 30 min the two become comparable. The profile of latent heating at 180 min indicates that ice nucleation and diffusional growth processes still take place near the cloud top, and also implies that the upper part of the cloud remains quite moist at the end of the simulation.

These two examples demonstrate that cirrus clouds developed in different atmospheric environments exhibit quite different radiative properties. Liu *et al.* (2003b) also show the profiles for warm stable and cold stable cirrus cases, and their properties are again quite different.

The above results are obtained under the assumption that ice crystals in the cirrus are hexagonal columns. It is well known that the shape (habit) of ice crystals has great impact on their interaction with radiation, so it is possible that cirrus clouds consisting of different types of ice crystals may exhibit very different radiative properties. To illustrate this point, Fig. 15.12 shows the time evolution of IR heating profiles of cirrus clouds developed under warm stable (four left panels) and cold stable (four right panels). The soundings used for warm stable and cold stable environmental conditions are the same as that of Fig. 13.14 except that the unstable layer is changed into a stable condition. In each type of environment, the developments of four different cirrus clouds are simulated; each is assumed to consist of a specific habit of ice crystals. These four ice habits are hexagonal column, hexagonal plate, spheres, and bullet rosette. We see not only that cirrus clouds formed in different atmospheric environments exhibit different radiative properties but also that, even in the same environment, clouds consisting of different habits of ice crystals also interact with radiation in very different ways. The largest contrast is between the heating profiles of rosettes and spheres. Taking the warm stable case, for example, the peak heating for cirrus of rosettes is ~ 14°C day^{-1} while that for cirrus of spheres is less than 1°C day^{-1}! The same can be said for the cold stable case.

The heating profiles due to short-wave radiation also exhibit substantial differences, although the contrasts are not as drastic as for IR.

To address this kind of uncertainty, it will be necessary to perform observations of the global distributions of cirrus clouds and investigate their microphysical contents (particle types, habits, concentrations, etc.), then use good cirrus models with radiative interaction physics to assess the uncertainties involved. Uncertainties associated with other types of clouds can be assessed in a similar way.

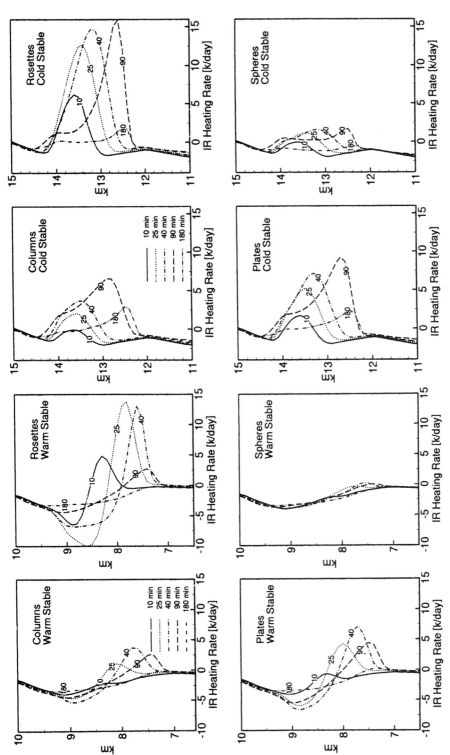

Fig. 15.12 Profiles of horizontally averaged IR heating rates for cirrus clouds that consist of four different ice crystal types. The two columns on the left are for the warm stable case, whereas the two columns on the right are for the cold stable case. From Liu *et al.* (2003b). Reproduced by permission of the American Meteorological Society.

There are many factors that contribute to the uncertainties mentioned in IPCC AR4 WG1, but the above discussions should make it clear that the aerosol–cloud factor can indeed generate very large uncertainties in climate predictions using models.

15.4 Impact of deep convective clouds on the stratosphere–troposphere exchange

So, we have seen above that the clouds exert a radiative forcing on the climate system, but this is envisioned as due to the change in the aerosol system. The aerosol factor is strongly tied with human industrial activity, which causes major changes in the physicochemical nature and concentration of the aerosol particles, which, in turn, cause changes in the clouds. In this scenario, the major ingredient of the clouds, water vapor, is assumed to play a passive role – it merely responds to environmental changes but is not itself a climate forcing. Even if the Earth–atmosphere system warms up, causing the water vapor concentration to increase due to evaporation, precipitation is assumed to increase so as to exactly balance out the additional evaporation on an annual basis. In other words, it is assumed that there is a global hydrological equilibrium.

But some recent observations indicate that this may not be entirely true. These observations are related to the fluctuations of water vapor in the stratosphere. Even though the stratosphere is very dry, with a typical water vapor concentration of ~ 3 ppmv (parts per million by volume), it has been observed that this concentration can fluctuate by as much as 50% on a decadal time scale. Fig. 15.13 shows the water vapor concentration in the lower stratosphere as measured over the past few decades by several different observational platforms, including a balloon-borne frostpoint

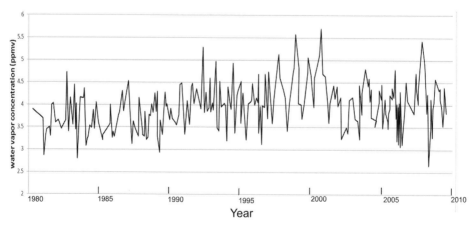

Fig. 15.13 Fluctuation of water vapor concentration in the lower stratosphere in the period 1980–2010. Data derived from Solomon *et al.* (2010).

hygrometer over Boulder, Colorado, the AURA MLS, UARS HALOE, and SAGE II satellites. The frostpoint hygrometer data cover the longest time period, but only at one location, whereas the satellite data have much wider spatial coverage.

It is seen that for the period from 1980 to 2000 there was an increasing trend peaking at 2000 with a concentration about 4.5 ppmv. Thereafter, the trend is decreasing. At present, it is not known why there is such a fluctuation. But the magnitude of the fluctuation can be as much as 50% between the mean maximum and minimum. Even though water vapor has very low concentration in the stratosphere, it is a strong greenhouse gas and, because of its altitude, it can exert strong radiative forcing there. Thus, the problem of changing vapor concentration in the stratosphere and its possible impact on the climate system should be assessed carefully (Forster and Shine, 2002). Although the residence time of water vapor in the stratosphere is still being studied, it is generally thought to be longer than a year. In this case, the changing vapor concentration in the stratosphere should be considered as an important climatic forcing.

This leads to the question of how water vapor enters the stratosphere. It is agreed by most researchers that the major water vapor source is the evaporation of water (liquid or solid) over the Earth's surface. Thus it is logical to consider that the vapor in the stratosphere comes from the troposphere through some vertical transport process. Holton *et al.* (1995) proposed a schematic of the global water vapor cycle, including the stratosphere, as shown in Fig. 15.14. In this scheme, water vapor is transported from the surface by large-scale ascent and some overshooting tropical deep convective clouds (often called "hot towers") that penetrate the tropopause (as represented by the 380 K isentropic surface). Then the vapor is pumped toward middle and high latitudes by the Rossby wave breaking labeled as "wave-driven extratropical pump" in the chart. There the large-scale descent brings the vapor downward into a "middle world" (the dark shaded region below the 380 K surface) and via isentropic mixing transports water vapor back into the troposphere. In principle, this scheme should also work for other trace chemicals as well.

Recent observations show that the actual situation appears to be much more complicated, and the details are being studied. On the one hand, it has been reported that tropical hot towers rarely exceed 14 km, which is about 3 km short of the cold-point tropical tropopause at 17 km, so exactly how the hot towers penetrate the tropopause to transport water vapor has been debated (see e.g. Highwood and Hoskins, 1998; Küpper *et al.*, 2004). The layer between ~14 and ~17 km is designated as the "tropical tropopause layer" (TTL), which is currently being studied actively. On the other hand, the deep convective clouds in mid-latitudes have been observed to penetrate through the local tropopause, indicating that the mid-latitudes do not simply behave as a receiving end of the transport. It is this latter point that we will describe in some detail in the following.

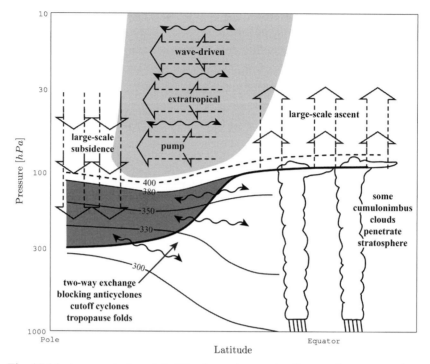

Fig. 15.14 A conceptual model of the dynamic aspects of stratosphere–troposphere exchange. Adapted from Holton *et al.* (1995).

Careful studies of the satellite images of some severe thunderstorms show that many of them have one or more chimney-plume-like cloud features above the anvils of these storms. For mid-latitude storms of such intensity, their anvils are usually at about the same level as the local tropopause. Thus these above-anvil cloud features indicate that they are in the stratosphere. Some of them have been estimated to be about 3 km above the anvil, so that they are located in the lower stratosphere. Setvák and Doswell (1991) and Levizzani and Setvák (1996) were the first to report these features definitively. Fig. 15.15 shows two examples of these above-anvil plumes.

These plumes were once thought to be a rarity, but later studies have shown that they are quite common and occur for many storms over various continents and oceans. Later, it was realized that many of them had also been recorded by ground-based observations but this was not recognized as such at the time (see Wang *et al.*, 2009).

These features are obviously clouds and, since the temperatures at the storm top are usually very cold, they must be a kind of cirrus cloud, i.e. they consist of ice particles. Since the stratosphere is usually very dry, it should not have enough water vapor to form these features. So where does the moisture come from? Wang (2003) used the simulated CCOPE supercell storm results (see Chapter 13) to demonstrate

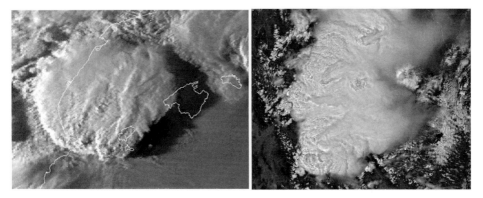

Fig. 15.15 (left) NOAA-12 AVHRR band 1 image, at 17:24 UTC on 11 September 1996, of a severe tornadic storm system occurring over Ibiza in the Balearic Islands of Spain (39°30'N 3°00'E). Image courtesy of Martin Setvák. Data source: NOAA/ CHMI. (right) MODIS Aqua band 1 image of a storm system in northern Mexico, at 20:05 UTC on 7 May 2007, showing multiple extensive above-anvil plumes. Image courtesy of Martin Setvák. Data source: NASA/LAADS. From Wang *et al.* (2011).

that the moisture comes from the storm itself via a mechanism called *storm-top gravity wave breaking*. This mechanism is explained in the following.

During a vigorous convection process such as thunderstorms, the local tropopause is strongly excited by the convective energy and is no longer a horizontal layer as during calm times. Instead, internal gravity waves are generated near the cloud top and propagate upwards into the stratosphere. A special characteristic of the internal gravity waves propagating in a stratified fluid like the atmosphere is that its group velocity (and hence the wave energy) propagates in a direction perpendicular to its phase. For a brief description of the internal gravity waves, see Holton (2004); for a more detailed discussion, see Pedlosky (2003).

The tropopause is generally regarded as an isentropic surface. Since normal propagation of internal gravity waves is adiabatic, it does not cause materials (such as air or water vapor) to be transported across an isentropic surface such as the tropopause. Thus, although the gravity waves excited at the storm top make the local tropopause wavy, there is no net cross-tropopause material transport occurring. The gravity waves merely distort the tropopause shape. In fact, under this adiabatic situation, even the storm overshooting phenomenon should be regarded as a temporary distortion of the tropopause, and so the sighting of an overshooting top does not necessarily indicate net transport of material into the stratosphere.

But under certain conditions, which are related to the stratification and wind shear, wave breaking can occur. This is often associated with the building up of a critical layer near the cloud top when the wind speed at that level and the wave speed become the same. Under this condition, the only mechanism by which the wave can dissipate energy is via wave breaking. When this occurs, materials can be transported across isentropic surfaces because wave breaking is not adiabatic and is irreversible in principle. The cloud model study shows that it is precisely the storm top gravity wave breaking that causes water vapor to be injected into the stratosphere and forms the above-anvil cirrus plumes.

Fig. 15.16 shows two frames of the central x–z cross-section of the simulated CCOPE storm as performed by Wang *et al.* (2010), who re-ran the WISCDYMM using a higher spatial resolution at 0.5 km × 0.5 km × 0.2 km.

Fig. 15.16(a) shows that wave breaking is occurring at $x \sim 45$ km, which is about 10 km behind the overshooting top, and a large patch of moisture (indicated by the high RH_i region) penetrates the isentropic surface to reach a level of $z \sim 14.2$ km. Careful inspection of the θ_e contours shows that the contours associated with this patch exhibit the "turning-over" signature of breaking waves, indicating that the breaking is

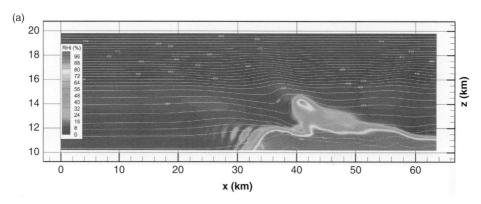

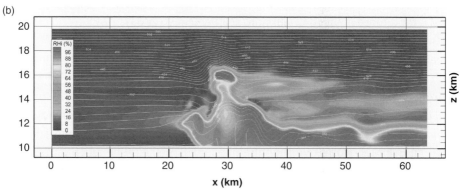

Fig. 15.16 Two snapshots of the gravity wave breaking process at the cloud top of the simulated CCOPE supercell (see Chapter 13). (a) Wave breaking in the anvil sheet. (b) Wave breaking in the overshooting area. Grayscale/color: RH_i. White contours: θ_e. See also color plates section.

due to dynamics and is not a numerical artifact. The patch would eventually become detached from the anvil and form an above-anvil plume. Fig. 15.16(b) shows another frame of a wave breaking event. This time the breaking occurs right at the over-shooting top and the θ_e contours also show a turning-over signature.

This phenomenon is now known as the "jumping cirrus", a term coined by Fujita (1982), who was the first to describe it based on his aircraft observations of severe storms. Fujita (1982) described what he saw about the jumping cirrus as follows:

One of the most striking features seen repeatedly above the anvil top is the formation of cirrus cloud which jumps upward from behind the overshooting dome as it collapses violently into the anvil cloud.

Fig. 15.17 shows the modeled sequence of the breaking at the top of the CCOPE supercell that fits precisely Fujita's statement. In this figure, the white arrow

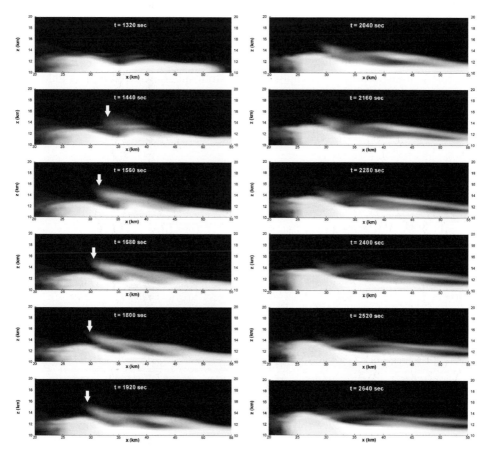

Fig. 15.17 Snapshots of the RH_i profiles in the central east–west cross-section of the simulated storm from $t = 1320$ s to 2640 s. From Wang (2004). Reproduced by permission of the American Geophysical Union.

indicates the front tip of the jumping cirrus. It is seen that, as time proceeds, the tip of the cirrus appears to go upstream relative to the storm (and remember that the storm is moving to the east, i.e. the $+x$ direction, at the same time) while it is jumping to a higher level. Both the vertical and horizontal speeds have a magnitude of about 10 m s^{-1}, indeed deserving the adjective "jumping".

Eventually the jump stops and the cirrus retreats, as seen in the right-hand side panels. At the same time the "tail end" of the cirrus layer becomes higher and detached from the anvil, forming an above-anvil plume. There are now videos showing the jumping cirrus phenomenon and some of them are available in Wang *et al.* (2009).

More recently, satellite vertical cross-sectional images from CloudSat and CALIPSO provide more direct evidence of the anvil top plumes as an active mechanism of injecting moisture into the stratosphere. Fig. 15.18 shows such an example. This is a severe storm system that occurred over eastern Angola. The overshooting top of the storm is indicated by the gray/red arrow. The jumping cirrus was downwind of the overshooting top and indicated by the black arrow. Note that this event occurred at latitude about 12°S, which is well within the tropical belt.

The above discussions demonstrate that deep convective storms play an important role in transporting water vapor into the stratosphere. Currently, it is unknown what percentage this mechanism contributes to the stratospheric water vapor, as no reliable global estimates exist.

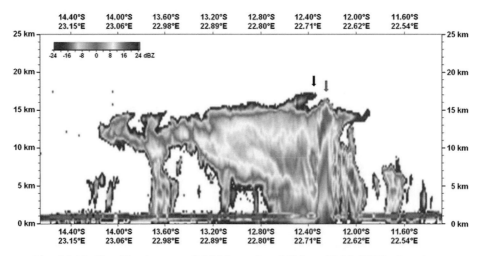

Fig. 15.18 CloudSat image, of 14 November 2009 at 12:20 UTC, showing a jumping cirrus occurrence (black arrow) at the top of a deep convective storm system in the Southern Hemisphere over eastern Angola in Southern/Central Africa. The overshooting top of the storm is indicated by the red arrow. The jumping cirrus was downwind of the overshooting top. Adapted from Wang *et al.* (2011). Data source: NASA/CloudSat. See also color plates section.

In addition to being a possible climatic forcing, water vapor in the stratosphere can also play a role in regulating the ozone concentration, because H_2O is the source material for making odd hydrogen species HO_x, which is known to cause ozone depletion in the stratosphere via some catalytic cycles.

15.5 Artificial weather modifications

One of the first motivations for studying cloud physics (including cloud dynamics) is to influence weather processes using artificial means. Although, strictly speaking, artificial weather modifications are not limited to modifying clouds, the modification of clouds represents the majority of such activities. Also, the modification of clouds mostly involves *cloud seeding*. This section will provide a brief description of the main ideas involved in such activities.

There are various reasons for performing cloud seeding, but the two main motivations are (1) to increase rainfall and (2) to alleviate severe weather. It is easy to understand the desire to increase rainfall because agricultural production is closely related to the amount of rain available during the growing seasons, and hence rain-making, if successful, can be highly beneficial to productivity. The alleviation of severe weather is also closely related to agricultural activity. One notable example is the attempt to suppress severe hail fall, which annually causes crop damage of multi-billions of dollars. Of course, hail can also damage houses, cars, and other property, so there are other sectors of society (such as the insurance industry) that are interested in hail alleviation if it can be achieved. Both rain-making and hail suppression involve seeding clouds.

Aside from these two main motivations, other minor (as yet) motivations include fog dissipation (e.g. in airports) and rain suppression (e.g. for major outdoor events such as the Olympic Games or the Presidential Inauguration Ceremony).

Cloud seeding can be divided into two categories: (1) warm cloud seeding and (2) cold cloud seeding:

- *Warm cloud seeding*. If the cloud temperature is relatively high, so that only the warm rain process can be expected to occur, then warm cloud seeding is implemented. For a given amount of water vapor available, one way to increase either the number concentration or the size of drops in the cloud is by seeding it with hygroscopic nuclei such as salt. It is hoped that the salt nuclei will produce not only more but larger drops. As we have seen in Chapter 11, the presence of large drops initially could stimulate the cloud to decrease its colloidal stability and increase the chance of rain.
- *Cold cloud seeding*. If the presence of mixed-phase clouds is expected, then a cold cloud seeding procedure can be implemented. As we have discussed in

Chapter 7, nature is relatively short of ice nuclei (IN) as compared to cloud condensation nuclei (CCN), and hence it is possible that human injection of artificial IN may cause significant changes in cloud precipitation probability. The IN that are most widely used are dry ice (solid CO_2) and silver iodide (AgI). Dry ice is broken into small pieces and seeded into clouds usually by aircraft. AgI, on the other hand, is dispersed either by incinerators located on the ground and allowing the AgI smoke to go into the cloud with updraft, or by incinerators carried by aircraft that disperse the smoke directly into the cloud. The hope is that this will increase the concentration of ice crystals and allow the Bergeron–Findeisen process to play out for increasing rainfall.

In the case of hail suppression, the main idea is to *overseed* the cloud with nuclei so that hopefully the large number but small size ice particles will form and hence alleviate the formation of large hailstones.

While some success has been claimed for such activities, there is as yet no overall acceptance by the scientific community that artificial weather modification leads to positive results. This is especially the case regarding the long-term effectiveness of rain-making, as one needs to have valid control cases to serve as a reference point to judge whether or not certain techniques indeed produce an increase in rainfall in the long run, and those control cases are very difficult to obtain. Consequently, the effectiveness of artificial weather modification is still somewhat controversial to date.

Problems

15.1. The collection kernel (15.12) of the flux model is obtained by calculating the flux of aerosol particles toward the surface of the drop, but the collection kernel of the trajectory model is obtained by calculating the collision cross-section of a particle. The two models are indeed very different, yet the computed K values can match very well. This implies that the two definitions of collection kernel are indeed equivalent. Can you make a physical argument why this should be so?

15.2. Assume that the boundary conditions of (15.9) are applicable for a drop with radius $a = 1$ and far-field aerosol concentration $n_\infty = 1$ (disregard the units). Based on this information, derive an explicit expression for the collection kernel K in (15.12).

15.3. Do the same as in problem 15.2 but for an ice crystal with capacitance C (refer to the electrostatic analogy in Chapter 9) collecting aerosol particles to derive an expression for the collection kernel.

References

Ackerman, S. A., and G. L. Stephens (1987) The absorption of solar radiation by cloud droplets: an application of anomalous diffraction theory. *J. Atmos. Sci.*, 44, 1574–1588.

Anderson, J. K., K. Droegemier, and R. B. Wilhelmson (1985) Simulation of the thunderstorm subcloud environment. *Preprints, 14th Conf. on Severe Local Storms, Indianapolis, IN*. Boston, MA: American Meteorological Society, pp. 147–150.

Andsager, K., K. V. Beard, and N. F. Laird (1999) Laboratory measurements of axis ratios for large raindrops. *J. Atmos. Sci.*, 56, 2673–2683.

Auer, A. H., Jr., and J. D. Marwitz (1972) Hail in the vicinity of organized updrafts. *J. Appl. Meteorol.*, 11, 748–752.

Auer, A. H., Jr., and D. L. Veal (1970) The dimension of ice crystals in natural clouds. *J. Atmos. Sci.*, 27, 919–926.

Baboolal, L. A., H. R. Pruppacher, and J. H. Topalian (1981) A sensitivity study of a theoretical model of SO_2 scavenging by water drops in air. *J. Atmos. Sci.*, 38, 856–870.

Bailey, M., and J. Hallett (2004) Growth rates and habits of ice crystals between $-20°$ and $-70°C$. *J. Atmos. Sci.*, 61, 514–544.

Bailey, M., and J. Hallett (2009) A comprehensive habit diagram for atmospheric ice crystals: confirmation from the laboratory, AIRS II, and other field studies. *J. Atmos. Sci.*, 66, 2888–2899.

Baker, M. B., H. J. Christian, and J. Latham (1995) A computational study of the relationships linking lightning frequency and other thundercloud parameters. *Q. J. R. Meteorol. Soc.*, 121, 1525–1548.

Beard, K. V. (1976) Terminal velocity and shape of cloud and precipitation drops aloft. *J. Atmos. Sci.*, 33, 851–864.

Beard, K. V., and C. H. Chuang (1987) A new model for the equilibrium shapes of raindrops. *J. Atmos. Sci.*, 44, 1509–1524.

Beard, K. V., and R. J. Kubesh (1991) Laboratory measurements of small raindrop distortion. Part 2: Oscillation frequencies and modes. *J. Atmos. Sci.*, 48, 2245–2264.

Beard, K. V., and H. T. Ochs (1983) Measured collection efficiencies for cloud drops. *J. Atmos. Sci.*, 40, 146–153.

Beard, K. V., and H. T. Ochs, III (1993) Warm-rain initiation: an overview of microphysical mechanisms. *J. Appl. Meteorol.*, 32, 608–625.

Beard, K. V., and H. T. Ochs (1995) Collisions between small precipitation drops. Part II: Formulas for coalescence, temporary coalescence, and satellites. *J. Atmos. Sci.*, 52, 3977–3996.

Beard, K. V., and H. R. Pruppacher (1969) A determination of the terminal velocity and drag of small water drops by means of a wind tunnel. *J. Atmos. Sci.*, 26, 1066–1072.

Beard, K. V., and H. R. Pruppacher (1971) A wind tunnel investigation of the rate of evaporation of small water drops falling at terminal velocity in air. *J. Atmos. Sci.*, 28, 1455–1464.

Beard, K. V., D. B. Johnson, and D. Baumgardner (1986) Aircraft observations of large raindrops in warm, shallow convective clouds. *Geophys. Res. Lett.*, 13, 991–994.

Beard, K. V., H. T. Ochs, and R. J. Kubesh (1989a) Natural oscillations of small raindrops. *Nature*, 342, 408–410.

Beard, K. V., J. Q. Feng, and C. Chuang (1989b) A simple perturbation model for the electrostatic shape of falling drops. *J. Atmos. Sci.*, 46, 2404–2418.

Beard, K. V., R. J. Kubesh, and H. T. Ochs (1991) Laboratory measurements of small raindrop distortion. Part 1: Axis ratio and fall behavior. *J. Atmos. Sci.*, 48, 698–710.

Beard, K. V., V. N. Bringi, and M. Thurai (2010) A new understanding of raindrop shape. *Atmos. Res.*, 97, 396–415.

Berry, E. X. (1968) Modification of the warm rain process. *Proc. First Natl. Conf. Weather Modification, State University of New York, Albany.* Boston, MA: American Meteorological Society, pp. 81–88.

Berry, E. X., and R. L. Reinhardt (1974) An analysis of cloud drop growth by collection. Part II. Single initial distributions. *J. Atmos. Sci.*, 31, 2127–2135.

Bigg, E. K., and S. C. Hopwood (1963) Ice nuclei in the Antarctic. *J. Atmos. Sci.*, 20, 185–188.

Bird, R. B., W. E. Stewart, and E. N. Lightfoot (1960) *Transport Phenomena.* New York, NY: Wiley.

Blanchard, D. C. (1969) The oceanic production rate of cloud nuclei. *J. Rech. Atmos.*, 4, 1–6.

von Blohn, N., K. Diehl, S. K. Mitra, and S. Borrmann (2009) Riming of graupel: wind tunnel investigations of collection kernels and growth regimes. *J. Atmos. Sci.*, 66, 2359–2366.

Blyth, A. M., W. A. Cooper, and J. B. Jensen (1988) A study of the source of entrained air in Montana cumuli. *J. Atmos. Sci.*, 45, 3944–3964.

Borovikov, A. M., I. I. Gaivoronskii, E. G. Zak, *et al.* (1963) *Cloud Physics.* Jerusalem, Israel: Israel Program of Scientific Translations.

Borrmann, S., R. Jaenicke, and P. Neumann (1993) On spatial distributions and inter-droplet distances measured in stratus clouds with in-line holography. *Atmos. Res.*, 29, 229–245.

Bowen, E. G. (1950) The formation of rain by coalescence. *Austral. J. Sci. Res.*, 3, 193–213.

Braham, R. R. (1974) Cloud physics of urban weather modification: a preliminary report. *Bull. Am. Meteorol. Soc.*, 55, 100–105.

Braham, R., Jr. (1990) Snow particle size spectra in lake effect snows. *J. Appl. Meteorol.*, 29, 200–207.

Brandes, E. A., G. Zhang, and J. Sun (2006) On the influence of assumed drop size distribution form on radar-retrieved thunderstorm microphysics. *J. Appl. Meteorol. Climatol.*, 45, 259–268.

Bringi, V. N., and V. Chandrasekar (2001) *Polarimetric Doppler Weather Radar: Principles and Applications.* Cambridge, UK: Cambridge University Press.

Brook, M. (1958) Studies of charge separation during ice–ice contact. *Recent Advances in Thunderstorm Electricity.* New York, NY: Pergamon, pp. 383–390.

Brown, P. S. (1988) The effects of filament, sheet, and disk breakup upon the drop spectrum. *J. Atmos. Sci.*, 45, 712–718.

Bruntjes, R. T., A. J. Heymsfield, and T. W. Krauss (1987) An examination of double-plate ice crystals and the initiation of precipitation in continental cumulus clouds. *J. Atmos. Sci.*, 44, 1331–1349.

Byers, H. R., and R. R. Braham, Jr. (1949) *The Thunderstorm: Final Report of the Thunderstorm Project*. Washington, DC: US Government Printing Office.

Cameron, A. G. W. (1981) In *Essays in Nuclear Astrophysics*, ed. C. A. Barnas, D. D. Clayton, and D. N. Schramm. Cambridge, UK: Cambridge University Press.

Carrier, G. F. (1953) *On Slow Viscous Flow*, Final Report, Office of Naval Research, Contract No. 653–00/1.

Castro, A., J. L. Marcos, J. Dessens, J. L. Sanchez, and R. Fraile (2004) Concentration of ice nuclei in continental and maritime air masses in Leon (Spain). *Atmos. Res.*, 47–48, 155–167.

Chagnon, C. W., and C. E. Junge (1961) The vertical distribution of submicron particles in the stratosphere. *J. Meteorol.*, 18, 746–752.

Chen, C. J., and P. K. Wang (2009) Diffusion growth of solid and hollow hexagonal ice columns. *Nuovo Cimento*, 124, 87–97.

Chen, J. P. (1994) Theory of deliquescence and modified Köhler curves. *J. Atmos. Sci.*, 51, 3505–3516.

Cheng, L., and M. English (1983) A relationship between hailstone concentration and size. *J. Atmos. Sci.*, 40, 204–213.

Chin, E. H. C., and M. Neiburger (1972) A numerical simulation of the gravitational coagulation process for cloud droplets. *J. Atmos. Sci.*, 29, 718–727.

Chiruta, M., and P. K. Wang (2003) On the capacitance of bullet rosette crystals. *J. Atmos. Sci.*, 60, 836–846.

Chiruta, M., and P. K. Wang (2005) The capacitance of solid and hollow hexagonal ice columns. *Geophys. Res. Lett.*, 32, L05803.

Chiu, C. S., and J. D. Klett (1976) Convective electrification of clouds. *J. Geophys. Res.*, 81, 1111–1124.

Chuang, C. C., and K. V. Beard (1990) A numerical model for the equilibrium shape of electrified raindrops. *J. Atmos. Sci.*, 47, 1374–1389.

Clough, S. A., Y. Beers, G. P. Klein, and L. S. Rothman (1973) Dipole moment of water from Stark measurements of H_2O, HDO, and D_2O. *J. Chem. Phys.*, 59, 2254–2259.

Cober, S. G., and R. List (1993) Measurements of the heat and mass transfer parameters characterizing conical graupel growth. *J. Atmos. Sci.*, 50, 1591–1609.

Connolly, P. J., C. Emersic, and P. R. Field (2012) A laboratory investigation into the aggregation efficiency of small ice crystals. *Atmos. Chem. Phys.*, 12, 2055–2076.

Cotton, W. R., and R. A. Anthes (1989) *Storm and Cloud Dynamics*. San Diego, CA: Academic Press.

Crutzen, P. J. (1976) The possible importance of CSO for the sulfate layer of the stratosphere. *Geophys. Res. Lett.*, 3(2), 73–76.

Curtius, J., R. Weigel, H.-J. Vössing, *et al.* (2005) Observations of meteoric material and implications for aerosol nucleation in the winter Arctic lower stratosphere derived from in situ particle measurements. *Atmos. Chem. Phys.*, 5, 3053–3069.

Czys, R. R. (1994) Preliminary laboratory results on the coalescence of small precipitation-size drops falling freely in a refrigerated environment. *J. Atmos. Sci.*, 51, 3209–3218.

Czys, R. R., and H. T. Ochs (1988) The influence of charge on the coalescence of water drops in free fall. *J. Atmos. Sci.*, 45, 3161–3168.

Damiani, R., G. Vali, and S. Haimov (2006) The structure of thermals in cumulus from airborne dual-Doppler radar observations. *J. Atmos. Sci.*, 63, 1432–1450.

Davies, C. N. (1945) Definitive equations for the fluid resistance of spheres. *Proc. Phys. Soc.*, 57, 259–270.

Dinger, J. E., and R. Gunn (1946) Electrical effects associated with a change of state of water. *Terr. Magn. Atmos. Electr.*, 51, 477–494.

Döppenschmidt, A., and H.-J. Butt (2000) Measuring the thickness of the liquid-like layer on ice surfaces with atomic force microscopy. *Langmuir*, 16, 6709–6714.

Dosch, H., A. Lied, and J. H. Bilgram (1995) Glancing-angle X-ray scattering studies of the premelting of ice surfaces. *Surf. Sci.*, 327, 145–164.

Doswell, C. A. (Ed.) (2001) *Severe Convective Storms*. Boston, MA: American Meteorological Society.

Drake, J. C., and B. J. Mason (1966) The melting of small ice spheres and cones. *Q. J. R. Meteorol. Soc.*, 92, 500–509.

Dufour, L., and R. Defay (1963) *Thermodynamics of Clouds*. New York, NY: Academic Press.

Dusek, U., G. P. Frank, L. Hildebrandt, *et al.* (2006) Size matters more than chemistry for cloud-nucleating ability of aerosol particles. *Science*, 312, 1375–1378.

Einstein, A. (1905) Über die von der molekularkinetischen Theorie der Wärme geforderte Bewegung von in ruhenden Flüssigkeiten suspendierten Teilchen. *Ann. Physik*, 17, 549–560.

Feingold, G., and Z. Levin (1986) The lognormal fit to raindrop spectra from frontal convective clouds in Israel. *J. Clim. Appl. Meteorol.*, 25, 1346–1363.

Few, A. A., A. J. Dessler, D. J. Latham, and M. Brook (1967) A dominant 200-Hertz peak in the acoustic spectrum of thunder. *J. Geophys. Res.*, 72, 6149–6154.

Fiebig, M., C. R. Lunder, and A. Stohl (2009) Tracing biomass burning aerosol from South America to Troll Research Station, Antarctica. *Geophys. Res. Lett.*, 36, L14815.

Field, P. R., and A. J. Heymsfield (2003) Aggregation and scaling of ice crystal size distributions. *J. Atmos. Sci.*, 60, 544–560.

Fletcher, N. H. (1962) *The Physics of Rainclouds*. Cambridge, UK: Cambridge University Press.

Flossmann, A. I. (1998) Interaction of aerosol particles and clouds. *J. Atmos. Sci.*, 55, 879–887.

Flossmann, A. I., and W. Wobrock (2010) A review of our understanding of the aerosol–cloud interaction from the perspective of a bin resolved cloud scale modeling. *Atmos. Res.*, 97, 478–497.

Forster, P. M. de F., and K. P. Shine (2002) Assessing the climate impact of trends in stratospheric water vapor. *Geophys. Res. Lett.*, 29(6), 1086.

Fujita, T. T. (1982) Principle of stereographic height computations and their application to stratospheric cirrus over severe thunderstorms. *J. Meteorol. Soc. Japan*, 60, 355–368.

Fujita, T. T. (1985) *The Downburst – Microburst and Macroburst. Satellite and Mesometeorology Research Project (SMRP)*, Research Paper 210, Department of Geophysical Sciences, University of Chicago, NTIS PB-148880, February.

Fukuta, N. (1963) Ice nucleation by metaldehyde. *Nature*, 199, 475–476.

Fukuta, N., and B. J. Mason (1963) Epitaxial growth of ice on organic crystals. *J. Phys. Chem. Solids*, 24, 715–718.

Furukawa, Y., M. Yamamoto, and T. Kuroda (1987) Ellipsometric study of the transition layer on the surface of an ice crystal. *J. Crystal Growth*, 82, 665–677.

Gillespie, D. T. (1975) An exact method for numerically simulating the stochastic coalescence process in a cloud. *J. Atmos. Sci.*, 32, 1977–1989.

Gish, O. H. (1944) Evaluation and interpretation of the columnar resistance of the atmosphere. *Terr. Magn. Atmos. Electr.*, 49, 159–168.

Goddard, J. W. F., and S. M. Cherry (1984) The ability of dual-polarization radar (copular linear) to predict rainfall rate and microwave attenuation. *Radio Sci.*, 19, 201–208.

Greenfield, S. (1957) Rain scavenging of radioactive particulate matter from the atmosphere. *J. Meteorol.*, 14, 115–125.

Gunn, K. L. S., and J. S. Marshall (1958) The distribution with size of snowflakes. *J. Meteorol.*, 15, 452–461.

Gunn, R., and G. D. Kinzer (1949) The terminal velocity of fall for water droplets in stagnant air. *J. Meteorol.*, 4, 243–248.

Hall, W. D., and H. R. Pruppacher (1976) The survival of ice particles falling from cirrus clouds in subsaturated air. *J. Atmos. Sci.*, 33, 1995–2006.

Hallett, J., and S. C. Mossop (1974) Production of secondary ice particles during the riming process. *Nature*, 249, 26–28.

Hallgren, R., and C. L. Hosler (1960) Preliminary results in the aggregation of ice crystals. In *Physics of Precipitation*, ed. H. Weickmann. AGU Geophysical Monograph, No. 5, AGU Publ. No. 746. Baltimore, MD: Waverly Press, pp. 257–263.

Hashino, T., M. Chiruta, and P. K. Wang (2010) A numerical study on the riming process in the transition from a pristine crystal to a graupel particle. In *The 13th Conference on Cloud Physics*, Portland, OR, 28 June–2 July 2010. Boston, MA: American Meteorological Society, Paper 1.86.

Hegg, D. A., L. F. Radke, and P. V. Hobbs (1990) Particle production associated with marine clouds. *J. Geophys. Res.*, 95, 13917–13926.

Heymsfield, A. J. (1972) Ice crystal terminal velocities. *J. Atmos. Sci.*, 29, 1348–1357.

Heymsfield, A. J. (1975) Cirrus uncinus generating cells and the evolution of cirriform clouds. Part III: Numerical computations of the growth of the ice phase. *J. Atmos. Sci.*, 32, 820–830.

Heymsfield, A. J., P. N. Johnson, and J. E. Dye (1978) Observations of moist adiabatic ascent in Northeast Colorado cumulus congestus clouds. *J. Atmos. Sci.*, 35, 1689–1703.

Heymsfield, A. J., A. Bansemer, P. R. Field, *et al.* (2002) Observations and parameterizations of particle size distributions in deep tropical cirrus and stratiform precipitating clouds: results from in situ observations in TRMM field campaigns. *J. Atmos. Sci.*, 59, 3457–3491.

Highwood, E. J., and B. J. Hoskins (1998) The tropical tropopause. *Q. J. R. Meteorol. Soc.*, 124, 1579–1604.

Hill, M. J. M. (1894) On a spherical vortex. *Phil. Trans. R. Soc. Lond. A*, 185, 213–245.

Hinds, W. C. (1982) *Aerosol Technology: Properties, Behavior, and Measurement of Airborne Particles*. New York, NY: Wiley-Interscience.

Hobbs, P. V. (1974) *Ice Physics*. Oxford, UK: Clarendon Press.

Hobbs, P. V., and A. L. Rangno (1990) Rapid development of high ice particle concentrations in small polar maritime cumuliform clouds. *J. Atmos. Sci.*, 47, 2710–2722.

Hobbs, P. V., L. F. Radke, and S. E. Shumway (1970) Cloud condensation nuclei from industrial sources and their apparent influence on precipitation in Washington State. *J. Atmos. Sci.*, 27, 81–89.

Hobbs, P. V., S. Chang, and J. D. Locatelli (1974) The dimensions and aggregation of ice crystals in natural clouds. *J. Geophys. Res.*, 79, 2199–2206.

Hobbs, P. V., D. A. Bowdle, and L. F. Radke (1985) Particles in the lower troposphere over the High Plains of the United States. Part I: Size distributions, elemental compositions, and morphologies. *J. Clim. Appl. Meteorol.*, 24, 1344–1356.

Hocking, L. M. (1959) The collision efficiency of small droplets. *Q. J. R. Meteorol. Soc.*, 85, 44–50.

Hocking, L. M., and P. R. Jonas (1970) The collision efficiency of small drops. *Q. J. R. Meteorol. Soc.*, 96, 722–729.

Hoffmann, C., R. Funk, M. Sommer, and Y. Li (2008) Temporal variations in PM10 and particle size distribution during Asian dust storms in Inner Mongolia. *Atmos. Environ.*, 42, 8422–8431.

Holton, J. R. (2004) *An Introduction to Dynamic Meteorology*, 4th edn. Burlington, MA: Academic Press.

Holton, J. R., P. H. Haynes, M. E. McIntyre, *et al.* (1995) Stratosphere–troposphere exchange. *Rev. Geophys.*, 33, 403–439.

Hosler, C. L., and R. E. Hallgren (1960) The aggregation of small ice crystals. *Disc. Faraday Soc.*, 30, 200–207.

Houze, R. A., Jr. (1993) *Cloud Dynamics*. San Diego, CA: Academic Press.

Huang, C., K. T. Wikfeldt, T. Tokushima, *et al.* (2009) The inhomogeneous structure of water at ambient conditions. *Proc. Natl. Acad. Sci.*, 106, 15214–15218.

Iribarne, J. V., and H. R. Cho (1980) *Atmospheric Physics*. Dordrecht, The Netherlands: Reidel.

Iribarne, J. V., and W. L. Godson (1973) *Atmospheric Thermodynamics*. Dordrecht, The Netherlands: Reidel.

Iwai, K. (1983) Three-dimensional structure of plate-like snow crystals. *J. Meteorol. Soc. Japan*, 61, 746–755.

Iwai, K. (1989) Three-dimensional structures of natural snow crystals by stereo-photomicrographs. *Atmos. Res.*, 24, 137–147.

Iwai, K. (1999) Three dimensional fine structures of bullet-type snow crystals and their growth conditions observed at Syowa Station, Antarctica [in Japanese]. *Seppyo [Snow and Ice]*, 61, 3–12.

Jaenicke, R. (1988) Properties of atmospheric aerosols. In *Landolt-Börnstein: Numerical Data and Functional Relationships in Science and Technology*, New Series, Vol. 4, *Meteorology*, Ser. Vol. 4b, *Physical and Chemical Properties of the Air*. Berlin, Germany: Springer, Chap. 9.3.

Janhäll, S., M. O. Andreae, and U. Pöschl (2010) Biomass burning aerosol emissions from vegetation fires: particle number and mass emission factors and size distributions. *Atmos. Chem. Phys.*, 10, 1427–1439.

Jayaratne, E. R., C. P. R. Saunders, and J. Hallett (1983) Laboratory studies of the charging of soft-hail during ice crystal interactions. *Q. J. R. Meteorol. Soc.*, 109, 609–630.

Jayaweera, K. O. L. F., and B. J. Mason (1965) The behavior of freely falling cylinders and cones in a viscous fluid. *J. Fluid Mech.*, 22, 709–720.

Jensen, E. J., P. Lawson, B. Baker, *et al.* (2009) On the importance of small ice crystals in tropical anvil cirrus. *Atmos. Chem. Phys.*, 9, 5519–5537.

Ji, W., and P. K. Wang (1998) On the ventilation coefficients of falling ice crystals at low–intermediate Reynolds numbers. *J. Atmos. Sci.*, 56, 829–836.

Johnson, D. E., P. K. Wang, and J. M. Straka (1995) A study of microphysical processes in the 2 August 1981 CCOPE supercell storm. *Atmos. Res.*, 33, 93–123.

Jost, W. (1960) *Diffusion in Solids, Liquids, Gases*, 3rd printing. New York, NY: Academic Press.

Junge, C. E. (1955) The size distribution and aging of natural aerosols as determined from electrical and optical data on the atmosphere. *J. Meteorol.*, 12, 13–25.

Junge, C. E. (1963) *Air Chemistry and Radioactivity*. New York, NY: Academic Press.

Junge, C. E., C. W. Chagnon, and J. E. Manson (1961) Stratospheric aerosols. *J. Meteorol.*, 18, 81–107.

Kajikawa, M. (1972) Measurement of falling velocity of individual snow crystals. *J. Meteorol. Soc. Japan*, 50, 577–583.

Kajikawa, M. (1982) Observation of the falling motion of early snow flakes. Part I. Relationship between the free-fall pattern and the number and shape of component snow crystals. *J. Meteorol. Soc. Japan*, 60, 797–803.

Keith, W. D., and C. P. R. Saunders (1989) The collection efficiency of a cylindrical target for ice crystals. *Atmos. Res.*, 23, 83–95.

Kessler, E. (1969) *On the Distribution and Continuity of Water Substance in Atmospheric Circulation*. Meteorological Monograph, Vol. 10, No. 32. Boston, MA: American Meteorological Society, pp. 1–84.

Kittel, C. (1996) *Introduction to Solid State Physics*, 7th edn. New York, NY: Wiley.

Klemp, J. B., and R. B. Wilhelmson (1978) The simulation of three-dimensional convective storm dynamics. *J. Atmos. Sci.*, 35, 1070–1096.

Klett, J. D., and M. H. Davis (1973) Theoretical collision efficiencies of cloud droplets at small Reynolds numbers. *J. Atmos. Sci.*, 30, 107–117.

Knight, C. A. (1979) Observations of the morphology of the melting snow. *J. Atmos. Sci.*, 36, 1123–1130.

Kogan, Y. L., Z. N. Kogan, and D. B. Mechem (2009) Fidelity of analytic drop size distributions in drizzling stratiform clouds based on large eddy simulations. *J. Atmos. Sci.*, 66, 2335–2348.

Kovetz, A., and B. Olund (1969) The effect of coalescence and condensation on rain formation in a cloud of finite vertical extent. *J. Atmos. Sci.*, 26, 1060–1065.

Krehbiel, P. R., R. J. Thomas, W. Rison, *et al.* (2000) GPS-based mapping system reveals lightning inside storms. *EOS, Trans. Am. Geophys. Union*, 81, 21–25.

Kubicek, A., and P. K. Wang (2012) A numerical study of the flow fields around a typical conical graupel falling at various inclination angles. *Atmos. Res.*, 118, 15–26.

Küpper, C., J. Thuburn, G. C. Craig, and T. Birner (2004) Mass and water transport into the tropical stratosphere: a cloud-resolving simulation. *J. Geophys. Res.*, 109, D10111.

Lamb, D., and J. Verlinde (2011) *Physics and Chemistry of Clouds*. Cambridge, UK: Cambridge University Press.

Lang, T. J., L. J. Miller, M. Weisman, *et al.* (2004) The severe thunderstorm electrification and precipitation study. *Bull. Am. Meteorol. Soc.*, 85, 1107–1125.

Langmuir, I. (1948) The production of rain by a chain reaction in cumulus clouds at temperatures above freezing. *J. Meteorol.*, 5, 175–192.

Latham, J. (1981) The electrification of thunderstorms. *Q. J. R. Meteorol. Soc.*, 107, 277–298.

Latham, J., and B. J. Mason (1961) Generation of electric charge associated with the formation of soft hail in thunderstorms. *Proc. R. Soc. A*, 260, 537–549.

Latham, J., and C. P. R. Saunders (1970) Experimental measurements of the collection efficiencies of ice crystals in electric fields. *Q. J. R. Meteorol. Soc.*, 96, 257–265.

Laws, J. O., and D. A. Parsons (1943) The relation of raindrop size to intensity. *Trans. Am. Geophys. Union*, 24, 452–460.

Le Clair, B. P., A. E. Hamielec, H. R. Pruppacher, and W. D. Hall (1972) A theoretical and experimental study of the internal circulation in water drops falling at terminal velocity in air. *J. Atmos. Sci.*, 29, 728–740.

Lee, R. E., Jr., M. R. Lee, and J. M. Strong-Gunderson (1993) Insect cold-hardiness and ice nucleating active microorganisms including their potential use for biological control. *J. Insect Physiol.*, 39, 1–12.

Levizzani, V. and M. Setvák (1996) Multispectral, high resolution satellite observations of plumes on top of convective storms. *J. Atmos. Sci.*, 53, 361–369.

Lew, J. K., D. C. Montague, H. R. Pruppacher, and R. M. Rasmussen (1986a) A wind tunnel investigation on the riming of snowflakes. Part I: Porous disks and large stellars. *J. Atmos. Sci.*, 43, 2392–2409.

Lew, J. K., D. C. Montague, H. R. Pruppacher, and R. M. Rasmussen (1986b) A wind tunnel investigation on the riming of snowflakes. Part II: Natural and synthetic aggregates. *J. Atmos. Sci.*, 43, 2410–2417.

Libbrecht, K. G. (2005) The physics of snow crystals. *Rep. Prog. Phys.*, 68, 855–895.

Lin, H. M., P. K. Wang, and R. E. Schlesinger (2005) Three-dimensional nonhydrostatic simulations of summer thunderstorms in the humid subtropics versus High Plains. *Atmos. Res.*, 78, 103–145.

Liou, K. N. (1992) *Radiation and Cloud Processes in the Atmosphere: Theory, Observation, and Modeling*. New York, NY: Oxford University Press.

List, R. J. (1963) *Smithsonian Meteorological Tables*, 6th edn. Washington, DC: Smithsonian Institution.

List, R. (1965) The mechanism of hailstone formation. In *Proceedings of the International Conference on Cloud Physics*, Tokyo and Sapporo, pp. 481–491.

List, R., and G. M. McFarquhar (1990) The role of breakup and coalescence in the three-peak equilibrium distribution of raindrops. *J. Atmos. Sci.*, 47, 2274–2292.

List, R., G. B. Lesins, F. Garcia-Garcia, and D. B. McDonald (1987) Pressurized icing tunnel for graupel, hail and secondary raindrop production. *J. Atmos. Sci.*, 44, 455–463.

List, R., R. Nissen, and C. Fung (2009a) Effects of pressure on collision, coalescence, and breakup of raindrops. Part I: Experiments at 50 kPa. *J. Atmos. Sci.*, 66, 2190–2203.

List, R., C. Fung, and R. Nissen (2009b) Effects of pressure on collision, coalescence, and breakup of raindrops. Part II: Parameterization and spectra evolution at 50 and 100 kPa. *J. Atmos. Sci.*, 66, 2204–2215.

Liu, C., E. R. Williams, E. J. Zipser, and G. Burns (2010) Diurnal variations of global thunderstorms and electrified shower clouds and their contribution to the global electrical circuit. *J. Atmos. Sci.*, 67, 309–323.

Liu, H. C., P. K. Wang, and R. E. Schlesinger (2003a) A numerical study of cirrus clouds. Part I: Model description. *J. Atmos. Sci.*, 60, 1075–1084.

Liu, H. C., P. K. Wang, and R. E. Schlesinger (2003b) A numerical study of cirrus clouds. Part II: Effects of ambient temperature and stability on cirrus evolution. *J. Atmos. Sci.*, 60, 1097–1119.

Locatelli, J. D., and P. V. Hobbs (1974) Fall speeds and masses of solid precipitation particles. *J. Geophys. Res.*, 79, 2185–2197.

Lohmann, U., L. Rotstayn, T. Storelvmo, *et al.* (2010) Total aerosol effect: radiative forcing or radiative flux perturbation? *Atmos. Chem. Phys.*, 10, 3235–3246.

Lorrain, P., and D. R. Corson (1970) *Electromagnetic Fields and Waves*, 2nd edn. San Francisco, CA: W. H. Freeman.

Low, T. B., and R. List (1982a) Collision, coalescence and breakup of raindrops. Part I: Experimentally established coalescence efficiencies and fragment size distributions in breakup. *J. Atmos. Sci.*, 39, 1591–1606.

Low, T. B., and R. List (1982b) Collision, coalescence and breakup of raindrops. Part II: Parameterizations of fragment size distributions. *J. Atmos. Sci.*, 39, 1607–1619.

Ludlam, F. H. (1958) The hail problem. *Nubia*, 1, 12–96.

Ludlam, F. H., and R. S. Scorer (1957) *Cloud Study: A Pictorial Guide*. London, UK: John Murray.

Lüönd, F., O. Stetzer, A. Welti, and U. Lohmann (2010) Experimental study on the ice nucleation ability of size-selected kaolinite particles in the immersion mode. *J. Geophys. Res.*, 115, D14201.

MacGorman, D. R., and W. D. Rust (1998) *The Electrical Nature of Storms*. New York, NY: Oxford University Press.

MacGorman, D. R., W. D. Rust, C. L. Ziegler, *et al.* (2008) TELEX the Thunderstorm Electrification and Lightning Experiment. *Bull. Am. Meteorol. Soc.*, 89, 997–1013.

Macklin, W. C. (1961) Accretion in mixed clouds. *Q. J. R. Meteorol. Soc.*, 87, 413–424.

Macklin, W. C. (1963) Heat transfer from hailstones. *Q. J. R. Meteorol. Soc.*, 89, 360–369.

Magono, C., and C. W. Lee (1966) Meteorological classification of natural snow crystals. *J. Fac. Sci., Hokkaido Univ., Ser.* 7, 2, 321–335.

Makkonen, L. (2012) Misinterpretation of the Shuttleworth equation. *Scr. Mater.*, 66, 627–629.

Malkin, T. L., B. J. Murray, A. V. Brukhno, J. Anwar, and C. G. Salzmann (2012) Structure of ice crystallized from supercooled water. *Proc. Natl. Acad. Sci.*, 109, 1041–1045.

Mansell, E. R., C. L. Ziegler, and E. C. Bruning (2010) Simulated electrification of a small thunderstorm with two-moment bulk microphysics. *J. Atmos. Sci.*, 67, 171–194.

Markson, R. (2007) The global circuit intensity: its measurement and variation over the last 50 years. *Bull. Am. Meteorol. Soc.*, 88(2), 223–241.

Marshall, J. S., and W. M. K. Palmer (1948) The distribution of raindrops with size. *J. Meteorol.*, 5, 165–166.

Martin, J. J., P. K. Wang, H. R. Pruppacher, and R. L. Pitter (1981) A numerical study of the effect of electric charges on the efficiency with which planar ice crystals collect supercooled water drops. *J. Atmos. Sci.*, 38, 2462–2469.

Martin, R. S., T. A. Mather, D. M. Pyle, *et al.* (2008) Composition-resolved size distributions of volcanic aerosols in the Mt. Etna plumes. *J. Geophys. Res.*, 113, D17211.

Mason, B. J. (1956) On the melting of hailstones. *Q. J. R. Meteorol. Soc.*, 82, 209–216.

Mason, B. J. (1971) *The Physics of Clouds*. Oxford, UK: Clarendon Press.

Matsumoto, M., S. Saito, and I. Ohmine (2002) Molecular dynamics simulation of the ice nucleation and growth process leading to water freezing. *Nature*, 416, 409–413.

Matsuo, T., and Y. Sasyo (1981) Empirical formula for the melting rate of snowflakes. *J. Meteorol. Soc. Japan*, 59, 1–9.

McDonald, J. E. (1963) Use of the electrostatic analogy in studies of ice crystal growth. *Z. Angew. Math. Phys.*, 14, 610–620.

McFarquhar, G. M., J. Um, M. Freer, *et al.* (2007) Importance of small ice crystals to cirrus properties: observations from the Tropical Warm Pool International Cloud Experiment (TWP-ICE). *Geophys. Res. Lett.*, 34, L13803.

Meszaros, A., and K. Vissy (1974) Concentration, size distribution and chemical nature of atmospheric aerosol particles in remote oceanic areas. *J. Aerosol Sci.*, 5, 101–109.

Milbrandt, J. A., and M. K. Yau (2005) A multimoment bulk microphysics parameterization. Part II: A proposed three-moment closure and scheme description. *J. Atmos. Sci.*, 62, 3065–3081.

Miller, N. L., and P. K. Wang (1989) A theoretical determination of the efficiency with which aerosol particles are collected by falling columnar ice crystals. *J. Atmos. Sci.*, 46, 1656–1663.

Mitchell, D. L. (2000) Parameterization of the Mie extinction and absorption coefficients for water clouds. *J. Atmos. Sci.*, 57, 1311–1326.

Mitchell, D. L., A. Huggins, and V. Grubisic (2006) A new snow growth model with application to radar precipitation estimates. *Atmos. Res.*, 82, 2–18.

Mitchell, D. L., S. K. Chai, Y. Liu, A. J. Heymsfield, and Y. Dong (1996) Modeling cirrus clouds. Part I: Treatment of bimodal size spectra and case study analysis. *J. Atmos. Sci.*, 53, 2952–2966.

Mitra, S. K., O. Vohl, M. Ahr, and H. R. Pruppacher (1990) A wind tunnel and theoretical study of the melting behavior of atmospheric ice particles. IV: Experiment and theory for snow flakes. *J. Atmos. Sci.*, 47, 584–591.

Murakami, M., and T. Matsuo (1990) Development of the hydrometeor videosonde. *J. Atmos. Ocean. Technol.*, 7, 613–620.

Murphy, D. M. (2003) Dehydration in cold clouds is enhanced by a transition from cubic to hexagonal ice. *Geophys. Res. Lett.*, 30, 2230.

Murphy, D. M., and T. Koop (2005) Review of the vapour pressures of ice and supercooled water for atmospheric applications. *Q. J. R. Meteorol. Soc.*, 131, 1539–1565.

Newton, C. W., and H. R. Newton (1959) Dynamical interactions between large convective clouds and environment with vertical shear. *J. Meteorol.*, 16, 483–496.

Ochs, H. T. (1978) Moment-conserving techniques for warm cloud microphysical computations. Part II. Model testing and results. *J. Atmos. Sci.*, 35, 1959–1973.

Ochs, H. T., R. R. Czys, and K. V. Beard (1986) Laboratory measurements of coalescence efficiencies for small precipitation drops. *J. Atmos. Sci.*, 43, 225–232.

Ochs, H. T., III, K. V. Beard, N. F. Laird, D. J. Holdridge, and D. E. Schaufelberger (1995) Effects of relative humidity on the coalescence of small precipitation drops in free fall. *J. Atmos. Sci.*, 52, 3673–3680.

Ogura, Y., and N. A. Phillips (1962) Scale analysis of deep and shallow convection in the atmosphere. *J. Atmos. Sci.*, 19, 173–179.

Ohtake, T. (1970) Factors affecting the size distribution of raindrops and snowflakes. *J. Atmos. Sci.*, 27, 804–813.

Orville, H. D., and F. J. Kopp (1977) Numerical simulation of the life history of a hailstorm. *J. Atmos. Sci.*, 34, 1596–1618.

Orville, R. E. (1994) Cloud-to-ground lightning flash characteristics in the contiguous United States: 1989–1991. *J. Geophys. Res.*, 99, 10833–10841.

Paluch, I. R. (1979) The entrainment mechanism in Colorado cumuli. *J. Atmos. Sci.*, 36, 2467–2478.

Park, R. W. (1970) *Behavior of water drops colliding in humid nitrogen.* Ph.D. thesis, University of Wisconsin-Madison.

Pedlosky, J. (2003) *Waves in the Ocean and Atmosphere.* Berlin, Germany: Springer.

Peppler, W. (1940) *Unterkühlte Wasserwolken und Eiswolken [Supercooled water clouds and ice clouds].* Forschungs- und Erfahrungsberichte des Reichswetterdienstes B, No. 1, pp. 3–68.

Pessi, A. T., and S. Businger (2009) Relationships among lightning, precipitation, and hydrometeor characteristics over the North Pacific Ocean. *J. Appl. Meteor. Climatol.*, 48, 833–848.

Pflaum, J. C., and H. R. Pruppacher (1979) A wind tunnel investigation of the growth of graupel initiated from frozen drops. *J. Atmos. Sci.*, 36, 680–689.

Pflaum, J. C., J. J. Martin, and H. R. Pruppacher (1978) A wind tunnel investigation of the hydrodynamic behavior of growing, freely falling graupel. *Q. J. R. Meteorol. Soc.*, 104, 179–187.

Pinsky, M. B., and A. P. Khain (2004) Collisions of small drops in a turbulent flow. Part II: Effects of flow accelerations. *J. Atmos. Sci.*, 61, 1926–1939.

Pinsky, M. B., A. P. Khain, and M. Shapiro (1999) Collisions of small drops in a turbulent flow. Part I: Collision efficiency. Problem formulation and preliminary results. *J. Atmos. Sci.*, 56, 2585–2600.

Pitter, R. L., and H. R. Pruppacher (1974) A numerical investigation of collision efficiencies of simple ice plates colliding with supercooled water drops. *J. Atmos. Sci.*, 31, 551–559.

Pitter, R. L., H. R. Pruppacher, and A. E. Hamielec (1973) A numerical study of viscous flow past a thin oblate spheroid at low and intermediate Reynolds numbers. *J. Atmos. Sci.*, 30, 125–134.

Pitter, R. L., H. R. Pruppacher, and A. E. Hamielec (1974) A numerical study of the effect of forced convection on mass transport from a thin oblate spheroid of ice in air. *J. Atmos. Sci.*, 31, 1058–1066.

Podzimek, J. (1966) Experimental determination of the "capacity" of ice crystals. *Stud. Geophys. Geodet.*, 10, 235–238.

Pöschl, U., S. T. Martin, B. Sinha, *et al.* (2010) Rainforest aerosols as biogenic nuclei of clouds and precipitation in the Amazon. *Science*, 329, 1513–1515.

Prenni, A. J., M. D. Petters, S. M. Kreidenweis, *et al.* (2009) Relative roles of biogenic emissions and Saharan dust as ice nuclei in the Amazon basin. *Nature Geosci.*, 2, 402–405.

Prodi, F. (1976) Scavenging of aerosol particles by growing ice crystals. In *International Conference on Cloud Physics*, Boulder, CO, 26–30 July 1976. Preprints, pp. 70–75.

Pruppacher, H. R., and K. V. Beard (1970) A wind tunnel investigation of the internal circulation and shape of water drops falling at terminal velocity in air. *Q. J. R. Meteorol. Soc.*, 96, 247–256.

Pruppacher, H. R., and J. D. Klett (1997) *Microphysics of Clouds and Precipitation*. Dordrecht, The Netherlands: Kluwer.

Rakov, V. A., and M. A. Uman (2003) *Lightning: Physics and Effects*. Cambridge, UK: Cambridge University Press.

Rasmussen, R. M., and A. J. Heymsfield (1987) Melting and shedding of graupel and hail. Part I: Model physics. *J. Atmos. Sci.*, 44, 2754–2763.

Rasmussen, R. M., and H. R. Pruppacher (1982) A wind tunnel and theoretical study of the melting behavior of atmospheric ice particles. I: A wind tunnel study of frozen drops of radius < 500 μm. *J. Atmos. Sci.*, 39, 152–158.

Rasmussen, R. M., V. Levizzani, and H. R. Pruppacher (1984a) A wind tunnel and theoretical study of the melting behavior of atmospheric ice particles. II: A theoretical study for frozen drops of radius < 500 μm. *J. Atmos. Sci.*, 41, 374–380.

Rasmussen, R. M., V. Levizzani, and H. R. Pruppacher (1984b) A wind tunnel and theoretical study of the melting behavior of atmospheric ice particles. III. Experiment and theory for spherical ice particles of radius > 500 μm. *J. Atmos. Sci.*, 41, 381–388.

Rasmussen, R., C. Walcek, H. R. Pruppacher, *et al.* (1985) A wind tunnel investigation of the effect of an external vertical electric field on the shape of electrically uncharged rain drops. *J. Atmos. Sci.*, 42, 1647–1652.

Reid, J. S., T. F. Eck, S. A. Christopher, *et al.* (2005) A review of biomass burning emissions. Part III: Intensive optical properties of biomass burning particles. *Atmos. Chem. Phys.*, 5, 827–849.

Reif, F. (1965) *Fundamentals of Statistical and Thermal Physics*. New York, NY: McGraw-Hill.

Reinking, R. (1979) The onset and steady growth of snow crystals by accretion of droplets. *J. Atmos. Sci.*, 36, 870–881.

Reynolds, O. (1876) On the manner in which raindrops and hailstones are formed. *Proc. Lit. Phil. Soc., Manchester*, 16, 23–33.

Reynolds, S. E., M. Brook, and M. F. Gourley (1957) Thunderstorm charge separation. *J. Meteorol.*, 14, 426–436.

Richards, C. N., and G. A. Dawson (1971) The hydrodynamic instability of water drops falling at terminal velocity in vertical electric fields. *J. Geophys. Res.*, 76, 3445–3455.

Rogers, D. C. (1974) *The Aggregation of Natural Ice Crystals*. Research Report AR110, Department of Atmospheric Resources, University of Wyoming, Laramie, WY.

Rosenfeld, D., U. Lohmann, G. B. Raga, *et al.* (2008) Flood or drought: How do aerosols affect precipitation? *Science*, 321, 1309–1313.

Ryan, B. T. (1974) Growth of drops by coalescence: the effect of different collection kernels and of additional growth by condensation. *J. Atmos. Sci.*, 31, 1942–1948.

Sambles, J. R., L. M. Skinner, and N. D. Listgarten (1970) An electron microscope study of evaporating small particles: the Kelvin equation for liquid lead and the mean surface energy of solid silver. *Proc. R. Soc. London, A,* 318, 507–522.

Sazaki, G., S. Zepeda, S. Nakatsubo, E. Yokoyama, and Y. Furukawa (2010) Elementary steps at the surface of ice crystals visualized by advanced optical microscopy. *Proc. Natl. Acad. Sci.,* 107, 19702–19707.

Sazaki, G., S. Zepeda, S. Nakatsubo, M. Yokomine, and Y. Furukawa (2012) Quasi-liquid layers on ice crystal surfaces are made up of two different phases. *Proc. Natl. Acad. Sci.,* 109, 1052–1055.

Schlamp, R. J., S. N. Grover, H. R. Pruppacher, and A. E. Hamielec (1976) A numerical investigation of the electric charges and vertical external electric fields on the collision efficiency of cloud drops. *J. Atmos. Sci.,* 33, 1747–1755.

Schlamp, R. J., S. N. Grover, H. R. Pruppacher, and A. E. Hamielec (1979) A numerical investigation of the effect of electric charges and vertical external electric fields on the collision efficiency of cloud drops: Part II. *J. Atmos. Sci.,* 36, 339–349.

Schlesinger, R. E. (1973) A numerical model of deep moist convection: Part I. Comparative experiments for variable ambient moisture and wind shear. *J. Atmos. Sci.,* 30, 835–856.

Schlottke, J., W. Straub, K. Beheng, H. Gomma, and B. Weigand (2010) Numerical investigation of collision-induced breakup of raindrops. Part I: Methodology and dependencies on collision energy and eccentricity. *J. Atmos. Sci.,* 67, 557–575.

Schmidt, R. A. (1984) Measuring particle size and snowfall intensity in drifting snow. *Cold Regions Sci. Technol.,* 9, 121–129.

Schuman, T. E. W. (1938) The theory of hailstone formation. *Q. J. R. Meteorol. Soc.,* 64, 3–21.

Schumann, W. O. (1952) Über die Dämpfung der elecktromagnetischen Eigenschwingungen des Systems Erde–Lufte–Ionosphare. *Z. Naturforsch. A,* 7, 250–252.

Schwarz, S. E. (1986) Mass-transport considerations pertinent to aqueous phase reactions of gases in liquid-water clouds. In *Chemistry of Multiphase Systems,* ed. W. Jaeschke. NATO ASI Series. Berlin, Germany: Springer, pp. 415–477.

Scorer, R. S. (1997) *Dynamics of Meteorology and Climate.* Chichester, UK: Wiley.

Scorer, R. S., and F. H. Ludlam (1953) Bubble theory of penetrative convection. *Q. J. R. Meteorol. Soc.,* 79, 94–103.

Seinfeld, J. H., and S. N. Pandis (2006) *Atmospheric Chemistry and Physics: From Air Pollution to Climate Change,* 2nd edn. New York, NY: Wiley-Interscience.

Sekhon, R. S., and R. C. Srivastava (1970) Snow size spectra and radar reflectivity. *J. Atmos. Sci.,* 27, 299–307.

Setvák, M., and C. A. Doswell (1991) The AVHRR channel 3 cloud top reflectivity of convective storms. *Mon. Weather Rev.,* 119, 841–847.

Shafrir, U., and M. Neiburger (1963) Collision efficiencies of two spheres falling in a viscous medium. *J. Geophys. Res.,* 68, 4141–4147.

Shuttleworth, R. (1950) The surface tension of solid. *Proc. Phys. Soc. A,* 63, 444–457.

Simpson, J. (1971) On cumulus entrainment and one-dimensional models. *J. Atmos. Sci.,* 28, 449–455.

Skatskii, V. I. (1965) Some results from experimental study of the liquid water content in cumulus clouds. *Izv. Atmos. Oceanic Phys.,* 1, 479–487.

Solomon, S., D. Qin, M. Manning, *et al.* (eds) (2007) *Climate Change 2007: The Physical Science Basis.* Contribution of Working Group I to the Fourth Assessment Report of the Intergovernmental Panel on Climate Change. Cambridge, UK: Cambridge University Press.

Solomon, S., K. Rosenlof, R. Portmann, *et al.* (2010) Contributions of stratospheric water vapor to decadal changes in the rate of global warming. *Science Express*, 28 January, pp. 1–6.

Squires, P. (1958a) The microstructure and colloidal stability of warm clouds. Part II. The causes of the variations in microstructure. *Tellus*, 10, 262–271.

Squires, P. (1958b) Penetrative downdraughts in cumuli. *Tellus*, 10, 381–389.

Srivastava, R. C. (1967) A study of the effect of precipitation on cumulus dynamics. *J. Atmos. Sci.*, 24, 36–45.

Stommel, H. (1947) Entrainment of air into a cumulus cloud. *J. Meteorol.*, 4, 91–94.

Straka. J. M. (1989) *Hail growth in a highly glaciated central High Plains multi-cellular hailstorm.* Ph.D. thesis, Department of Meteorology, University of Wisconsin, Madison, WI.

Straka, J. M. (2010) *Cloud and Precipitation Microphysics: Principles and Parameterizations.* Cambridge, UK: Cambridge University Press.

Straub, W., K. D. Beheng, A. Seifert, J. Schlottke, and B. Weigand (2010) Numerical investigation of collision-induced breakup of raindrops. Part II: Parameterizations of coalescence efficiencies and fragment size distributions. *J. Atmos. Sci.*, 67, 576–588.

Sturniolo, O., A. Mugnai, and F. Prodi (1995) A numerical sensitivity study on the back-scattering at 35.8 GHz from precipitation-sized hydrometeors. *Radio Sci.*, 30(4), 903–919.

Sun, F. L. (1993) *On the bimodal size distribution of hydrometeors in clouds.* M.Sc. thesis, Department of Atmospheric and Oceanic Sciences, University of Wisconsin-Madison, Madison, WI.

Szakáll, M., K. Diehl, S. K. Mitra, and S. Borrmann (2009) A wind tunnel study on the shape, oscillation, and internal circulation of large raindrops with sizes between 2.5 and 7.5 mm. *J. Atmos. Sci.*, 66, 755–765.

Szakáll, M., S. K. Mitra, K. Diehl, and S. Borrmann (2010) Shapes and oscillations of falling raindrops – a review. *Atmos. Res.*, 97, 416–425.

Szyrmer, W., and I. Zawadzki (1997) Biogenic and anthropogenic sources of ice-forming nuclei: a review. *Bull. Am. Meteorol. Soc.*, 78, 209–228.

Takahashi, T. (1973) Measurement of electric charge of cloud droplets, drizzle, and rain-drops. *Rev. Geophys. Space Phys.*, 11, 903–924.

Takahashi, T. (1978) Riming electrification as a charge generation mechanism in thunder-storms. *J. Atmos. Sci.*, 35, 1536–1548.

Takahashi, T., and K. Miyawaki (2002) Reexamination of riming electrification in a wind tunnel. *J. Atmos. Sci.*, 59, 1018–1025.

Takahashi, T., T. Endoh, and G. Wakahama (1991) Vapor diffusional growth of free-falling snow crystals between −3 and −23°C. *J. Meteorol. Soc. Japan*, 69, 15–30.

Takeda, T. (1971) Numerical simulation of a precipitating convective cloud: the formation of a "long-lasting" cloud. *J. Atmos. Sci.*, 28, 350–376.

Takeuti, T., M. Nakano, M. Brook, D. J. Raymond, and P. Krehbiel (1978) The anomalous winter thunderstorms of the Hokuriku coast. *J. Geophys. Res.*, 83, 2385–2394.

Taneda, S. (1956) Experimental investigation of the wake behind a sphere at low Reynolds numbers. *J. Phys. Soc. Japan*, 11, 1101–1108.

Telford, J. W. (1955) A new aspect of coalescence theory. *J. Meteorol.*, 12, 436–444.

Thorpe, A. D., and B. J. Mason (1966) The evaporation of ice spheres and ice crystals. *Br. J. Appl. Phys.*, 17, 541–548.

Thurai, M., G. J. Huang, V. N. Bringi, W. L. Randeu, and M. Schönhuber (2007) Drop shapes, model comparisons, and calculations of polarimetric radar parameters in rain. *J. Atmos. Ocean. Technol.*, 24, 1019–1032.

Thurai, M., V. N. Bringi, M. Szakáll, *et al.* (2009) Drop shapes and axis ratio distributions: comparison between 2D video disdrometer and wind-tunnel measurements. *J. Atmos. Ocean. Technol.*, 26, 1427–1432.

Tinsley, B. A., G. B. Burns, and L. Zhou (2007) The role of the global electric circuit in solar and internal forcing of clouds and climate. *Adv. Space Res.*, 40, 1126–1139.

Tokay, A., and K. V. Beard (1996) A field study of raindrop oscillations. Part I: Observation of size spectra and evaluation of oscillation causes. *J. Appl. Meteorol.*, 35, 1671–1687.

Trentmann, J., G. Luderer, T. Winterrath, *et al.* (2006) Modeling of biomass smoke injection into the lower stratosphere by a large forest fire (Part I): reference simulation. *Atmos. Chem. Phys.*, 6, 5247–5260.

Turner, D. D. (2005) Arctic mixed-phase cloud properties from AERI lidar observations: algorithm and results from SHEBA. *J. Appl. Meteorol.*, 44, 427–443.

Turner, J. S. (1969) Buoyant plumes and thermals. *Annu. Rev. Fluid Mech.*, 1, 29–44.

Twomey, S. (1964) Statistical effects in the evolution of a distribution of cloud droplets by coalescence. *J. Atmos. Sci.*, 21, 553–557.

Twomey, S., and T. A. Wojciechowski (1969) Observations of the geographical variation of cloud nuclei. *J. Atmos. Sci.*, 26, 684–688.

Ulbrich, C. W. (1983) Natural variations in the analytical form of raindrop size distribution. *J. Clim. Appl. Meteorol.*, 22, 1764–1775.

Um, J., and G. M. McFarquhar (2009) Single-scattering properties of aggregates of bullet rosettes in cirrus. *J. Appl. Meteorol. Climatol.*, 46, 757–775.

Uman, M. A. (1969) *Lightning*. New York, NY: McGraw-Hill.

Uman, M. A. (1987) *The Lightning Discharge*. Orlando, FL: Academic Press.

Valdez, M. P., and K. C. Young (1985) Number fluxes in equilibrium raindrop populations: a Markov chain analysis. *J. Atmos. Sci.*, 42, 1024–1036.

Volmer, M. (1939) *Kinetik der Phasenbildung*. Dresden, Germany: Verlag Th. Steinkopff.

Vonnegut, B., and C. B. Moore (1958) Giant electrical storms. In *Recent Advances in Atmospheric Electricity*, ed. L. G. Smith. New York, NY: Pergamon, pp. 399–410.

Vrbka, L., and P. Jungwirth (2006) Homogeneous freezing of water starts in the subsurface. *J. Phys. Chem. B*, 110, 18126–18129.

Walcek, C. J., and H. R. Pruppacher (1984) On the scavenging of SO_2 by cloud and rain-drops: I. A theoretical study of SO_2 absorption and desorption for water drops in air. *J. Atmos. Chem.*, 1, 269–289.

Waldvogel, A. (1974) The N_0 jump of raindrop spectra. *J. Atmos. Sci.*, 31, 1067–1078.

Walker, J. C. G. (1977) *Evolution of the Atmosphere*. New York, NY: Macmillan.

Wang, P. K. (1982) Mathematical description of the shape of conical hydrometeors. *J. Atmos. Sci.*, 39, 2615–2622.

Wang, P. K. (1983) On the definition of collision efficiency of atmospheric particles. *J. Atmos. Sci.*, 40, 1051–1052.

Wang, P. K. (1997) Characterization of ice particles in clouds by simple mathematical expressions based on successive modification of simple shapes. *J. Atmos. Sci.*, 54, 2035–2041.

Wang, P. K. (1999) Three-dimensional representations of hexagonal ice crystals and hail particles of elliptical cross-sections. *J. Atmos. Sci.*, 56, 1089–1093.

Wang, P. K. (2002) *Ice Microdynamics*. San Diego, CA: Academic Press.

Wang, P. K. (2003) Acid rain and precipitation chemistry. In *Encyclopedia of Water Science*. New York, NY: Marcel Dekker.

Wang, P. K. (2004) A cloud model interpretation of jumping cirrus above storm top. *Geophys. Res. Lett.*, 31, L18106.

Wang, P. K., and S. M. Denzer (1983) Mathematical description of the shape of plane hexagonal snow crystals. *J. Atmos. Sci.*, 40, 1024–1028.

Wang, P. K., and W. Ji (1997) Simulation of three-dimensional unsteady flow past ice crystals. *J. Atmos. Sci.*, 54, 2261–2274.

Wang, P. K., and W. Ji (2000) Collision efficiencies of ice crystals at low–intermediate Reynolds numbers colliding with supercooled cloud droplets: a numerical study. *J. Atmos. Sci.*, 57, 1001–1009.

Wang, P. K., and H. R. Pruppacher (1977) Acceleration to terminal velocity of cloud and rain drops. *J. Appl. Meteorol.*, 16, 275–280.

Wang, P. K., S. N. Grover, and H. R. Pruppacher (1978) On the effect of electric charges on the scavenging of aerosol particles by cloud and small rain drops. *J. Atmos. Sci.*, 35, 1735–1743.

Wang, P. K., R. Rasmussen, C. C. Yang, H. R. Pruppacher and C. R. Viswanathan (1980) Heterogeneous nucleation of water and ice on a p–n junction. In *Symposium on Nucleation, 180th National American Chemical Society Meeting*, Las Vegas, 26–28 August 1980. Abstracts.

Wang, P. K., T. J. Greenwald, and J. Wang (1987) A three parameter representation of the shape and size distributions of hailstones – a case study. *J. Atmos. Sci.*, 44, 1062–1070.

Wang, P. K., M. Setvák, W. Lyons, W. Schmid, and H. Lin (2009) Further evidence of deep convective vertical transport of water vapor through the tropopause. *Atmos. Res.*, 94, 400–408.

Wang, P. K., S.-H. Su, M. Setvák, H.-M. Lin, and R. Rabin (2010) Ship wave signature at the cloud top of deep convective storms. *Atmos. Res.*, 97, 294–302.

Wang, P. K., S.-H. Su, Z. Charvát, J. Štástka, and H.-M. Lin (2011) Cross tropopause transport of water by mid-latitude deep convective storms: a review. *Terr. Atmos. Ocean. Sci.*, 22, 447–462.

Wang, Z., K. Sassen, D. N. Whiteman, and B. B. Demoz (2004) Studying altocumulus with ice virga using ground-based active and passive remote sensors. *J. Appl. Meteor.*, 43, 449–460.

Warneck, P. (1988) *Chemistry of the Natural Atmosphere*. San Diego, CA: Academic Press.

Warner, J. (1969) The microstructure of cumulus cloud. Part I. General features of the droplet spectrum. *J. Atmos. Sci.*, 26, 1272–1282.

Warner, J. (1970) On steady-state one-dimensional models of cumulus convection. *J. Atmos. Sci.*, 27, 1035–1040.

Westbrook, C. D., R. J. Hogan, and A. J. Illingworth (2008) The capacitance of pristine ice crystals and aggregate snowflakes. *J. Atmos. Sci.*, 65, 206–219.

Whitby, K. T. (1978) The physical characteristics of sulfur aerosols. *Atmos. Environ.*, 12, 135–159.

Wilhelmson, R. (1974) The life cycle of a thunderstorm in three dimensions. *J. Atmos. Sci.*, 31, 1629–1651.

Wilkins, R. D., and A. H. Auer, Jr. (1970) Riming properties of hexagonal ice crystals. *Preprints, Conf. on Cloud Physics, Fort Collins, CO*, American Meteorological Society, pp. 81–82.

Williams, E. R. (2005) Lightning and climate: a review. *Atmos. Res.*, 76, 272–287.

Williams, E. R., M. Weber, and R. Orville (1989) The relationship between lightning type and convective state of thunderclouds. *J. Geophys. Res.*, 94, 13213–13220.

Willis, P. T. (1984) Functional fits to some observed drop size distributions and parameterization of rain. *J. Atmos. Sci.*, 41, 1648–1661.

Willis, P. T., and P. Tuttleman (1989) Drop-size distributions associated with intense rainfall. *J. Appl. Meteorol.*, 28, 3–15.

Willmarth, W. W., Hawk, N. E., and R. L. Harvey (1964) Steady and unsteady motions and wakes of freely falling disks. *Phys. Fluids*, 7, 197–208.

Wilson, C. T. R. (1903) Atmospheric electricity. *Nature*, 68, 101–104.

Wilson, C. T. R. (1920) Investigations on lightning discharges and on the electric field of thunderstorms. *Phil. Trans. R. Soc. London A*, 221, 73–115.

Wilson, C. T. R. (1929) Some thundercloud problems. *J. Franklin Inst.*, 208, 1–12.

Wurden, G., and D. Whiteson (1996) High-speed plasma imaging: a lightning bolt. *IEEE Trans. Plasma Sci.*, 24, 83–84.

Zawadzki, I., W. Szyrmer, C. Bell, and F. Fabry (2005) Modeling of the melting layer. Part III: The density effect. *J. Atmos. Sci.*, 62, 3705–3723.

Index